Encyclopedia of Wireless Communications and Networks

Encyclopedia of Wireless Communications and Networks

Edited by **Bernhard Ekman**

NY RESEARCH
P R E S S

New York

Published by NY Research Press,
23 West, 55th Street, Suite 816,
New York, NY 10019, USA
www.nyresearchpress.com

Encyclopedia of Wireless Communications and Networks
Edited by Bernhard Ekman

International Standard Book Number: 978-1-63238-177-4 (Hardback)

Contents

Preface   VII

Part 1   **Wireless Communication Antennas**   1

Chapter 1   **Review of the Wireless Capsule
Transmitting and Receiving Antennas**   3
Zhao Wang, Eng Gee Lim, Tammam Tillo
and Fangzhou Yu

Chapter 2   **Latest Progress in MIMO Antennas Design**   23
Yue Li, Jianfeng Zheng and Zhenghe Feng

Chapter 3   **Travelling Planar Wave Antenna
for Wireless Communications**   47
Onofrio Losito and Vincenzo Dimiccoli

Chapter 4   **Superstrate Antennas for Wide
Bandwidth and High Efficiency
for 60 GHz Indoor Communications**   93
Hamsakutty Vettikalladi, Olivier Lafond
and Mohamed Himdi

Part 2   **Wireless Communication Hardware**   123

Chapter 5   **Hardware Implementation of Wireless
Communications Algorithms: A Practical Approach**   125
Antonio F. Mondragon-Torres

Chapter 6   **Analysis of Platform Noise Effect on Performance
of Wireless Communication Devices**   157
Han-Nien Lin

Chapter 7   **Gallium Nitride-Based Power Amplifiers for Future
Wireless Communication Infrastructure**   207
Suramate Chalermwisutkul

Part 3   Channel Estimation and Capacity   227

Chapter 8   **Indoor Channel Measurement**
             for Wireless Communication   229
             Hui Yu and Xi Chen

Chapter 9   **Channel Capacity Analysis Under Various Adaptation**
             Policies and Diversity Techniques over Fading Channels   255
             Mihajlo Stefanović, Jelena Anastasov, Stefan Panić, Petar Spalević
             and Ćemal Dolićanin

Chapter 10   **Superimposed Training-Aided Channel**
             Estimation for Multiple Input Multiple
             Output-Orthogonal Frequency Division Multiplexing
             Systems over High-Mobility Environment   277
             Han Zhang, Xianhua Dai, Daru Pan and Shan Gao

             Permissions

             List of Contributors

Preface

Every book is initially just a concept; it takes months of research and hard work to give it the final shape in which the readers receive it. In its early stages, this book also went through rigorous reviewing. The notable contributions made by experts from across the globe were first molded into patterned chapters and then arranged in a sensibly sequential manner to bring out the best results.

The book highlights recent advancements starting from the lowest layers to the upper layers of wireless communication networks and consists of "real-time" research development on the related issues. The information in this book has been systematically organized in order to make it easily accessible to the readers of all levels. It also preserves the balance between the recent research results and their theoretical support. A huge variety of new techniques in this field are investigated in this book. The authors attempt to present these topics in detail by discussing wireless communication antennas, hardware and channel estimation and capacity. Intelligent and reader-friendly elucidations are provided in this book to serve the readers of all levels, ranging from knowledgeable and practicing communication engineers to beginners or professional researchers.

It has been my immense pleasure to be a part of this project and to contribute my years of learning in such a meaningful form. I would like to take this opportunity to thank all the people who have been associated with the completion of this book at any step.

Editor

Part 1

Wireless Communication Antennas

Review of the Wireless Capsule Transmitting and Receiving Antennas

Zhao Wang, Eng Gee Lim, Tammam Tillo and Fangzhou Yu
Xi'an Jiaotong - Liverpool University
P.R. China

1. Introduction

The organization of American Cancer Society reported that the total number of cancer related to GI track is about 149,530 in the United State only for 2010 (American Cancer Society, 2010). Timely detection and diagnoses are extremely important since the majority of the GI related cancers at early-stage are curable.

However, the particularity of the alimentary track restricts the utilization of the current available examine techniques. The upper gastrointestinal tract can be examined by Gastroscopy. The bottom 2 meters makes up the colon and rectum, and can be examined by Colonoscopy. In between, lays the rest of the digestive tract, which is the small intestine characterised by being very long (average 7 meters) and very convoluted. However, this part of the digestive tract lies beyond the reach of the two previously indicated techniques. To diagnose the small intestine diseases, the special imaging techniques like CT scan or MRI are less useful in this circumstance.

Therefore, the non-invasive technique Wireless Capsule Endoscopy (WCE) has been proposed to enable the visualisation of the whole GI track cable freely. The WCE is a sensor device that contains a colour video camera and wireless radiofrequency transmitter, and battery to take nearly 55,000 colour images during an 8-hour journey through the digestive tract.

The most popular WCEs, are developed and manufactured by Olympus (Olympus, 2010), IntroMedic (IntroMedic, 2010) and Given Imaging (Given Imaging, 2010). However, there are still several drawbacks limiting the application of WCE. Recently, there are two main directions to develop the WCE. One is for enlarging the advantages of current wireless capsule, for example they are trying to make the capsule smaller and smaller, to enhance the propagation efficiency of the antenna or to reduce the radiated effects on human body. While, others are working on minimizing the disadvantages of capsule endoscope, for instance, they use internal and external magnetic field to control the capsule and use technology to reduce the power consumption.

The role of the WCE embedded antenna is for sending out the detected signals; hence the signal transmission efficiency of the antenna will directly decide the quality of received real-time images and the rate of power consumption (proportional to battery life). The human body as a lossy dielectric material absorbs a number of waves and decreases the power of receiving signals, presenting strong negative effects on the microwave propagation. Therefore, the antenna elements should ideally possess these features: first, the ideal antenna for the wireless capsule endoscope should be less sensitive to human tissue influence; second, the antenna should have enough bandwidth to transmit high resolution images and huge number of data; third, the enhancement of the antenna efficiency would facilitate the battery power saving and high data rate transmission.

In this chapter the WCE system and antenna specifications is first introduced and described. Next, the special consideration of body characteristics for antenna design (in body) is summarized. State-of-the-art WCE transmitting and receiving antennas are also reviewed. Finally, concise statements with a conclusion will summarize the chapter.

2. Wireless Capsule Endoscopy (WCE) system

In May of 2000, a short paper appeared in the journal Nature describing a new form of gastrointestinal endoscopy that was performed with a miniaturized, swallowable camera that was able to transmit color, high-fidelity images of the gastrointestinal tract to a portable recording device (Iddan et al., 2000). The newer technology that expands the diagnostic capabilities in the GI tract is capsule endoscopes also known as wireless capsule endoscopy. One example of the capsule is shown in Figure 1.

Fig. 1. Physical layout of the WCE (Olympus, 2010).

The capsule endoscopy system is composed of several key parts (shown in Figure 2): image sensor and lighting, control unit, wireless communication unit, power source, and mechanical actuator. The imaging capsule is pill-shaped and contains these miniaturized elements: a battery, a lens, LEDs and an antenna/transmitter. The physical layout and conceptual diagram of the WCE are depicted in Figure 1 and Figure 2, respectively. The capsule is activated on removal from a holding assembly, which contains a magnet that keeps the capsule inactive until use. When it is used, capsule record images and transmit them to the belt-pack receiver. The capsule continues to record images at a rate over the course of the 7 to 8 hour image acquisition period, yielding a total of approximately 55,000 images per examination. Receiver/Recorder Unit receives and records the images through an antenna array consisting of several leads that connected by wires to the recording unit, worn in standard locations over the abdomen, as dictated by a template for lead placement. The antenna array and battery pack can be worn under regular clothing. The recording device to which the leads are attached is capable of recording the thousands of images transmitted by the capsule and received by the antenna array. Once the patient has completed the endoscopy examination, the antenna array and image recording device are returned to the health care provider. The recording device is then attached to a specially modified computer workstation (Gavriel, 2000). The software shows the viewer to watch the video at varying rates of speed, to view it in both forward and reverse directions, and to capture and label individual frames as well as brief video clips.

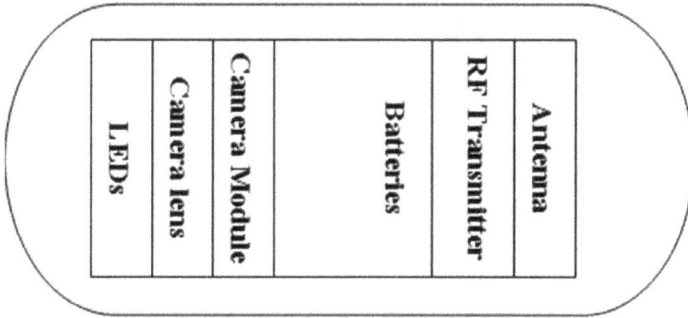

Fig. 2. Conceptual diagram of the WCE.

Since the device received FDA (American Food and Drug Administration) clearance in August 2001, over 1,000,000 examinations have been conducted globally. The 11mm by 26mm M2A capsule is propelled passively, one end of the capsule contains an optical dome with six white Light Emitting Diodes and a CMOS camera that captures 2 images a second (Given Imaging, 2010). These images relayed via a transmitter using a radio frequency signal to an array of aerials from where they are transferred over the wires to a data-recorder. The sensor array allows for continues triangulation of the position of the capsule inside the body of the patient. The accuracy of the capsule location provide by this method was reported to be +/-3 cm (Ravens & Swain, 2002). In December 2004, FDA approved a second type of capsule developed by Given Imaging-the PillCam ESO, which allows the evaluation of esophageal disease. The response to this demand materialized in the development of the pillCam ESO which has the higher frame rate and CMOS cameras positioned at both ends of the capsule. This capsule acquires and transmits seven frames per second from each camera, giving a total of 14 frames per second (Mishkin et al. 2006). Due to the increased frame rate, the capsule battery life is only 20 minutes. In October 2005, Olympus launched a competitor system called EndoCapsule in Europe. The difference lies in the use of a different imaging technology-CCD, which the manufacturers claim is of higher quality (Fuyono, I. 2005). Another feature of EndoCapsule is the Automatic Brightness Control (ABC), which provides an automatic illumination adjustment as the conditions in the GI tract vary. In October 2006, Given Imaging received the CE Mark to market a third capsule-the PillCam COLON though out the European Union. This capsule measures 11mm by 31mm, that is slightly larger than previous products. It captures 4 images a second for up to 10hours. A new feature in Given Imaging capsules is an automatic lighting control (Eliakim et al. 2006; Schoofs et al., 2006). In 2007, PillCam SB2 was cleared for marketing in the US. According to the manufacturers, it offers advanced optics and a wider field of view. PillCam SB2 also captures nearly twice the mucosal area per image. It also provides Automatic Light Control for optimal illumination of each image. In 2009, the second-generation capsule, PillCam COLON2, was cleared by the European Union. The capsule has the ability to adjust the frame rate in real time to maximize colon tissue coverage. To present, Olympus is working on the development of a new generation capsule endoscope, which features magnetic propulsion. Apart from the novel propulsion and guidance system, the capsule designers aim to provide a drug delivery system, a body fluid sampling system and also the ultrasound scan capability. RF System Lab Company announced the design of the new Sayaka capsule (RF System Lab, 2010), which acquires images at a rate of 30 frames per second and generate about 870,000 over an eight hour period of operation. Also, further applications of magnetic fields are presented (Lenaertes & puers, 2006).

3. Antenna specifications for WCE

Wireless capsule transmitting and receiving antennas belong to wireless communication unit. The transceiver in conjugation with an antenna was utilised. A bidirectional communication between the capsule and the external communication unit at recommended frequency for industrial, scientific and medical usage was established. Wireless capsule endoscopy transmitting antenna is for sending out the detected signal and receiving antenna receive the signal outside human body. The signal transmission efficiency of the antenna will directly decide the quality of the received real-images and rate of power consumption. Because a lossy dielectric material absorbs a number of waves and decreases the power of receiving signal, it presents strong negative effects on the microwave propagation (Johnson, & Guy, 1972). Therefore, some features to ideally possess are required. The WCE antenna should be less sensitive to human tissue influence. Enough bandwidth to transmit high resolution images and huge number of data is a requirement for antenna. Also, power saving and high data rate transmission can be obtained with enhancement of antenna efficiency.

In addition to the standard constraints in electronic design, a number of main challenges arise for systems that operate inside the human body. The size of the capsule endoscope system should be small because small-sized capsules are easier to swallow. Therefore, the foremost challenge is miniaturization to obtain an ingestible device (the volume should be smaller than endoscopy). The availability of small-scale devices can place severe constraints on a design, and the interconnection between them must be optimized. The size constraints lead to another challenge, noise. The coexistence of digital integrated circuits, switching converters for the power supply, and communication circuits in close vicinity of the analog signal conditioning could result in a high level of noise affecting the input signal. Therefore, capsule designers must take great care when selecting and placing components, to optimize the isolation of the front end.

The next vital challenge is to reduce power consumption. In particular, the generated wireless signal must not interfere with standard hospital equipment but still be sufficiently robust to overcome external interferences. On the basis of Friis's formula, the total loss between transmitter and receiver increases with the distance between the transmitting and the receiving antennas increasing. As the result of the dispersive properties of human body materials, the transmitting power absorbed by body varies according to the antenna's operating frequency. The radiated field intensity inside and outside the torso or gut area is determined for FCC regulated medical and Industrial Scientific Medical (ISM) bands, including the 402MHz to 405MHz for Medical Implant Communications Service (MICS), 608MHz to 614 MHz for Wireless Medical Telemetry Service (WMTS), and the 902MHz-928MHz ISM frequency band. Moreover FCC has allocated new bands at higher frequencies such as 1395MHz–1400 MHz wireless medical telemetry services (WMTS) band. Carefully selection of target frequency is important during the antenna design.

The effective data rate was estimated to be about 500 Kbps (Rasouli et al. 2010). The transmit power must be low enough to minimize interference with users of the same band while being strong enough to ensure a reliable link with the receiver module. Lower frequencies are used for ultrasound (100 kHz to 5 MHz) and inductive coupling (125 kHz to 20 MHz). The human body is no place for operational obscurity, so the control software must enforce specific rules to ensure that all devices operate as expected. For that reason, key programs must be developed in a low-level (often assembly) language. The last challenge concern encapsulating the circuitry in appropriate biocompatible materials is to protect the patient from potentially harmful substances and to protect the device from the GI's hostile environment. The encapsulation of contactless sensors (image, temperature, and so on) is relatively simple compared to the packaging of chemical sensors that need direct access to the GI fluids. Obtaining FDA (Food and Drug Administration) approval for the US market or CE (European Conformity) marking in Europe involves additional requirements. Capsules must undergo extensive material-toxicity and reliability tests to ensure that ingesting them causes no harm. The maximal data rate of this transmitter is limited by the RC time constant of the Rdata resistor and the capacitance seen at the base. It is clear that formal frequency higher than 1/(Rdata*Cbase), the modulation index decreases, because the injected base current is shorted in the base capacitance. Although the occupied bandwidth decreases, the S/N ratio decreases too, and robust demodulation becomes more difficult at faster modulation rates. From experiments, the limit was found to be at 2Mbps [22]. Considering the sensitivity of small receivers for biotelemetry, the designed antenna should have a gain that exceeds −20 dB (Chi et al. 2007; Zhou et al. 2009).

4. Special consideration of body characteristics for antenna design

The antenna designed for biomedical telemetry is based on the study of the materials and the propagation characteristics in the body. Because of the different environment, the wave radio propagation becomes different in free space. The human body consists of many tissues with different permittivity and conductivity, which leads to different dielectric properties.

The same radio wave propagating through different media may exhibit different features. From an electromagnetic point of view, materials can be classified as conductive, semi conductive or dielectric media. The electromagnetic properties of materials are normally functions of the frequency, so are the propagation characteristic. Loss tangent defined as the ratio of the imaginary to the real parts of the permittivity, which is equation (Kraus & Fleisch, 1999).

$$tan\delta = \frac{\sigma}{\omega\varepsilon} \tag{1}$$

With the specific classification are given in (Kraus & Fleisch, 1999), the body material is dielectric material. The loss tangent is just a term in the bracket. The attenuation constant is actually proportional to the frequency if the loss tangent is fixed; where the attenuation constant is

$$a = \omega\sqrt{\mu\varepsilon}\left[\frac{1}{2}\left(\sqrt{1+\frac{\sigma^2}{\varepsilon^2\omega^2}}-1\right)\right]^{1/2}. \tag{2}$$

The dominant feature of radio wave propagation in media is that the attenuation increases with the frequency. With the formula

$$\gamma = \sqrt{j\omega\mu(\sigma + j\omega\varepsilon)}, \tag{3}$$

$$E = E_0 e^{j\omega t - \gamma z}, \tag{4}$$

$$H = \frac{j}{\omega\mu}\nabla \times E, \tag{5}$$

It can find out that the power of E plane and H plane reduce with high dielectric constant and conductivity. The total power is consumed easily in human body. The efficiency of antenna becomes lower than free space. With the formula

$$v = \frac{1}{\sqrt{\mu\varepsilon}} \text{ and } \beta = \omega\sqrt{\mu\varepsilon}\left[\frac{1}{2}\left(\sqrt{1+\frac{\sigma^2}{\varepsilon^2\omega^2}}+1\right)\right]^{1/2} = \frac{2\pi}{\lambda}, \tag{6}$$

in a high dielectric material, the electrical length of the antenna is elongated. Compare dipole antenna in the air and in the body material, they have same physical length but electrical lengths are not same. Because of the high permittivity, the antenna in the body material has longer electrical length. The time-averaged power density of an EM wave is

$$S_{av} = \frac{1}{2}\sqrt{\frac{\varepsilon}{\mu}}E_0^2, \tag{7}$$

which leads to high power density in human body. The intrinsic impedance of the material and is determined by ratio of the electric field to the magnetic field (Huang & Boyle, 2008).

$$\eta = \sqrt{\frac{j\omega\mu}{\sigma + j\omega\varepsilon}}. \tag{8}$$

Based on wave equation $\nabla^2 E - \gamma^2 E = 0$, A and B in the wave propagating trigonometric form $E = xA\cos(\omega t - \beta z) + yB\sin(\omega t - \beta z)$ can be determined. With the relationship of A and B, it can confirm shape of polarization.

The multi-layered human body characteristic can be simplified as one equivalent layer with dielectric constant of 56 and the conductivity of 0.8 (Kim & Rahmat-Samii, 2004). So, with the change from free space to body materials, dielectric constant changes from 1 to 56 and conductivity changes from 0 to 0.8. What's more, to detect the transmitted signal independent of transmitter a position, the antenna is required the omni-directional radiation pattern (Kim & Rahmat-Samii, 2004; Chirwa et al., 2003). To investigate the characteristics of antennas for capsule endoscope, the human body is considered as an averaged homogeneous medium as described by the Federal Communications Commission (FCC) and measured using a human phantom (Kwak et al., 2005; Haga et al., 2009).

5. State-of-the-art WCE transmitting and receiving antennas

An antenna plays a very crucial role in WCE systems. Wireless capsule transmitting and receiving antennas belong to wireless communication unit, which provides a bidirectional communication between the capsule and the external communication unit at recommended frequency at which industrial, scientific and medical band was established. Wireless capsule endoscopy transmitting antenna is for sending out the detected signal and receiving antenna receive the signal outside human body. This section is to discuss the current performance of both WCE transmitting and receiving antennas.

5.1 Transmitting antennas

The capsule camera system is shown in Figure 2. One of the key challenges for ingestible devices is to find an efficient way to achieve RF signal transmission with minimum power consumption. This requires the use of an ultra-low power transmitter with a miniaturized antenna that is optimized for signal transmission through the body. The design of an antenna for such a system is a challenging task (Norris et al., 2007). The design must fulfill several requirements to be an effective capsule antenna, including: miniaturization to achieve matching at the desired bio-telemetric frequency; omni-directional pattern very congruent to that of a dipole in order to provide transmission regardless of the location of the capsule or receiver; polarization diversity that enables the capsule to transmit efficiently regardless of its orientation in the body; easy and understandable tuning adjustment to compensate for body effects. Types of transmitting antenna are used such as the spiral antennas, the printed microstrip antennas, and conformal antennas as shown in following subsections.

5.1.1 Spiral antennas

A research group from Yonsei University, South Korea, proposed a series of spiral and helical antennas providing ultra-wide bandwidth at hundreds of megahertz.

Single arm spiral antenna

The first design is a miniaturized normal mode helical antenna with the conical structure (Kwak et al., 2005). To encase in the capsule module, the conical helical antenna is reduced only in height with the maintenance of the ultra-wide band characteristics. Thus, the spiral shaped antenna is designed with the total spiral arm length of a quarter-wavelength. The configuration of the designed antenna is shown in Figure 3(a). It is composed of a radiator and probe feeding structure. The proposed antenna is fabricated on the substrate with 0.5-oz copper, 3 mm substrate height, and dielectric constant of 2.17. The diameter of the antenna is 10.5 mm and 0.5 mm width conductor.

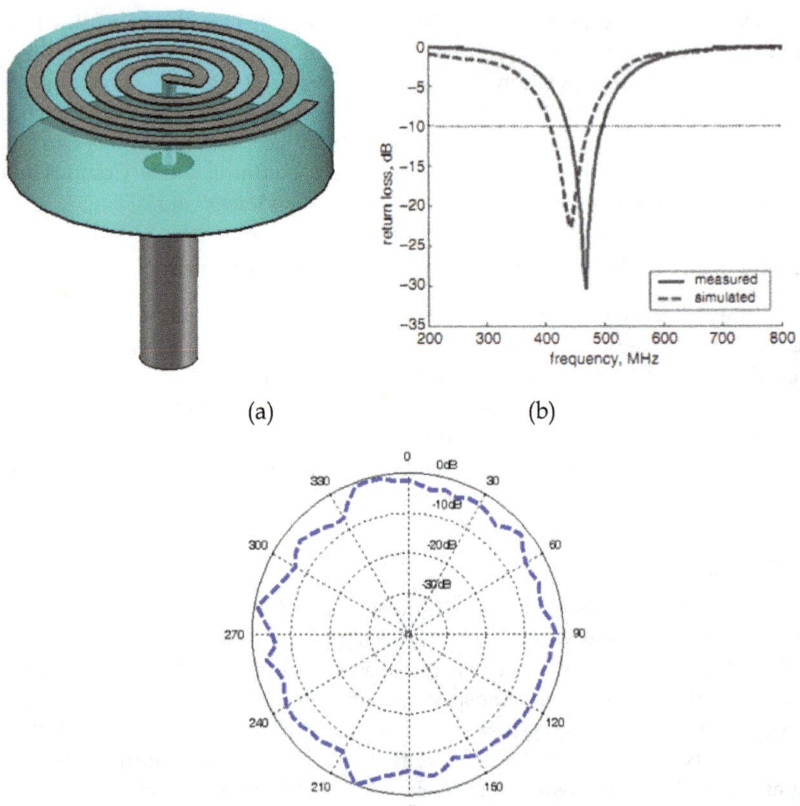

(a)   (b)

(c)

Fig. 3. Single arm spiral antenna (Kwak et al. 2005): (a) the geometric structure; (b) simulated and measured return losses; (c) azimuth pattern at 430MHz.

The simulated and the measured return losses of the antenna surrounded by human body equivalent material are shown in Figure 3(b). It can be observed that the bandwidth of the proposed spiral shaped antenna for S_{11}<-10dB is 110 MHz of 400-510 MHz and the fractional bandwidth is 24.1 %, which is larger than 20%, the reference of the UWB fractional bandwidth. The measurement result of the azimuth radiation pattern is shown in Figure 3(c). The normalized received power level is varying between 0dB to -7dB, which can be considered as an omni-directional radiation pattern.

Dual arm spiral antenna

The dispersive properties of human body suggested that signals are less vulnerable when they are transmitted at lower frequency range. Therefore, a modified design is proposed to provide ultra-wide bandwidth at lower frequency range (Lee et al. 2007). Figure 4(a) shows the geometry of a dual spiral antenna. The newly proposed antenna is composed of two spirals connected by the single feeding line. The radius of designed antenna is 10.1mm and its height is about 3.5mm. To design a dual spiral antenna, two substrate layers are used. The upper and lower substrate layers have the same dielectric constant of 3.5 and the thicknesses of them are both 1.524mm. Two spirals with the same width of 0.5mm and the same gap of 0.25mm have different overall length. The lower spiral antenna is a 5.25 turn structure and the upper spiral is 5 turns.

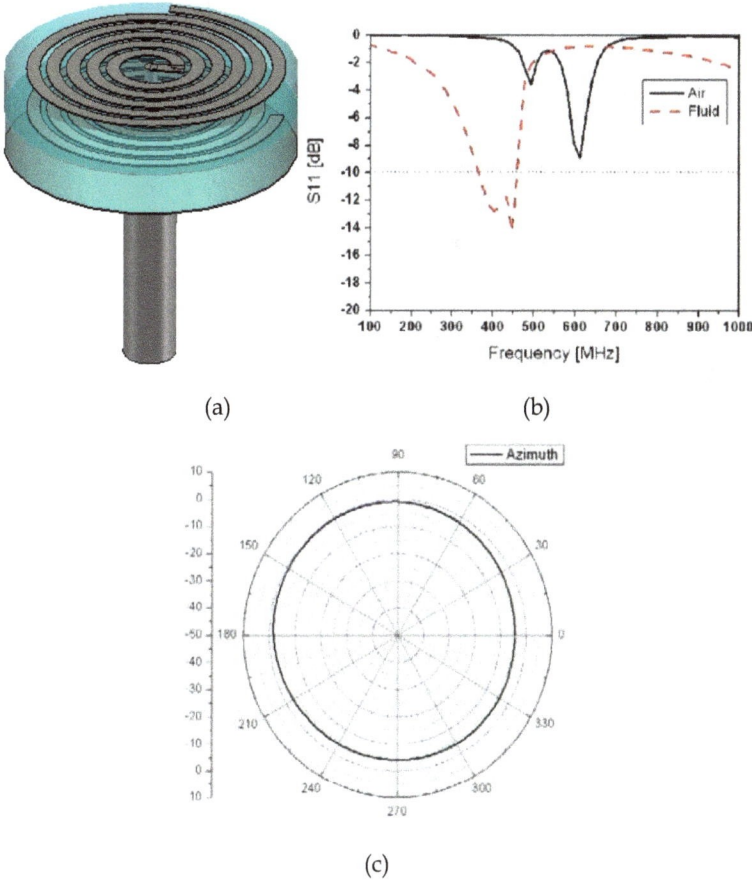

(a)

(b)

(c)

Fig. 4. Dual arm spiral antenna (Lee et al. 2007): (a) the geometric structure; (b) measured return losses; (c) azimuth pattern at 400MHz.

The return loss of the proposed antenna was measured in the air and in the simulating fluid of the human tissue as shown in Figure 4(b). Because of considering electrical properties of equivalent material of human body, return loss characteristic in the air is not good but dual resonant characteristic is shown in the air. However, the proposed antenna has low return loss value at operating frequency in the fluid and its bandwidth is 98MHz (from 360MHz to 458MHz) in the fluid, with the fractional bandwidth of about 25%. The simulated radiation pattern as shown in Figure 4(c) is omni-directional at the azimuth plane with 5dB variation.

Conical helix antenna

Extensive studies of the helical and spiral antennas were conducted with modified geometric structures. For example, a conical helix antenna fed through a 50 ohm coaxial cable is shown in Figure 5. Compared to small spiral antenna, conical spiral takes up much space. However, additional space is not necessary because a conical spiral can use the end space of the capsule as shown in Figure 5(a). The radius of the designed antenna is 10mm and the total height is 5 mm. This size is enough to be encased in small capsule.

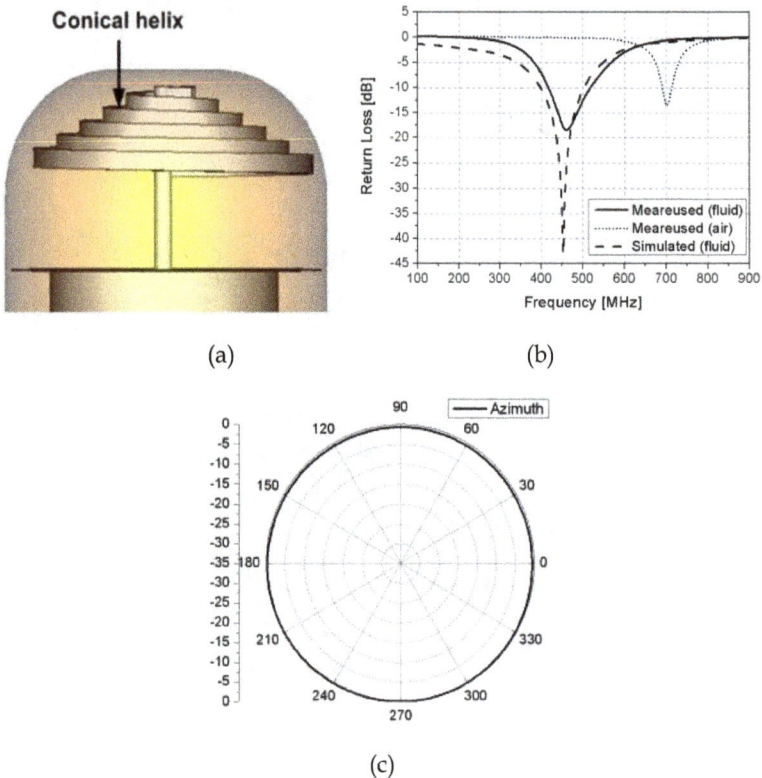

(a)   (b)

(c)

Fig. 5. Conical helix antenna (Lee et al. 2008): (a) the geometric structure; (b) simulated and measured return losses; (c) azimuth pattern at 450MHz.

The proposed antenna provides a bandwidth of 101MHz (from 418MHz to 519MHz) in the human body equivalent material as shown in Figure 5(b). Its center frequency is 450MHz, so the fractional bandwidth is about 22%. The normalized simulated radiation pattern is shown in Figure 5(c). The proposed antenna has omni-directional radiation pattern with less than 1dB variation.

Fat arm spiral antenna

Another modified design is the fat arm spiral antenna as shown in Figure 6(a). The spiral arm is 3mm wide and separated from ground plane with a 1mm air gap. The antenna is simulationally investigated in the air, in the air with capsule shell and in the human body equivalent material.

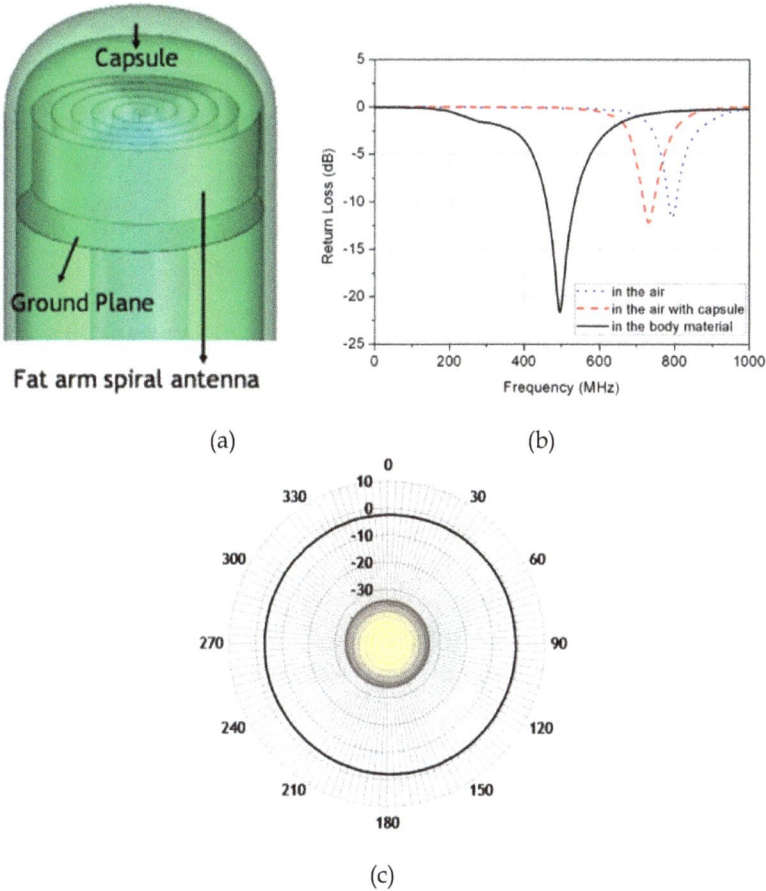

(a)   (b)

(c)

Fig. 6. Fat arm spiral antenna (Lee et al. 2010): (a) the geometric structure; (b) return losses; (c) azimuth pattern at 450MHz.

The return losses of the antenna in free space, with dielectric capsule shell and in the liquid tissue phantom are plotted in Figure 6(b). The resonant frequency is observed about 800 MHz in the air, and reduced to 730 MHz due to the capsule effects on the effective dielectric constant and matching characteristic. When the proposed antenna is emerged in the equivalent liquid, it shows good matching at a resonant frequency and its bandwidth is 75 MHz (460 ~ 535 MHz) for S_{11} less than -10dB. The radiation pattern illustrated in Figure 6(c) presents that this antenna also provides omni-directional feature at azimuth plane.

Square microstrip loop antenna

A square microstrip loop antenna (Shirvante et al. 2010) is designed to operate on the Medical Implant Communication Service (MICS) band (402MHz -405MHz). The antenna is patterned on a Duroid 5880 substrate with a relative permittivity ε_r of 2.2 and a thickness of 500μm as shown in Figure 7(a). The area of the antenna is approximately 25 mm² which is smaller enough to be encased in a swallowable capsule for children.

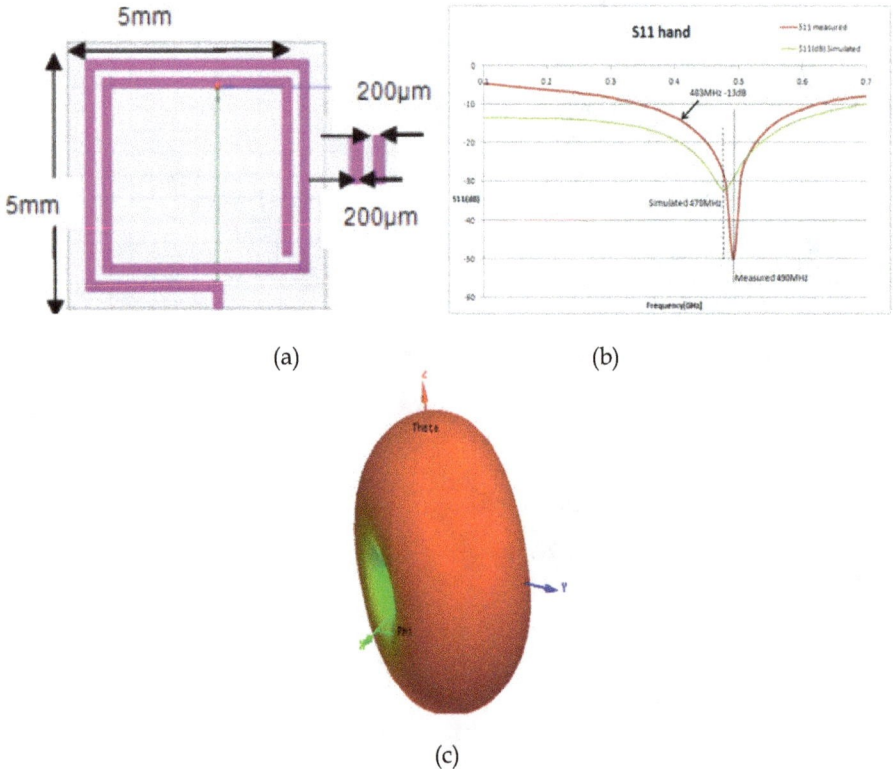

(a)   (b)

(c)

Fig. 7. square microstrip loop antenna (Shirvante et al. 2010): (a) the geometric structure; (b) simulated and measured return losses; (c) azimuth pattern at 403MHz.

The simulated and measured return losses as shown in Figure 7(b) presents that the antenna provides enough bandwidth to cover the 402MHz to 405MHz band. At the FSK operating frequency 403MHz, the measured return loss is -13dB. Moreover, the designed antenna shows a large tolerance to impedance variation at the MICS band, in correspondance to ε_r variation. The designed antenna also has an omni-directional radiation pattern at azimuth plane.

5.1.2 Conformal antennas

A conformal geometry exploits the surface of the capsule and leaves the interior open for electrical components including the camera system. Several designs made efficient usage of the capsule shell area are selected as examples and introduced in this subsection.

Conformal chandelier meandered dipole antenna

The conformal chandelier meandered dipole antenna is investigated as a suitable candidate for wireless capsule endoscopy (Izdebski et al., 2009). The uniqueness of the design is its miniaturization process, conformal structure, polarization diversity, dipole-like omni-directional pattern and simple tunable parameters (as shown in Figure 8(a)). The antenna is offset fed in such a way that there is an additional series resonance excited in addition to the parallel resonance (as shown in Figure 8(b)). The two arms with different lengths generate the dual resonances. This additional series resonance provides better matching at the frequency of interest. This antenna is designed to operate around 1395MHz – 1400 MHz wireless medical telemetry services (WMTS) band.

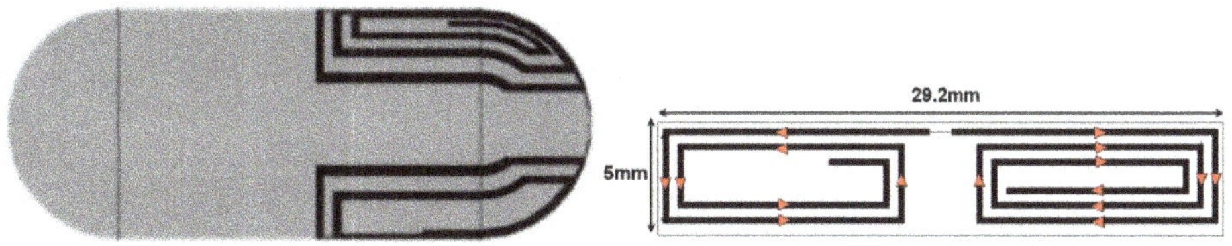

(a)   (b)

Fig. 8. Conformal chandelier meandered dipole antenna (Izdebski et al., 2009): (a) the geometric structure of the conformal chandelier meandered dipole antenna; (b) Offset Planar Meandered Dipole Antenna with current alignment vectors.

The offset planar meandered dipole antenna is simulated on a 0.127 mm thick substrate with a dielectric constant of 2.2. The antenna is placed in the small intestine and it is observed that there is a lot of detuning due to the body conductivity and the dielectric constant (average body composition has a relative permittivity of 58.8 and a conductivity of 0.84S/m). The series resonance shifts closer to 600 MHz. The antenna is then retuned to the operational frequency of 1.4 GHz by reducing the length of the dipole antenna. The return losses of both the detuned and tuned antenna are shown Figure 9(a). Figure 9(b) shows the radiation pattern of the tuned antenna inside the human body at 1.4 GHz.

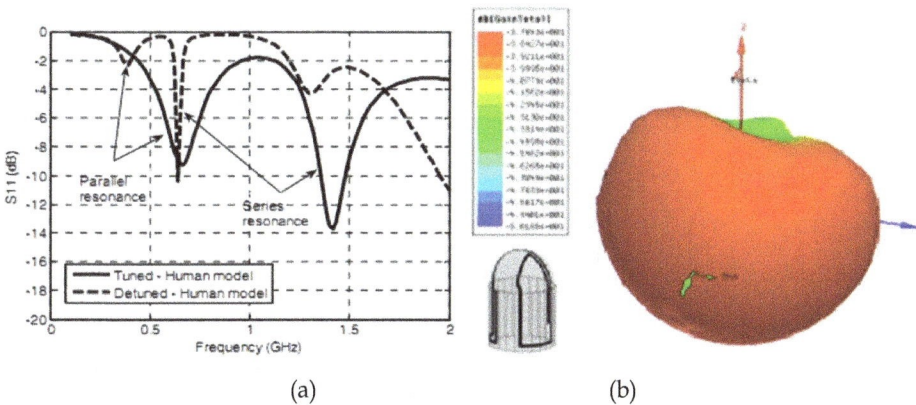

(a)   (b)

Fig. 9. Conformal chandelier meandered dipole antenna (Izdebski et al., 2009): (a) the return losses of detuned and tuned structure in human model; (b) azimuth pattern at 1.4GHz.

The radiation pattern is dipole-like but tilted due to the conformity of the structure. The axial ratio (dB) for the conformal chandelier meandered dipole antenna is about 7dB (elliptical polarization). It possesses all the characteristics of planar structure along with polarization diversity.

Outer-wall loop antenna

The proposed outer-wall loop antenna (Yun et al., 2010.) makes maximal use of the capsule's outer surface, enabling the antenna to be larger than inner antennas. As shown in Figure 10(a), the antenna is part of the outer wall of the capsule, thus decreasing volume and increasing performance, and uses a meandered line for resonance in an electrically small area. The capsule shell with the relative permittivity of 3.15 has the outer and the inner radius of the capsule as 5.5mm and 5mm, respectively. Its length is 24 mm. The height of the meander line and gap between meander patterns are set to 7mm and 2.8mm, respectively. The opposite side of the loop line is meandered in the same way. Although capsule size is reduced, the radius of sphere enclosing the entire structure of the antenna is increased.

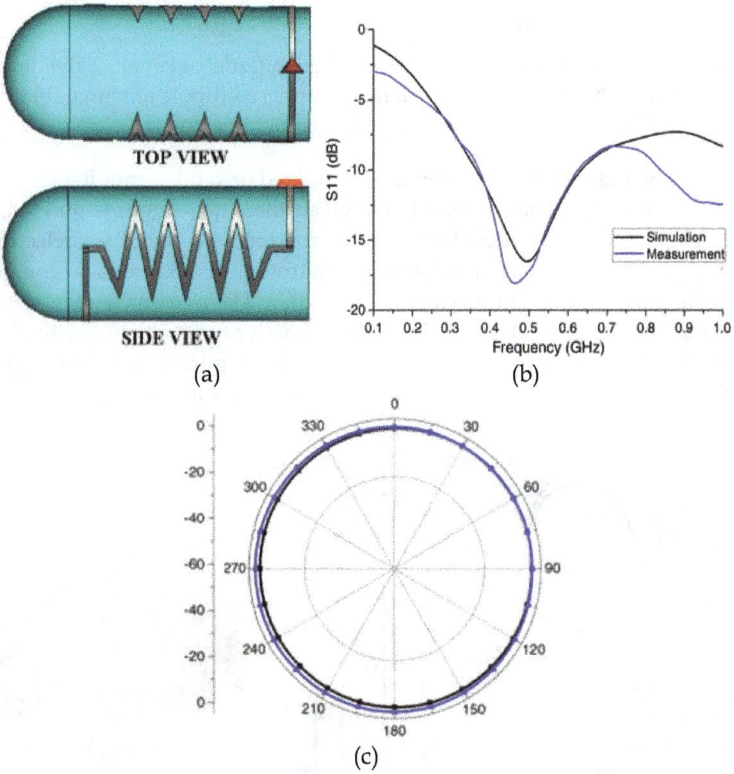

Fig. 10. Outer-wall loop antenna (Yun et al., 2010.): (a) the geometric structure; (b) simulated and measured return losses; (c) azimuth pattern at 500MHz.

Figure 10(b) shows that the proposed antenna has an ultra wide bandwidth of 260 MHz (from 370MHz to 630 MHz) for VSWR<2 and an omnidirectional radiation pattern at azimuth plane (as shown in Figure 10(c)). Using identical antenna pairs in the equivalent body phantom fluid, antenna efficiency is measured to 43.7% (3.6 dB).

5.2 Receiving antennas

The receiving antennas are operating outside of human body, which is no longer limited by its size. Therefore, the design of receiving antennas is less challenge than the design of transmitting antennas. In this subsection, several types of receiving antenna are selected as examples.

Narrow bandwidth antenna for receiver

A narrow bandwidth receiving antenna is designed using microstrip loop structure (Shirvante et al. 2010). The antenna is patterned using a milling machine on a Duroid 5880 substrate with a relative permittivity ε_r of 2.2 and a thickness of 500μm as shown in Figure 11(a). The overall length of the wire is approximately a quarter wavelengths: λair /4 = 187mm at 402MHz for air medium.

(a)   (b)

(c)

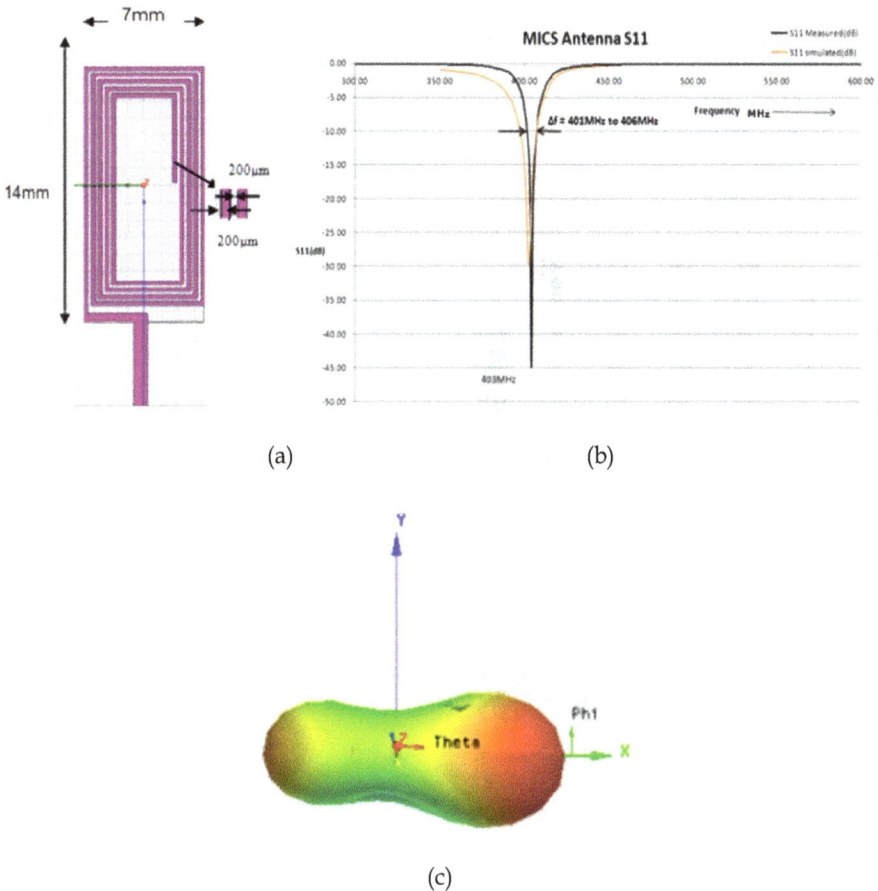

Fig. 11. Rectangular microstrip loop antenna (Shirvante et al. 2010): (a) the geometric structure; (b) simulated and measured return losses; (c) azimuth pattern at 403MHz.

Figure 11(b) shows the simulated and measured return losses of the proposed antenna. The return loss shows a deep null of -30dB at 403MHz. The directional rational pattern as shown in Figure 11(c) provides the possibility to aim the receiver to human body area, where the transmitter sends signals from. Therefore, for narrow bandwidth applications, such as the ASK or FSK modulation, the line loop antenna is a good choice.

Miniaturized microstrip planar antenna

To accommodate the antenna in a small communication unit, a meander line style structure is used (Babar et el., 2009). The antenna's radiating part is shorted with the ground plane, to further decrease the size of the antenna structure. The reduction of the size of the antenna by shortening also reduces the gain of the antenna, as decreasing the size of the antenna more than its wavelength affects the efficiency of the antenna.

The antenna was fabricated on a double sided copper FR4 – printed circuit board, with 1.6mm thickness as shown in Figure 12(a). The excitation is given through an SMA connector from the opposite direction of the PCB to the antenna structure. The total size of the antenna structure is 20mm x 37mm. There is no ground plane present on the opposite side of the PCB, where the antenna structure is present, which helps in getting an omni-directional radiation pattern.

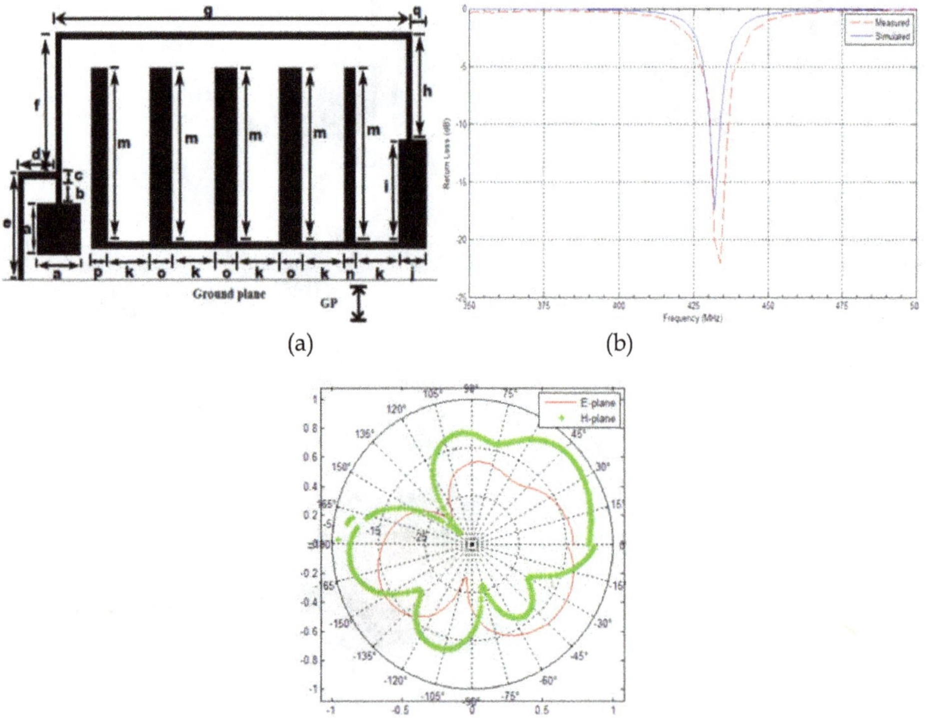

(a)

(b)

(c)

Fig. 12. Microstrip planar antenna (Babar et al. 2009): (a) the geometric structure; (b) simulated and measured return losses; (c) radiation patterns at 433MHz.

Figure 12(b) presents that the operating frequency of the antenna is 433 MHz with the bandwidth of 4MHz. Figure 12(c) shows the radiation pattern of the antenna's E and H-plane. The achieved max gain from the antenna was around -6.1 dBi.

Receiver antenna with buffer layer

The dual pentagon loop antenna having circularly polarization is proposed (Park, S. et al., 2008). The configuration of the proposed dual pentagon loop antenna is shown in Figure 13(a). The proposed antenna and the feeding structure were etched on the front and the back of a substrate (Figure 13(b)). And a-a' are b-b' are shorted as follows. The proposed antenna was designed a dual loop type to enhanced H-field since the current direction of each of loops is different. And there is a gap on each of loops to make a CP wave (Morishita & Hirasawa 1994; Sumi et al., 2004 as cited in Park, S. et al., 2008). The strip widths of the primary loop and of the CPW are 0.80 mm; the used substrate is R/flex 3850; $L1 = 12.93$ mm, $L2 = 10.97$ mm, $L3 = 10.21$ mm, $G = 0.49$ mm, $S1 = 26.01$ mm, $S2 = 1.65$ mm, $W1 = 5.80$ mm, $W2 = 1.70$ mm. The CPW feeding line on the back of substrate is used to efficiently excite balanced signal power which makes to have a broadband.

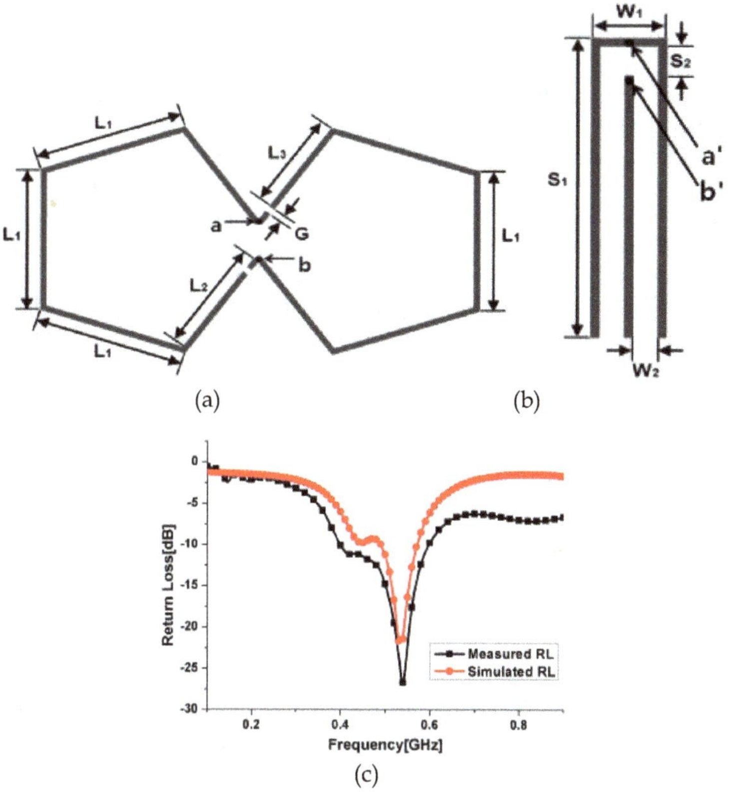

Fig. 13. Receiver antenna with buffer layer (Park, S. et al., 2008): (a) the pentagon dual loop antenna; (b) feeding structure; (c) simulated and measured return losses.

Figure 13(c) presents that the bandwidth of the receiver antenna is from 400 MHz to 600 MHz for VSWR≤2. As a wave in air meets a medium of which relative permittivity is very high over air, much reflection is inevitably generated. So we designed the buffer layer having ε_r between air and human body for reducing the reflection, artificially. The buffer layer which is added a little bit loss is attached on the back of the proposed antenna for reducing a size of antenna and back lobe power.

6. Conclusions

Because of the requirement of medical test for GI tract, WCE came to the world. It solves many restrictions on exploring GI tract. With the development from 2001, WCE has become a promising device with suitable requirement. It has image sensor and lighting, control unit, wireless communication unit, power source, and mechanical actuator. The system can be operated outside the human body, the size of the capsule endoscope system is smaller, and the interconnection between devices was optimized, power consumption also reduced with technology optimized. Some companies and individual are still studying on new functions and optimization.

For wireless capsule endoscopy antenna, several basic standards and situation of operation in human body were discussed. The signal transmission efficiency of the antenna will directly decide the quality of the received real-images and rate of power consumption. Because of the lossy material absorbs a number of waves and decreasing the power of receiving signal, human body presenting strong negative effects on the microwave propagation. Wireless capsule endoscopy transmitting antenna is for sending out the detected signal inside human body and receiving antenna receive the signal outside human body. Several transmitting antennas are introduced in this article. The two fundamental types of transmitting antenna are the spiral antennas and conformal antennas both feature as the small physical size, relatively large bandwidth, omni-directional pattern and polarization diversity. The receiving antennas operating outside of human body are also discussed, such as the narrow bandwidth antenna for receiver, microstrip meandered planar antenna and the receiver antenna with buffer layer. All of them operate well outside the human body.

7. Acknowledgement

This work is supported by the Natural Science Foundation of Jiangsu province (No. BK2010251 and BK2011352), Suzhou Science and Technology Bureau (No. SYG201011), and XJTLU Research Develop Fund (No. 10-03-16.).

8. References

American Cancer Society, (2010). Key statistics about cancers, *Official website of American Cancer Society*, access at Oct. 1st, 2010. <http://www.cancer.org/Cancer/index>

Babar, A. et al., (2009). Miniaturized 433 MHz antenna for card size wireless systems, *Proceeding of Antennas and Propagation Society International Symposium (APSURSI), 2009 IEEE*, Charleston, June, 2009.

Chi, B. et al., (2007). Low-power transceiver analog front-end circuits for bidirectional high data rate wireless telemetry in medical endoscopy applications, *IEEE Trans. Biomed. Eng.,* Vol. 54, No. 7, 2007, pp. 1291–1299.

Chirwa, L.C. et al., (2003). Radiation from ingested wireless devices in biomedical telemetry, *Electronic Letters,* Vol.39, No.2, 2003, pp.178-179.

Eliakim, R. et al., (2006). Evaluation of the PillCam Colon capsule in the detection of colonic pathology: results of the first multicenter, prospective, comparative study. *Endoscopy 2006,* Vol.38, No.10, 2006, pp. 963-970.

Fuyono, I., (2005). Olympus finds market rival hard to swallow, *Nature,* Vol. 438, 2005, p.913.

Gavriel, D. M., (2000). The development of the swallowable video capsule (M2A), *Gastrointestinal Endoscopy,* Vol. 52, No. 6, 2000, pp. 817-819.

Given Imaging, (2010). Overview of product, *Official website of Given Imaging,* access at Sep. 30th, 2011.
<http://www.givenimaging.com/en-int/HealthCareProfessionals/Pages/pageHCP.aspx>

Haga, N. et al., (2009). Characteristics of cavity slot antenna for body-area networks, *IEEE Trans. Antennas Propag.,* Vol. 57, No. 4, 2009, pp. 837–843.

Huang, Y. & Boyle, K., (2008). Radio Wave Propagation Characteristic in Media, *Antennas from Theory to Practice,* pp.93-95.

Iddan, G. G. et al., (2000). Wireless capsule endoscopy, *Nature,* Vol. 405, 2000, pp. 417-418.

IntroMedic, (2010). MicroCam Info, *Official website of IntroMedic,* access at Sep. 30th, 2011.
<http://www.intromedic.com/en/product/productInfo.asp>

Izdebski, P. et al., (2009). Ingestible Capsule Antenna for Bio-Telemetry, *Proceeding of IEEE International Workshop on Antenna Technology (iWAT) 2009,* Santa Monica, March, 2009.

Johnson, C. C. & Guy, A. W., (1972). Nonionizing electromagnetic wave effects in biological materials and systems, *Proceeding of IEEE,* Vol. 60, No. 6, 1972, pp.692–720.

Kim, J. & Rahmat-Samii, Y., (2004). Implanted antennas inside a human body: simulations, designs and characterizations, *IEEE transaction of Microwave theory and techniques, August,* Vol. 52. No. 8, 2004, pp. 1934-1943.

Kraus ,J. D. & Fleisch, D. A., (1999). Electromagnetics with Application, 5th edition, McGraw-Hill, 1999.

Kwak, S. I. et al., (2005). Ultra-wide band spiral shaped small antenna for the biomedical telemetry, *2005 Asia-Pacific Conference Proceedings (APMC),* Suzhou, December, 2005.

Lee, S. H. et al., (2007). A dual spiral antenna for wideband capsule endoscope system, *2007 Asia-Pacific Conference Proceedings (APMC),* Bangkok, December, 2007.

Lee, S. H. et al., (2008). A conical spiral antenna for wideband capsule endoscope system, *Proceeding of Antennas and Propagation Society International Symposium (AP-S) 2008,* San Diego, June, 2008.

Lee, S. H. et al., (2010). Fat arm spiral antenna for wideband capsule endoscope systems, *Proceeding of Radio and Wireless Symposium (RWS) 2010,* New Orleans, LA, January, 2010.

Lenaertes, B. & Puers, R., (2006). An omnidirectional transcutaneous power link for capsule endoscopy, *in Proceedings of International Workshop on Wearable and Implantable Body Sensor Networks,* 2006, pp.46-49.

Mishkin, D. S. et al., (2006). ASGE Technology Status Report, Wireless Capsule Endoscopy, *Gastrointestinal Endoscopy*, Vol. 63, No. 4, 2006, pp. 539-545.

Morishita, H. & Hirasawa, K., (1994). Wideband circularly-polarized loop antenna, *Proceeding of Antennas and Propagation Society International Symposium (AP-S) 1994*, Seattle, 1994.

Norris, M. et al., (2007). Sub miniature antenna design for wireless implants, *Proceedings of the IET Seminar on Antennas and Propagation for Body-Centric Wireless Communication*, London, 2007.

Olympas, (2010). EndoCapsule – Taking capsule endoscopy to next level, *Official website of Olympus*, access at Sep. 30th, 2011.
< http://www.olympus-europa.com/endoscopy/2001_5491.htm>

Park, S. et al., (2008). A New Receiver Antenna with Buffer Layer for Wireless Capsule Endoscopy in human body, *Proceeding of Antennas and Propagation Society International Symposium (AP-S) 2008*, San Diego, June, 2008.

Rasouli, M. et al., (2010). Wireless Capsule Endoscopes for Enhanced Diagnostic Inspection of Gastrointestinal Tract, *Proceeding of Robotics Automation and Mechatronics (RAM) 2010*, Singapore, June, 2010.

Ravens, A. F. & Swain, P., (2002). The wireless capsule: new light in the darkness, *Digestive Diseases*, Vol. 20, No. 2, 2002, pp.127-133.

RF System Lab, (2010), The next generation of capsule endoscopy - Sayaka, *Official website of RF System Lab*, access at Sep. 30th, 2011,
< http://www.rfamerica.com/sayaka/index.html>

Schoofs, N. et al., (2006). PillCam colon capsule endoscopy compared with colonoscopy for colorectal tumor diagnosis: a prospective pilot study. *Endoscopy 2006*, Vol. 38, No.10, 2006, pp. 971-977.

Shirvante, V. et al., (2010). Compact spiral antennas for MICS band wireless endoscope toward pediatric applications, *Proceeding of Antennas and Propagation Society International Symposium (APSURSI), 2010 IEEE*, Toronto, July, 2010.

Sumi, M. et al., (2004). Two rectangular loops fed in series for broad-band circular polarization and impedance matching, *IEEE Transaction on Antennas and Propagation*, Vol. 52, No. 2, pp. 551-554, 2004.

Yu, X. et al., (2006). Microstrip antennas for the wireless capsule endoscope system, *Patent CN 1851982A*, October. 2006.

Yun, S. et al., (2010). Outer-Wall Loop Antenna for Ultrawideband Capsule Endoscope System, *IEEE Antennas and Wireless Propagation Letters*, Vol. 9, pp.1135-1138, 2010.

Zhou,Y. et al., (2009). A wideband OOK receiver for wireless capsule endoscope, *European Microwave Conference 2009*, Rome, October, 2009.

2

Latest Progress in MIMO Antennas Design

Yue Li, Jianfeng Zheng and Zhenghe Feng
Tsinghua University
China

1. Introduction

Multiple-Input Multiple-Output (MIMO) wireless communication system, which is also called Multiple-Antenna system, is well known as one of the most important technologies and widely studied nowadays (Winters, 1987; Foschini & Gans, 1998; Marzetta & Hochwald, 1999; Raleigh & Cioffi, 1998). The main idea of MIMO wireless communication is to utilize the spatial degree of freedom of the wireless multi-path channel by adopting multiple antennas at both transmit and receive ends to improve spectrum efficiency and transmission quality of the wireless communication systems. MIMO technology is able to extremely improve the transmission data rates and alleviate the conflict between the increasing demand of wireless services and the scarce of electromagnetic spectrum. Two famous techniques of the MIMO systems are spatial multiplexing (SM) and transmit diversity (TD) (Nabar et al, 2002). In the scheme of SM, multiple data pipes between transmit and receive ends provide multiplexing gain to dramatically increase the channel capacity linearly with the number of antennas (Telatar, 1999; Bolcskei et al, 2002). The TD technologies, such as space-time coding, are adopted to improve the link reliability of wireless communication, especially in the multi-path fading channels (Marzetta & Hochwald, 1999; Tarokh et al, 1998; Bolcskei et al, 2001). The channel knowledge is not required in the transmit end for TD technologies. MIMO is the key technology for future wireless communication systems, such as 3GPP LTE, WiMAX 802.16, IEEE 802.20, IMT-Advanced and so on.

Although the spatial degree of freedom is important and has the potential to extremely increase the capacity of the MIMO systems, how to utilize the space resources is still needed to be studied. Physical layer design is the most important issue of wireless communication systems. Among all the components, the antenna is the interface of the MIMO wireless communication systems to the channel, which is the most sensitive part for the spatial degree of freedom. The system performance is directly dictated by the number of antennas adopted in transmit and receive end. The key issue to achieve high channel capacity of the MIMO system is the mutual coupling between antenna elements. In traditional MIMO systems, space-separated antenna array is adopted at the base station or mobile terminal. Nearly half of the wavelength is required to achieve acceptable isolation, about -15 dB for most of the situations. However, for the space is limited in both the base station and the mobile terminal, the mutual coupling between the adjacent antenna elements becomes more and more serious, restricts the performance of MIMO systems (Wallace & Jensen, 2004; Morris & Jensen, 2005). The design of antenna in space-limited MIMO system is still need further discussed. This chapter will focus on this topic.

In this chapter, we provide a comprehensive discussion on the latest technologies of antenna design for space-limited MIMO applications, such as minimized base stations, portable access points and mobile terminals. To solve the contradiction of system volume and antenna performance, two basic methods are proposed to maintain the channel capacity in a reduced system volume, as illustrated in Fig. 1. The first one is to reduce to volume each antenna occupied without decreasing the number of antenna elements. The polarization resource is one of the important space resources. Different from the space-separated antennas, the polarization antenna array can utilize the multiple field components to improve the spatial degree of freedom of MIMO systems within a limited space. And the antennas with different polarizations can locate in the same place to save the space occupied. The ports isolation is the challenge for antenna design. Another one is to enhance the antenna performance in the space-limited MIMO system, without increasing the antenna volume. Using switching mechanism, one more polarization or radiation pattern can be selected due to the channel conditions. Based on the adaptive antenna selection, suitable signal processing methods can be adopted alternatively to achieve better performance. The design of switching mechanism is the key issue for carefully consideration.

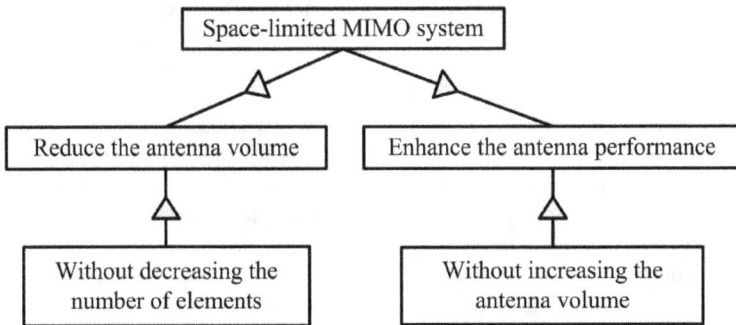

Fig. 1. Technical diagram for antenna design in space-limited MIMO system.

This chapter is organized as following. In Section 2, dual-polarized antenna solution is proposed as an example of 2-element polarization antenna array. Two practical designs are present to show the isolation enhancement between ports. Section 3 describes polarization reconfigurable antenna element based on the Section 2. Channel capacity benefit has been validated by experiment. In Section 4, another type of reconfigurable antenna, pattern reconfigurable antenna element is proposed. Section 5 will give a summery of this chapter.

2. Dual-polarized antenna

In this section, we talk about the polarization resource of antenna. The polarization antenna array has been studied in mobile communications for decades. In 1970s, the polarization characteristics of mobile wireless channel had been widely measured and discussed. The results illustrated that the correlation between feeding ports of different polarization antenna elements must be low to satisfy the requirements of diversity, and the volume occupied is much smaller than the space-separated antennas. Thus, more uncorrelated sub-channels can be obtained by using polarization antenna array. Further, the orientations of the mobile terminals are commonly not perpendicular to the ground. Polarization antenna array is an effective solution to reduce the polarization mismatch. In traditional cellular mobile communication systems, the system with polarization diversity antennas has a 7 dB gain than the one with space diversity antennas in Line-of-Sight scenarios, and a 1 dB gain in Non-Line-of-Sight scenarios (Nakano et al, 2002).

In MIMO systems, the channel capacity of MIMO system with polarization antenna array is approximately 10%~20% higher than that with space-separated co-polarized antenna array, though the system SNR of polarization antenna array is lower (Kyritsi et al, 2002; Wallace et al, 2003). Another measurement results in micro- and pico-cell show the channel capacity of MIMO systems with dual-polarized antenna elements are 14% higher than that with twice-numbered single-polarized antennas (Sulonen et al, 2003). Similar results are also obtained (Erceg et al, 2006). Of course, the dual-polarized antenna element can be treated as a 2-element single-polarized antenna array. For this application, two important issues must be considered: one is the ports isolation, the other one is the antenna dimension. High-isolated compact-volume dual-polarized antenna is our goal of design.

In resent research, different methods of isolation enhancement are introduced. An air bridge, which is utilized in the cross part of two feedings for high isolation, was proposed in (Barba, 2008; Mak et al, 2007). Different feed mechanisms, feed by probe and coupling through aperture, were used in (Guo et al, 2002). Another isosceles triangular slot antenna is proposed for wideband dual-polarization applications in (Lee et al, 2009). TE10 and TE01 modes are excited by two orthogonally arranged microstrips. The above mentioned methods are difficult to be realized in a compact structure and unable to be adopted in space-limited multiple antenna systems. In this section, we introduce two compact antenna designs with good ports isolation.

2.1 Dual-polarized slot antenna

For the purpose to realize dual orthogonal polarizations, slot structure is selected as the main radiator. As shown in Fig. 2, both vertical and horizontal polarizations can exist simultaneously in a rectangular slot. The operating frequency is dictated by the widths of the slot. The slot also has the advantages of wide bandwidth, bi-directional radiation pattern and high efficiency (Lee et al, 2009). However, how to excite these two polarizations is still a question. The traditional method is to feed both polarizations in the same way through two adjacent sides of the slot. Thus, the feeding structure is simple but with large dimension, which isn't able to fulfil our requirement of compact size.

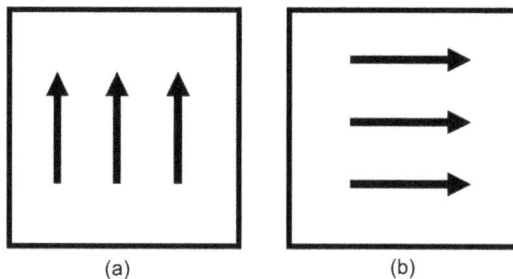

Fig. 2. Polarization mode in slot: (a) vertical polarization, (b) horizontal polarization.

In order to excite dual orthogonal polarizations in a compact structure, we utilized the dual modes of co-planar waveguide (CPW). Fig. 3 shows the geometry of the proposed antenna with CPW feeding structure. The overall dimensions of the antenna are 100x80 mm². The antenna is made of the substrate of FR4 (ε_r=4.4, tanδ=0.01), whose thickness is 1 mm. A 52x50 mm² slot, etched in the front side of light region, serves as the main radiator. In the back side of dark region, an L-shaped microstrip line is fed through port 1. The CPW is fed through port 2 in the front side. As shown in Fig. 4(a), when feeding through port 1, a normal odd mode of CPW is excited to feed the vertical polarization mode. When feeding through port 2, as shown in Fig. 4(b), the mode in the CPW is the even mode as a slot line, which can excite the horizontal polarization mode.

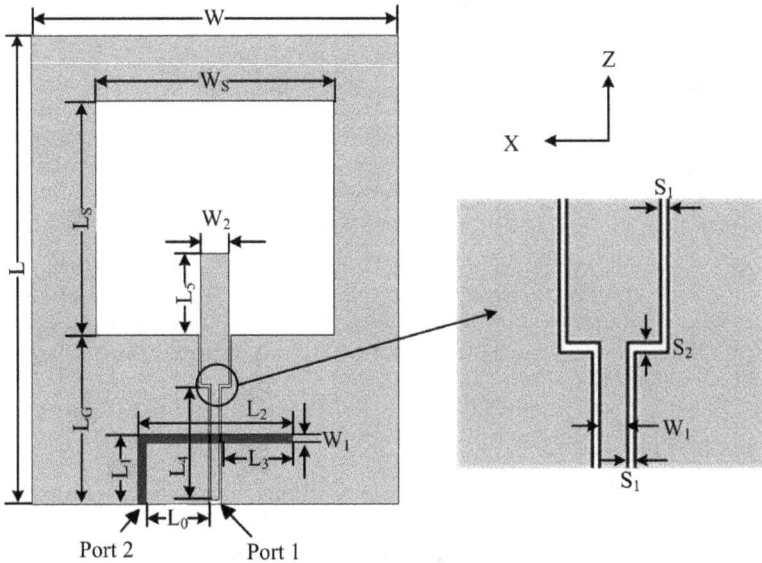

Fig. 3. The geometry of the proposed antenna. (L=100 mm, L_S=50 mm, L_G=36 mm, L_0=15 mm; L_1=15 mm, L_2=32 mm, L_3=12.5 mm, L_4=25.5 mm; L_5=19 mm, W_1=1.9 mm, W_2=6 mm, W_S=52 mm, W=80 mm, S_1=0.35 mm, S_2=0.5 mm. Reprinted from (Li et al, 2010) by the permission of IEEE).

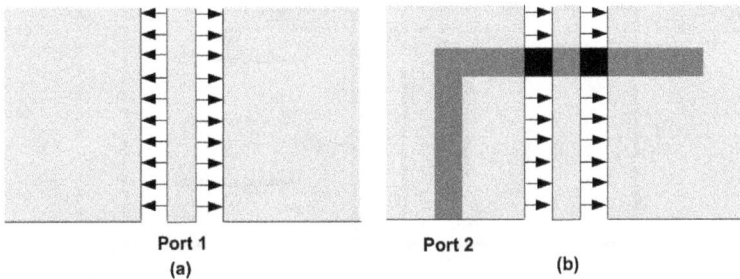

Fig. 4. Feeding modes in CPW: (a) odd mode, (b) even mode.

The current distributions of both polarizations are shown in Fig. 5 for better explanation. A half wavelength distribution appears on each side of slot. Dimensions of L_S and W_S determine the resonant frequencies of the vertical mode and horizontal mode respectively. The L_3 is the tuning parameter for matching port 1. To match port 2, dimensions of W_2, L_5 and L_6 need to be optimized. Due to the symmetric and anti-symmetric characteristics of the two modes in CPW, high isolation can be achieved between two ports. As a result, the feeding structure can excite both polarization modes simultaneously and independently.

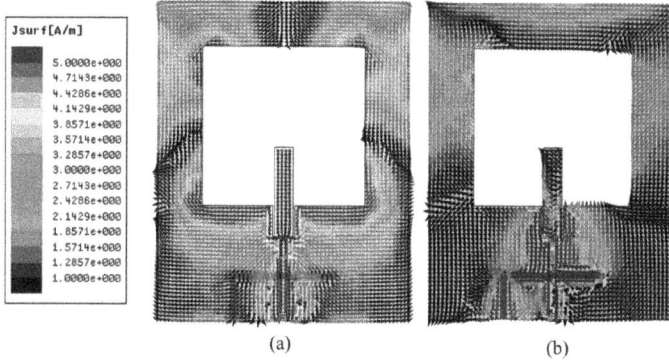

(a)   (b)

Fig. 5. Current distributions of (a) vertical polarization and (b) horizontal polarization.

To validate the design, the S parameters of the proposed antenna are simulated using Ansoft high frequency structure simulator (HFSS). The antenna has also been fabricated and measured. Fig. 6 shows the measured S parameter of the proposed antenna in solid lines, compared with the simulated ones in dash lines. The centre frequencies of the dual polarizations are both 2.4GHz. The bandwidths of -10dB reflection coefficient are 670MHz (1.96-2.63GHz, 27.9%) and 850MHz (1.93-2.75GHz, 35.4%) for horizontal polarization and vertical polarization, respectively. Throughout the WLAN frequency band (2.4-2.484GHz), the isolation between two ports in the required band is lower than -32.6dB. These results show that the proposed antenna is simpler, more compact than the references (Barba, 2008; Mak et al, 2007; Lee et al, 2009).

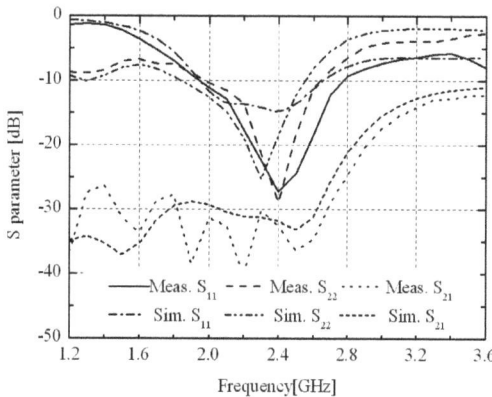

Fig. 6. Simulated and measured S parameters of the proposed antenna.

The radiation patterns of the proposed antenna when feeding through port 1 and 2 are shown in Fig. 7 and Fig. 8. For port 1, the vertical polarization case, the 3dB beam widths are

Fig. 7. Measured and simulated radiation patterns when feeding from port 1 at 2.4 GHz: (a) X-Y plane (b) Y-Z plane.

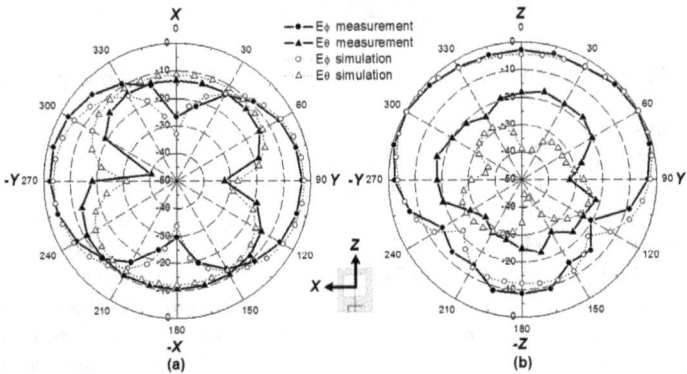

Fig. 8. Measured and simulated radiation patterns when feeding from port 2 at 2.4 GHz: (a) X-Y plane (b) Y-Z plane.

100° and 70° in E-plane (Y-Z plane) and H-plane (X-Y plane). From these results it may be noted that the cross polarization in X-Y plane is worse than what was achieved in earlier designs as values for cross polarization are not lower than -15dB. From the radiation patterns, however, we can observe that the poles of E_φ and E_θ are almost corresponding to the maximum of each other, which means the integration of the two patterns is close to zero. In other words, the signals of co and cross polarizations are almost uncorrelated. In the Y-Z plane, the cross polarization level is sufficiently low to be ignored. For port 2, the horizontal polarization is the dominant polarization. The 3dB beam widths are 60° and 180° in E-plane (X-Y plane) and H-plane (Y-Z plane). From the above discussion, we may conclude that the signals received by the two ports are uncorrelated, so dual-polarization in single antennas can be treated as two independent antennas. The radiation efficiency and gain of the proposed antenna are also measured. In the WLAN band of 2.4-2.484GHz, the efficiency is better than 91.2% and 84.4% for port 1 and 2; and the gain is better than 3.85 dBi and 5.21

dBi for port 1 and 2. The proposed antenna is a candidate for compact volume dual-polarized antenna application.

2.2 Dual-polarized loop antenna

The half wavelength resonant structure, such as the patch and the slot, is able to be adopted in dual-polarized antenna design. In order to realize even more compact dimension, we choose the loop antenna, whose circumference is one wavelength. The radiation patterns of the slot and the loop are almost the same. Also, the loop element can support two orthogonal polarizations using the same structure, shown in Fig. 9. Seen from these two modes, the current distribution is 90° rotated from one to another one. Good orthogonality is illustrated with high isolation. The current distribution of its one–wavelength mode is dictated by the feeding position, and feed should not be arranged at the position of the current null. However, the maximum point of one mode is the null of the other mode. It is difficult to feed the dual polarizations in one side of loop.

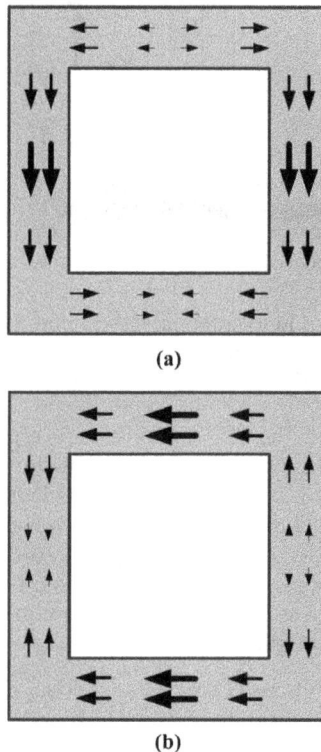

(a)

(b)

Fig. 9. Modes in loop antenna: (a) vertical polarization, (b) horizontal polarization.

The feeding method should be considered carefully. In order to excite two orthogonal one-wavelength modes, it is common to arrange two feeds at two orthogonal positions, which will make the overall dimension much larger. A compact size could be realized if such two modes of operation are fed at only one position. The compact CPW feed backed with microstrip line adopted in the last design is an effective solution to feed the dual-mode of loop antenna. Fig. 10 shows the geometry of the loop antenna, which is quite similar as the slot design. This antenna consists of a rectangular loop, a CPW feeding and a microstrip line, and supported by the same FR4 board as last design with the thickness of 1 mm. The loop has width of 4 mm; narrower than the slot design. The loop and CPW are etched on the front side and the microstrip line is printed on the back side.

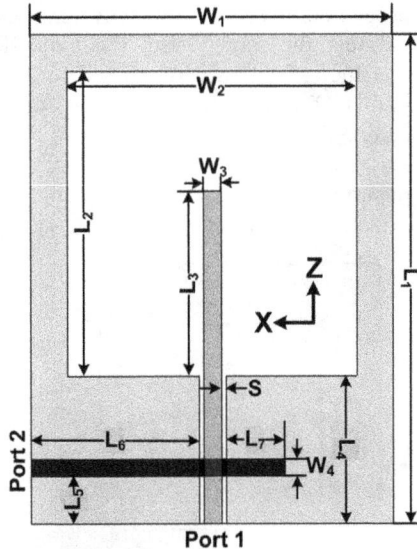

Port 1

Fig. 10. Geometry of the proposed loop antenna. (L_1=53 mm, L_2=33 mm, L_3=20 mm, L_4=16 mm; L_5=5.1 mm, L_6=18.5 mm, L_7=6.5 mm, W_1=40 mm, W_2=32 mm, W_3=2 mm, W_4=1.9 mm, S=0.5 mm. Reprinted from (Li et al, 2011a) by the permission of IEEE).

When the loop fed through port 1, the CPW operates at its typical symmetrical mode. In this mode the vertical polarization is excited. The inner conductor works as a monopole with the vertical polarization. The energy is coupled from monopole to the loop, exciting the vertical polarization mode. It is a good solution to feed the one-wavelength mode at the position of current null. The radiation consists of two modes, the one-wavelength mode of the loop and a monopole mode. When the loop is fed through port 2, the horizontal polarization of the loop antenna is excited. The feed is exactly at the maximum of current, and the horizontal mode is clearly excited in this configuration.

Fig. 11 shows the current distributions of two polarizations, which are totally different from the slot antenna. For the same application of 2.4 GHz WLAN in last design, the rectangular slot is etched in a large ground. The slot's length and width are approximately half wavelength. For a typical slot mode, the width of extended ground is a quarter of wavelength or smaller. If the size of surrounded ground decreases to some level, the slot turns to be a loop mode with the frequency shift. What's more, a loop has four edges with the overall dimension of the loop antenna is 40x53 mm², including the feeding structure. The slot antenna is with the dimension of 100x80 mm². It is clear that the area of the proposed antenna is only 26.5% of the slot one. Fig. 12 (a) and (b) show the loop antenna, in front and back views, respectively. Fig 12(c) shows the slot antenna design, which also operates in the same band. A significant size reduction is achieved using the loop design.

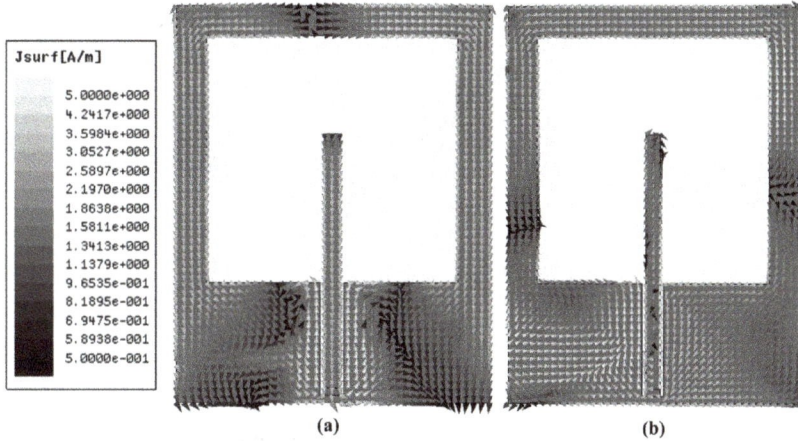

Fig. 11. Current distributions of (a) vertical polarization and (b) horizontal polarization.

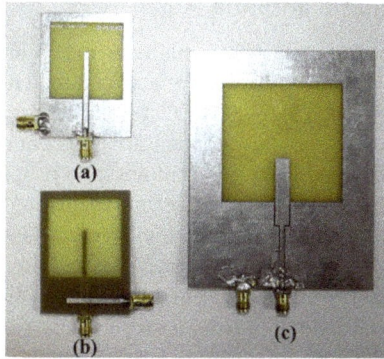

Fig. 12. Photograph of the loop antenna (a) front side, (b) back side and (c) the slot antenna.the total length is one wavelength. Therefore, the dimension of a rectangular loop antenna is much smaller than the slot design with large ground. However, the slot antenna can be adopted in the array design in the same ground for special requirements.

The measured and simulated S parameters are illustrated in Fig. 13. The -10 dB bandwidth of the reflection coefficients are 770 MHz (1.98-2.75 GHz, 32.1%) for the vertical polarization and 730 MHz (1.96-2.69 GHz, 30.4%) for the horizontal polarization, both covering the of 2.4 GHz WLAN band. The isolation in this band is better than -21.3 dB, which is lower than the slot design, as a cost of dimension reduction. The isolation deterioration is mainly contributed to the feeding structure of the vertical polarization. The feeding monopole is located at the current maximum point of the horizontal mode. The energy couples between two modes. But it still fulfils the -15 dB industrial requirement.The radiation pattern s of the loop antenna is quite similar to the slot antenna, but with a lower level of cross polarization. In the 2.4 GHz WLAN band, the measured gains are better than 2.9 dBi and 4.1 dBi.

Considering the compact structure of loop, this antenna is suitable for the space-limited systems.

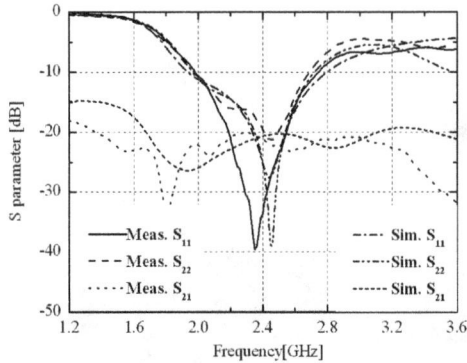

Fig. 13. Simulated and measured S parameters of the loop antenna.

3. Polarization reconfigurable antenna

As described in the introduction, reconfigurable antenna is an effective solution for the space-limited MIMO systems by adaptive antenna selection. This kind of systems is called adaptive MIMO system. The adaptive MIMO system takes the advantage of varying channel characteristics to make the best use of the improvement of channel capacity (Cetiner et al, 2004). Due to the channel condition, different antenna properties, such as polarizations and radiation patterns, are selected for better transmitting or receiving. Also, different data processing algorithms are used depending on the antenna selection. For this reason, the reconfigurable antenna is very important to the MIMO system, especially for the space-limited system. In this section, we will introduce the polarization reconfigurable antenna, based on the dual-polarized slot antenna described in the last section. In order to validate the benefit of polarization selecting, the channel capacity of a 2x2 MIMO system using the polarization reconfigurable antenna has been measured in a typical indoor scenario.

3.1 Reconfigurable mechanism

The geometry of the proposed reconfigurable slot antenna element is shown in Fig. 14, based on the design of (Li et al, 2010). The port 1 and port 2 are combined together and controlled by two PIN diodes. The port 1 is connected the microstrip line on the back side through a via hole, and controlled by PIN 1. The port 2 is connected directly to the microstrip line on the back side, and controlled by PIN 2. When PIN1 is ON and PIN2 is OFF, the antenna is fed through the port 1. The vertical polarization of the slot is excited. When PIN1 is OFF and PIN 2 is ON, the antenna is fed through the port 2, and the horizontal polarization of the slot is excited. Therefore, two ports are fed alternatively and controlled by the PIN diodes. The two PIN diodes need the bias circuit to control. Due to compact feed design, the two PIN diodes share the same bias circuit, saving the space of the antenna system and using less lumped components.

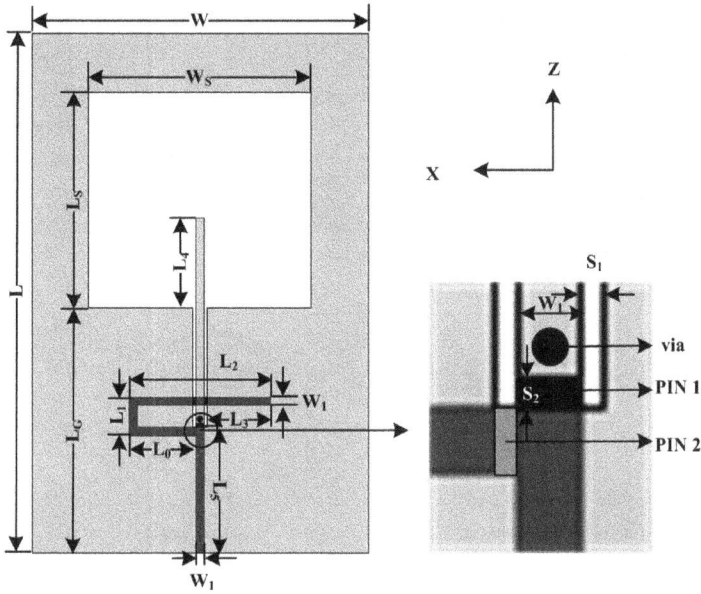

Fig. 14. Geometry of the proposed loop antenna. (L=120 mm, L_S=50 mm, L_G=36 mm, L_0=16 mm; L_1=8.9 mm, L_2=33.9 mm, L_3=15.3 mm, L_4=20.1 mm; L_5=30 mm, W_1=1.9 mm, W_S=53 mm, W=80 mm, S_1=0.7 mm, S_2=1 mm. Reprinted from (Li et al, 2011b) by the permission of John Wiley & Sons, Inc.).

A prototype of the dual-polarized slot antenna with switching mechanism is fabricated, and shown in Fig. 15. The PIN diodes with bias circuit are on the back side of the antenna. The detailed bias circuits of two PIN diodes (D1 and D2, Philips BAP64-03) are shown in Fig. 15 (c). The 'ON' and 'OFF' states of the two PIN diodes are controlled by a 1-bit single-pole 2-throw (SP2T) switch on the front side. The bias circuit consists of three RF choke inductors (L_{b1} L_{b2} and L_{b3}, 12 nH), a DC block capacitor (C_b, 120 pF), three RF shorted capacitors (C_{s1}

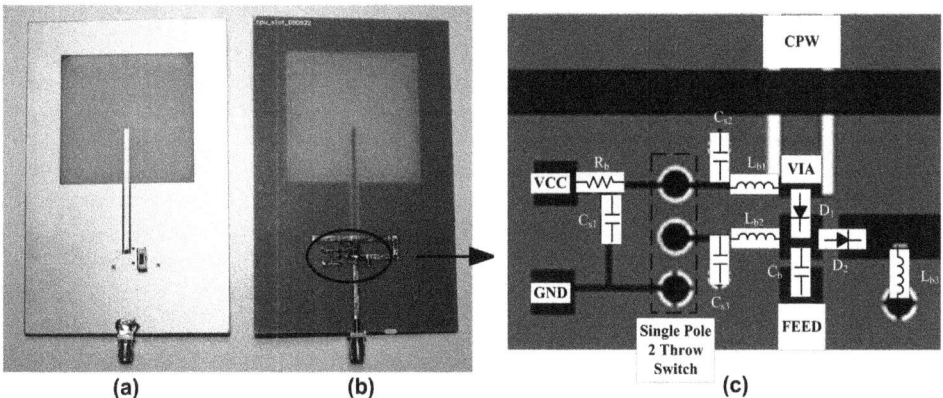

Fig. 15. Photograph of the antenna prototype (a) front side, (b) back side; (c) bias circuit of the PIN diodes.

C_{s2} and C_{s3}, 470 pF) and a bias resistor (R, 46 Ω). The bias resistor is selected depend on the value of VCC and the operating current of the PIN diode. In this application, the VCC is 3 V.

The measured reflection coefficients for both polarizations are shown in Fig. 16. Compared with results of the dual-polarized slot antenna in Fig.13, the difference is mainly contributed from the parasitic parameters of PIN diodes and the bias circuit. The -10dB bandwidths are 700MHz (2.02-2.72 GHz, 29.2%) and 940MHz (1.84-2.78 GHz, 40%) for vertical and horizontal polarizations, both covering the WLAN band (2.4-2.484 GHz). The gain decreases approximately 0.5 dB due to the insertion loss of PIN diodes.

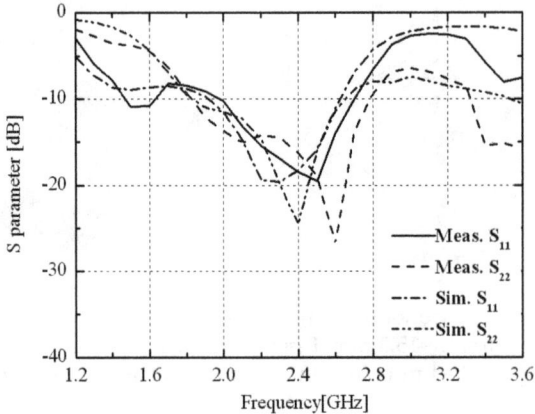

Fig. 16. Simulated and measured S parameters of the reconfigurable antenna.

3.2 Channel capacity measurement

In this section, we measured the channel capacity of a 2x2 MIMO system in a typical indoor scenario by using the proposed polarization reconfigurable antenna. The measurement setup is shown in Fig.17. The measurement system consists of an Agilent E5071B Vector network analyzer (VNA), which has 4 ports for simultaneous measurement, transmit and receive antennas, a computer and RF cables. Two standard omni-direcitional dipoles are utilized as the transmit antennas (TX), and arranged perpendicular to XY plane along Z axis. Two proposed reconfigurable antennas are used as the receive antennas (RX). The 2x2 antennas are connected to the 4 ports of the VNA. The computer is used to control the measurement procedures and record the measured channel responses. In order to validate the improvement in channel capacity by using reconfigurable antennas, another two reference dipoles are adopted as receive antennas for comparison. The measurement was carried out in a room of the Weiqing building, Tsinghua University, illustrated in Fig. 18. The framework of the room is reinforced concrete, the walls are mainly built by brick, and the ceiling is made with plaster plates with aluminium alloy framework. The heights of desk partition and wood cabinet are 1.4 m and 2.1 m. The transmit antennas are fixed in the middle of room (TX). The receive antennas are arranged in several typical locales which are noted as RX1-5 in Fig. 20. Here, the scenarios when the receive antennas are arranged in RX1 and RX2 are line-of sight (LOS), while that is NLOS when the receive antennas are arranged in RX3, RX4 and RX5. In this measured, the antennas used are fixed at the height of 0.8 m. The space of antenna elements in TX or RX is 0.5λ, with the mutual coupling less than -25dB.

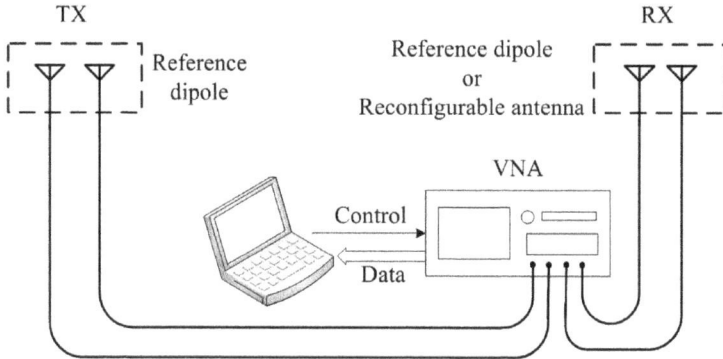

Fig. 17. Experiment setup of the measurement.

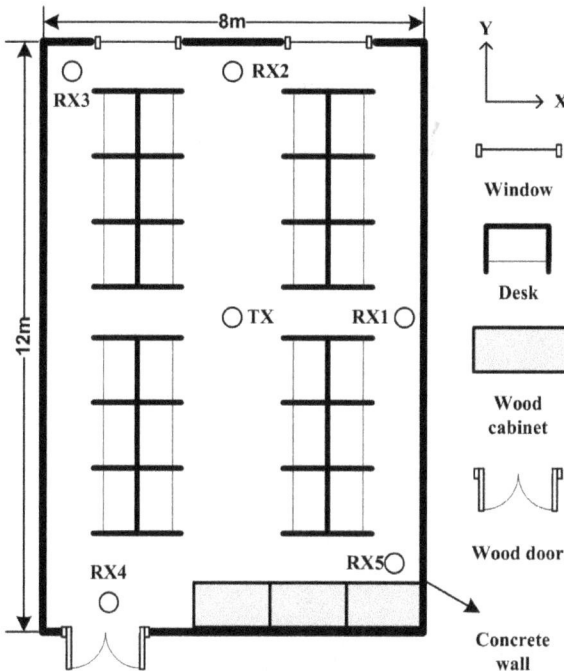

Fig. 18. Layout of measurement environment.

The measurement was carried out in the band of 2.2-2.6 GHz, with a step of 2 MHz. Three different orientations (ZZ, YY, and XX) of RX antennas were measured to simulate different operational poses of the mobile terminals. For two horizontal (H) and vertical (V) polarizations reconfigurable antennas, 4 configurations (HH, HV, VH, VV) were switched manually for each channel capacity measurement in a quasi-static environment, and the result with the biggest value was chosen for statistics. Given the small-scale fading effect, 4x4 grid locations for each RX position were measured. Therefore, a total 201x3x16x2=19296 measured channel capacity for LOS condition was obtained, and 201x3x16x3=28944 was the measured results for NLOS condition.

The channel capacity can be calculated through following formula (Foschini & Gans, 1998):

$$C = \log_2 \det[I_{N_r} + \frac{SNR}{N_t} H_n H_n^H]$$

(1)

where N_r and N_t are the numbers of RX and TX antennas. I_{Nr} is a $N_r \times N_r$ identity matrix, SNR is the signal-to-noise ratio at RX position, H_n is the normalized H, and $()^H$ is the Hermitian transpose. H is normalized by the received power in the 1x1 reference dipole with identical polarization. We selected the SNR when the average channel capacity is 5 bit/s/Hz in a 1x1 reference dipole system in LOS or NLOS scenario.

The measured Complementary Cumulative Distribution Functions (CCDF) of the channel capacity for the 2x2 MIMO system using polarization reconfigurable antennas in both LOS and NLOS conditions are shown in Fig. 19 and 20. As summarized in Table 1, the average and 95% outage channel capacities are both improved, especially in NLOS scenario. For NLOS, the received signal is mainly contributed from reflection and diffraction, which vary the polarization property of the wave. However, the path loss is higher in NLOS scenario. The transmit power should be enhanced to guarantee the system performance. Considering the insertion loss introduced from non-ideal PIN diodes, better capacity can be obtained by using high quality components. The measurement results prove the benefit by using polarization reconfigurable antennas.

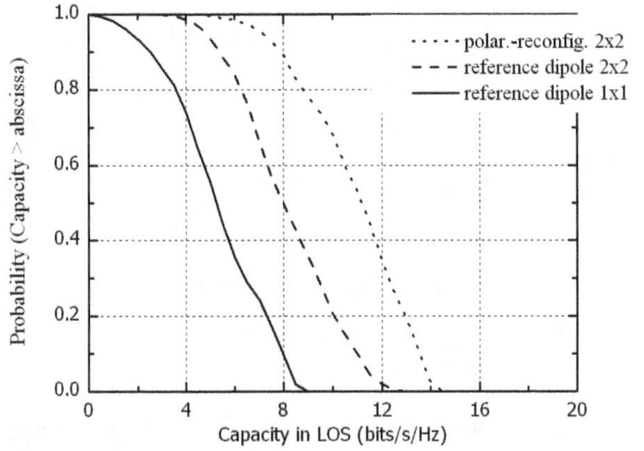

Fig. 19. CCDFs of channel capacity in LOS condition.

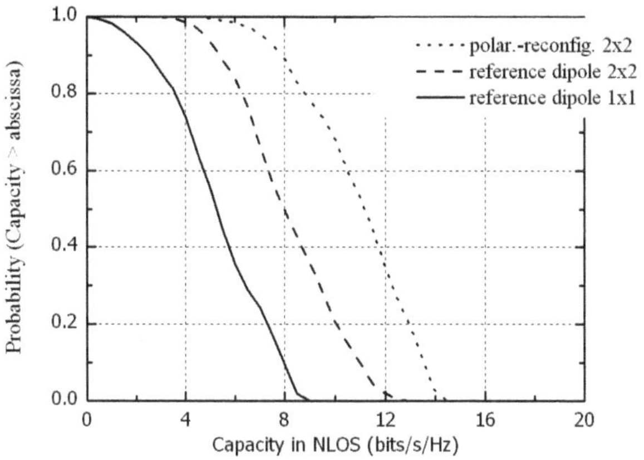

Fig. 20. CCDFs of channel capacity in NLOS condition.

Channel capacity	Condition	1x1 dipole	2x2 dipole	2x2 polar.-reconfig.
Average	LOS	5	7.86	10.62
	NLOS	5	9.9	13.18
95% outage	LOS	1.75	4.91	7.11
	NLOS	1.94	6.87	11.32

Table 1. Average and 95% Outage Channel Capacity (bit/s/Hz).

4. Pattern reconfigurable antenna

Pattern reconfigurable antenna is another type of reconfigurable antenna. Such antenna provides dynamic radiation coverage and mitigates multi-path fading. In this section, we introduce a design of pattern reconfigurable antenna with compact feeding structure. The benefit by using pattern reconfigurable antennas in the MIMO system is also proved by experiment of channel capacity measurement.

The configuration of the pattern reconfigurable antenna is shown in Fig. 21 (a). It is composed of an elliptical topped monopole, two Vivaldi notched slots and a typical CPW feed with 2 PIN diodes. The antenna is printed on the both sides of a 50 x 50 mm² Teflon substrate, with ε_r=2.65, tanδ=0.001 and thickness is 1.5 mm. The CPW is connected to the microstrip at the back side through several via holes. A 0.2 mm wide slit is cut from the ground on the front side for DC isolation. Three curves are used to define the shape of antenna, fitted to the coordinates in Fig. 21 (a). Curve 1 is defined by equation (2) and curve 2 is defined by equation (3). Curve 3 and curve 2 are symmetrical along X axis.

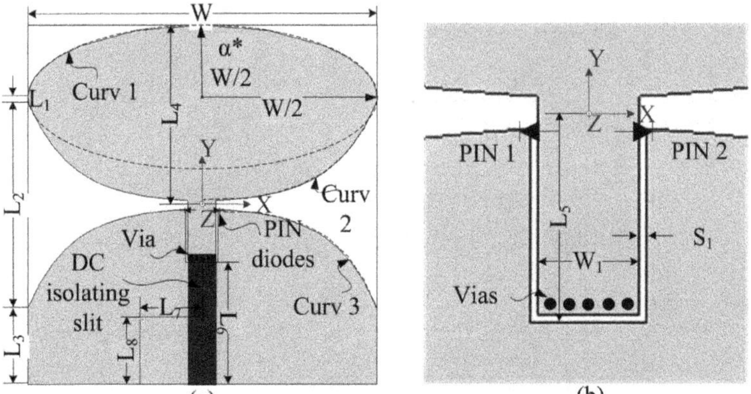

(a)   (b)

Fig. 21. Geometry of the proposed loop antenna. (L_1=1.74 mm, L_2=28.52 mm, L_3=10.74 mm, L_4=25 mm; L_5=8 mm, L_6=16.8 mm, L_7=10mm, L_8=10mm, W=50 mm, L_p=2 mm, W_1=4 mm, S_1=0.3 mm. Reprinted from (Li et al, 2010c) by the permission of IEEE).

$$\left(\frac{x}{W/2}\right)^2 + \left[\frac{y-(L_4-\alpha*W/2)}{\alpha*W/2}\right]^2 = 1 \tag{2}$$

where $L_4 - \alpha*W/2 \le y \le L_4$, and $\alpha = 0.4$.

$$y = C_1 e^{c \cdot x} + C_2 \tag{3}$$

where C_1=14, C_2=0.26, c=0.16.

4.1 CPW-slot transition design

Different radiation patterns are provided by different work states of the same antenna. In order to achieve different work states, a switchable CPW-to-slotline transition with two PIN diodes is proposed and sown in Fig. 21 (b). Three feeding modes are achieved in this structure by varying the states of PIN diodes. When both PIN diodes are OFF, the elliptical topped monopole is fed through a typical CPW and a nearly omni-directional radiation pattern is achieved in XZ plane. When PIN 1 is OFF and PIN 2 is ON, the right slotline is shorted. The left Vivaldi notched slot is fed through the left slotline (LS) of the CPW, and a unidirectional radiation pattern is formed along the –X axis. In the same way, when PIN 1 is ON and PIN 2 is OFF, a unidirectional beam along the +X axis is obtained in the right Vivaldi notched slot through the right slot (RS). The proposed CPW-to-slotline transition is able to achieve good switching from the CPW to slotline with any other extra structures. Compared with this design, the CPW-to-slotline transition reported in (Wu et al, 2008; Kim et al, 2007; Ma et al, 1999) all required extra structures for mode convergence, including $\lambda/2$ phase shifter (Ma et al, 1999)and $\lambda/4$ matching structures (Wu et al, 2008; Kim et al, 2007), which occupy considerable space in the feed network. Such structures are not suitable for the space-limited systems. The proposed CPW-to-slotline transition here is designed to reduce the overall dimensions of the antenna.

In order to explain work principle of the feed transition, the equivalent transmission line model is utilized, illustrated in Fig.22 and 23. The PIN diode is expressed as perfect conductor for 'ON' state and open circuit for the 'OFF' state. Fig. 22 (a) shows the normal CPW structure. By tuning the L_5, the radiation resistance $R_{monopole}$ of monopole is matched to 50Ω at the feed port. When the right slot is shorted by PIN diode, the antenna is fed through the RS mode. The diagram and equivalent transmission line model are depicted in Fig. 22 (b). The right slotline is used to feed the Vivaldi notched slot, and the shorted left slotline works as a matching branch. The shorted branch which is less than a quarter of wavelength serves as a shunt inductance and its value is determined by its length L_5. The position of the PIN diode is not fixed, and it is another freedom for impedance matching of RS feed. As shown in Fig. 23, the value of shunt inductance is $jZ_{slot}\tan[\beta slot(L_5-L_p)]$ and used to match the radiation resistance $R_{vivaldi}$. The advantage of this switchable feeding structure is that no extra structure is used in the CPW and slotline transition.

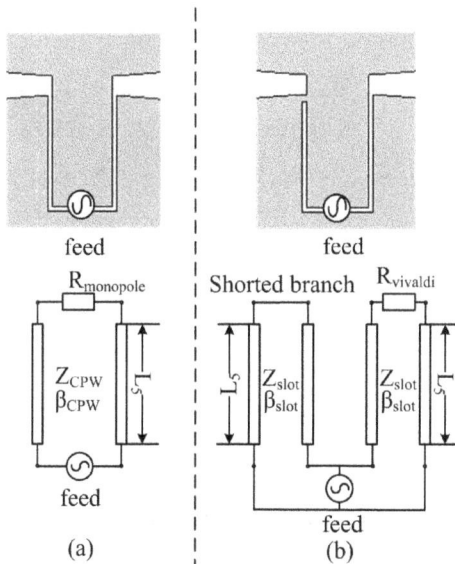

Fig. 22. Feed diagram and equivalent transmission line model. (a) CPW feed; (b) RS feed.

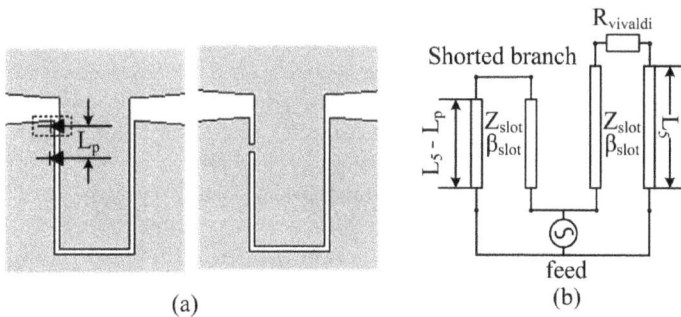

Fig. 23. Matching strategy of RS feed. (a) Feed diagram; (b) Transmission line model.

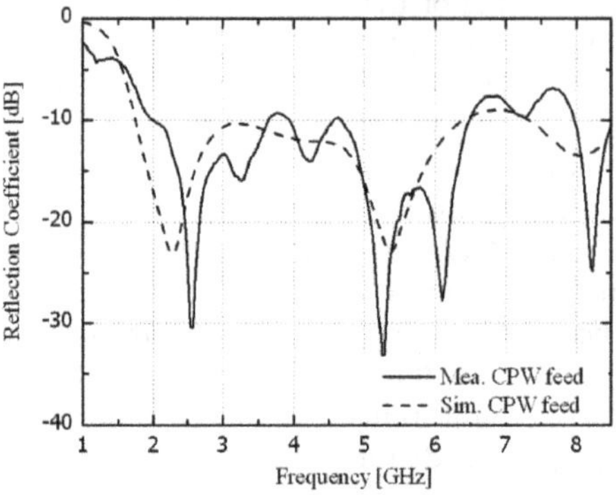

Fig. 24. Simulated and measured reflection coefficient of the reconfigurable antenna.

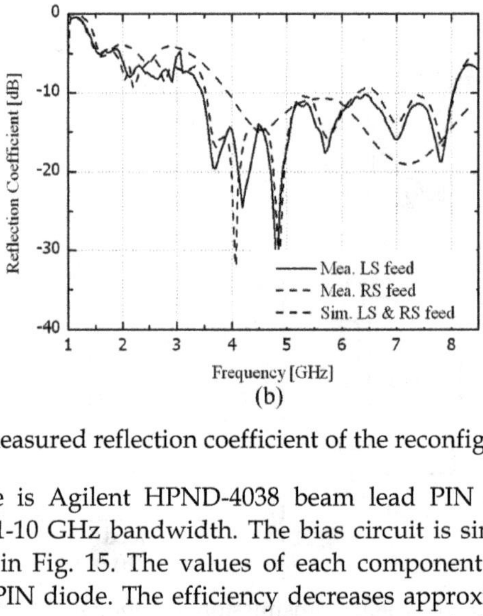

The selected PIN diode is Agilent HPND-4038 beam lead PIN diode, with acceptable performance in a wide 1-10 GHz bandwidth. The bias circuit is similar as the PIN diodes used in the last design in Fig. 15. The values of each component are determined by the working current of the PIN diode. The efficiency decreases approximately 0.3 dB by using this PIN diode. All the measurements were taken using an Agilent E5071B VNA. The simulated and measured reflection coefficients of CPW feed, LS and RS feeds are shown in Fig. 24. The measured -10dB bandwidths are 2.02-6.49 GHz, 3.47-8.03 GHz and 3.53-8.05 GHz for CPW feed, LS feed and RS feed, respectively. The overlap band from 3.53 GHz to 6.49 GHz is treated as the operation frequency for the reconfigurable patterns. The measured normalized radiation pattern in XZ and XY planes for CPW, LS and RS feed at 4, 5, 6 GHz are shown in Fig. 25. For the CPW feed, a nearly omni-directional radiation pattern appears in XZ plane and a doughnut shape in XY plane. For the LS or RS feed, a unidirectional beam appears along –X or +X axis, with acceptable front-to-back ratio better than 9.5dB. For the CPW feed, an average gain in the desired frequency range is 2.92 dBi. For the LS and RS feed, the average gains in the 4-6 GHz band are 4.29 dBi and 4.32 dBi. The improved gain is mainly contributed to the directivity of the slotline feed mode, and the diversity gain is achieved by switching the patterns.

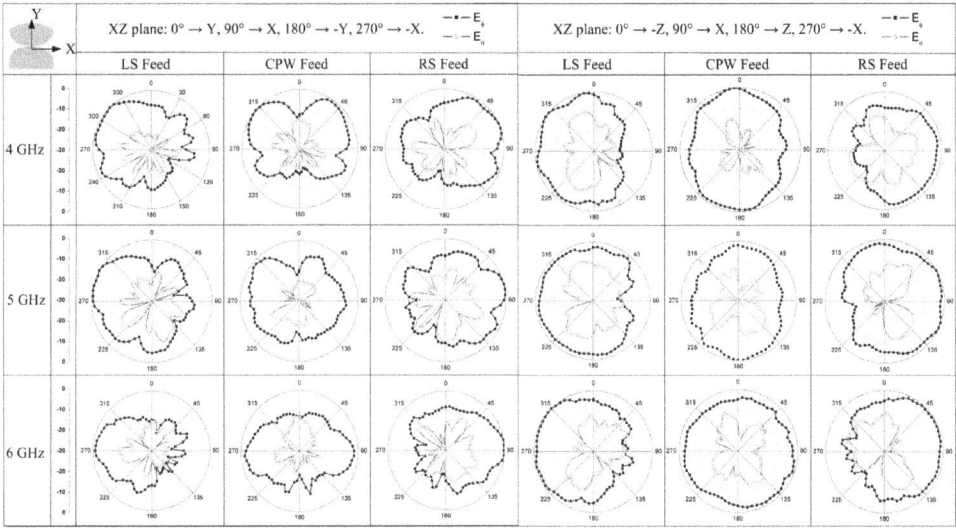

Fig. 25. Radiation patterns of the reconfigurable antenna.

4.2 Channel capacity measurement

The channel capacity of a 2x2 MIMO system by using the proposed pattern reconfigurable antenna is measured in this section. The experiment setup is as same as Fig. 17. At the TX end, two reference dipoles are arranged perpendicular to XZ plane along Y axis. Another two reference dipoles and two proposed pattern reconfigurable antennas are adopted at the RX end alternatively for comparison. Each port of the two wire dipoles has a bandwidth of 3.9-5.9 GHz with reflection coefficient better than –6 dB, and mutual coupling between the two ports is lower than –25 dB over the frequency band which is achieved by tuning the distance between two elements. Also, the isolation between two proposed pattern reconfigurable antennas is lower than –25 dB.

The measurement was also taken in the Weiqing building of Tsinghua University of Fig. 18. The locations of RX are different from last experiment. The position of RX4 is not measured. Therefore, the LOS scenario includes the RX1 and RX2, and the NLOS scenario includes the RX3 and RX5. The frequency range of measurement is 4-6 GHz, with a step of 10 MHz. A total number of 201 data points/results are obtained as samples. Three configurations (CPW, LS and RS) of each reconfigurable element of the receive end were switched together manually and the highest value signal was selected as the receiving signal. Also, considering the small-scale fading effect, 5x5 grid locations for each RX position were arranged. A total number of 2x201x25=10050 results were measured for LOS and NLOS scenarios respectively. In the measurement, a 2x2 channel matrix H is obtained. The channel capacity is calculated by formula (1) in the last section. We also selected the SNR when the average channel capacity is 5 bit/s/Hz in a 1x1 reference dipole system in LOS or NLOS scenario.

The measured CCDFs of channel capacity of LOS and NLOS scenarios are illustrated in Fig. 26 and 27. The results consist of the channel capacity information of 2x2 multiple antenna system using the proposed pattern reconfigurable antennas, compared with 1x1 and 2x2 systems using reference dipoles. As listed in Table 2, 2.28 bit/s/Hz and 4.13 bit/s/Hz of the average capacity enhancement are achieved in LOS and NLOS scenarios, and 2.51 bit/s/Hz and 3.75 bit/s/Hz enhancement for 95% outage capacities. In the NLOS scenario, the received signal is mainly contributed from reflection and diffraction of the concrete walls and the desk partitions, arriving at the direction of endfire. The diversity gain in the endfire increases the channel capacity. Considering the insertion loss introduced from the non-ideal PIN diodes, better performance of the proposed antenna can be achieved by using high quality switches, such as micro-electro-mechanical systems (MEMS) type switches with less insertion loss and parasitic parameters.

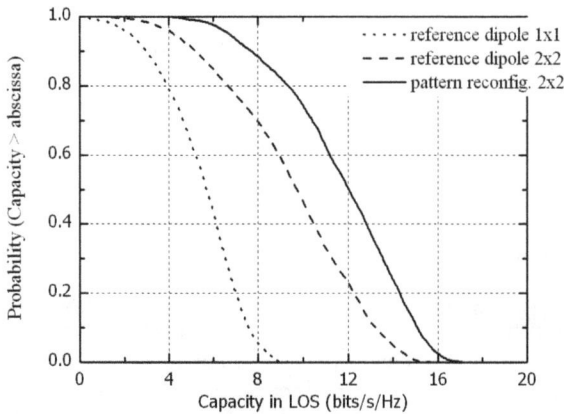

Fig. 26. CCDFs of channel capacity in LOS condition.

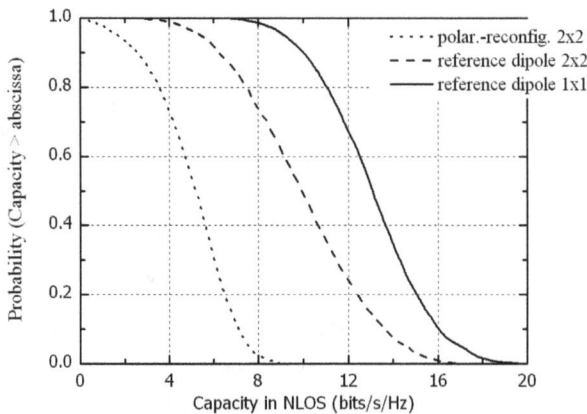

Fig. 27. CCDFs of channel capacity in NLOS condition.

Channel Capacity	Scenario	1x1 Dipole	2x2 Dipole	2x2 Pattern Reconfig.
Average	LOS	5	9.46	11.74
	NLOS	5	9.93	13.06
95% Outage	LOS	2.29	4.21	6.72
	NLOS	1.68	5.41	9.16

Table 2. Average and 95% Outage Channel Capacity (bit/s/Hz).

5. Conclusion

This chapter has introduced the now trend of antenna design in MIMO systems. From a view of antenna design, it is difficult to achieve good performance by using traditional antennas in space-limited MIMO systems. The mutual coupling between the antenna elements deteriorates the independence among multiple channels in MIMO systems. For both techniques of TD and SM, the benefit of multiple channels is difficult to be obtained due to the space limitation. However, the usage of miniaturized mobile terminals is popular and their size is getting much smaller. Here comes the contradiction between the antenna performance and the ministration of mobile handsets.

In this chapter, we are aimed to solve the space problem of antennas in MIMO systems. Two effective solutions are introduced here. The first one is to use polarization, an important spatial resource, to take the place of antenna element. Two orthogonal polarized antenna elements can be arranged together with acceptable isolation. In this way, the space between antenna elements is saved, making the overall antenna system more compact. As an important practical application, two types of dual-polarized antennas are presented and analyzed. Isolation enhancement methods are proposed, such as the feed design and operation modes design. The proposed antennas show the advantages of compact structure, high ports isolation and easy fabrication, and are suitable to be adopted in the space-limited MIMO systems.

Considering in the opposite way, the antennas with better performance in the original space is another solution in space-limited MIMO systems. The reconfigurable antenna is a prevalent type of antenna nowadays. Switching mechanism is added to achieve selectable polarizations, radiation patterns and other property. Different antenna configurations and corresponding signal processing methods are selected due to the channel information. The switching mechanism is the most important issue. Based on the dual-polarized slot antenna design, the PIN diodes are used to achieve polarization selection. A pattern reconfigurable antenna is design by using a switchable CPW-to-slotline feeding structure. In order to prove the benefit of the antenna selection, we design an experiment of channel capacity in a typical indoor environment. The results show that the channel capacity improves in both LOS and NLOS scenarios, especially in NLOS scenario. The reconfigurable antenna shows the potential application in space-limited MIMO systems.

6. Acknowledgment

This work is supported by the National Basic Research Program of China under Contract 2009CB320205, in part by the National High Technology Research and Development

Program of China (863 Program) under Contract 2009AA011503, the National Science and Technology Major Project of the Ministry of Science and Technology of China 2010ZX03007-001-01, and Qualcomm Inc..

7. References

Barba, M. (2008). A High-Isolation, Wideband and Dual-Linear Polarization Patch Antenna. *IEEE Transactions on Antenna and propagation,* vol.56, No.5, (May 2008), pp. 1472-1476, ISSN 0018-926X.

Bolcskei, H.; Nabar, R.; Erceg, V.; Gesbert, D. & Paulraj A. (2001). Performance of spatial multiplexing in the presence of polarization diversity. *Proceedings of IEEE International Conference on Acoustics, Speech, and Signal Processing,* pp. 2437-2440, ISBN 0-7803-7041-4, Salt Lake City, Utah, USA, May 7-11, 2001

Bolcskei, H.; Gesbert, D. & Paulraj A. (2002). On the capacity of OFDM based spatial multiplexing systems. *IEEE Transactions on Communications,* vol.50, No.2, (February 2002), pp. 225-234, ISSN 0090-6778.

Cetiner, B..; Jafarkhani, H.; Qian, J.; Hui, J.; Grau, A. & De Flaviis, F. (2004). Multifunctional reconfigurable MEMS integrated antennas for adaptive MIMO systems. *IEEE Communications Magazine,* vol.42, (December 2004), pp. 62-70, ISSN 0163-6804.

Erceg, V.; Sampath, H. & Catreux-Erceg, S. (2006). Dual-Polarization versus single-polarization MIMO channel measurement results and modeling. *IEEE Transactions on Wireless Communication,* vol.5, No.1, (January 2006), pp. 28-33, ISSN 1536-1276.

Foschini, G. & Gans, M. (1998). On limits of wireless communications in a fading environment when using multiple antennas. *Wireless Personal Communications,* vol.6, No.3, (Mar 1998), pp. 311-335, ISBN 0201634708.

Guo, Y.; Luk, K. & Lee K. (2002). Broadband dual polarization patch element for cellular-phone base stations. *IEEE Transactions on Antenna and propagations,* vol.50, No.2, (February 2002), pp. 251-253, ISSN 0018-926X.

Kim, H.; Chung, D.; Erceg, V.; Anagnostou, D. & Papapolymerou, J. (2007). Hardwired Design of Ultra-Wideband Reconfigurable MEMS Antenna. *Proceedings of IEEE 18th International Symposium on Personal, Indoor and Mobile Radio Communications,* pp. 1-4, ISBN 1-4244-1144-0, Athens, Greece, September 3-7, 2007.

Kyritsi, P.; Cox, D.; Valenzuela, R. and Wolniansky, P. (2002). Effect of antenna polarization on the capacity of a multiple element system in an indoor environment. *IEEE Journal on Selected Areas in Communications,* vol.20, No.6, (August 2002), pp. 1227-1239, ISSN 0733-8716.

Lee, C.; Chen, S. & Hsu, P. (2009). Isosceles Triangular Slot Antenna for Broadband Dual Polarization Applications. *IEEE Transactions on Antenna and propagations,* vol.57, No.10, (October 2009), pp. 3347-3351, ISSN 0018-926X.

Li, Y.; Zhang, Z.; Chen, W.; Feng, Z. & Iskander, M. (2010). A dual-polarization slot antenna using a compact CPW feeding structure. *IEEE Antennas and Wireless Propagations Letter,* vol.9, (December 2009), pp. 191-194, ISSN 1536-1225.

Li, Y.; Zhang, Z.; Feng, Z. & Iskander, M. (2011). Dual-mode Loop Antenna with Compact Feed for Polarization Diversity. *IEEE Antennas and Wireless Propagations Letter,* vol.10, (December 2010), pp. 95-98, ISSN 1536-1225.

Li, Y.; Zhang, Z.; Zheng, J. & Feng, Z. (2011). Channel capacity study of polarization reconfigurable slot antenna for indoor MIMO system. *Microwave and Optical Technology Letters*, vol.53, No.6, (March 2011), pp. 1029-1213, ISSN 1098-2760.

Li, Y.; Zhang, Z.; Zheng, J.; Feng, Z. & Iskander, M. (2011). Experimental Analysis of a Wideband Pattern Diversity Antenna with Compact Reconfigurable CPW-to-Slotline Transition Feed. *IEEE Transactions on Antenna and propagations*, accepted for publication, ISSN 0018-926X.

Lin, Y. & Chen, C. (2000). Analysis and applications of a new CPW-slotline transition. *IEEE Transactions on Microwave Theory and Techniques*, vol.48, No.3, (March 2000), pp. 463-466, ISSN 0018-9480.

Ma, K.; Qian, Y. & Itoh, T. (1999). Analysis and applications of a new CPW-slotline transition. *IEEE Transactions on Microwave Theory and Techniques*, vol.47, No.4, (April 1999), pp. 426-432, ISSN 0018-9480.

Mak, K.; Wong, H. & Luk, K. (2007). A Shorted Bowtie Patch Antenna with a Cross Dipole for Dual Polarization. *IEEE Antennas and Wireless Propagations Letter*, vol.6, (June 2007), pp. 126-129, ISSN 1536-1225.

Marzetta, T. & Hochwald, B. (1999). Capacity of a mobile multiple-antenna communication link in Rayleigh flat fading. *IEEE Transactions on Information Theory*, vol.45, No.1, (January 1999), pp. 139-157, ISSN 0018-9448.

Morris, M. & Jensen, M. (2005). Superdirectivity in MIMO systems. *IEEE Transactions on Antenna and propagation*, vol.53, No.9, (September 2005), pp. 2850-2857, ISSN 0018-926X.

Nabar, R.; Bolcskei, H.; Erceg, V.; Gesbert, D. & Paulraj A. (2002). Performance of multi-antenna signaling techniques in the presence of polarization diversity. *IEEE Transactions on Signal Processing*, vol.50, No.10, (October 2002), pp. 2553-2562, ISSN 1053-587X.

Nakano, M.; Satoh, T.; & Arai, H. (2002). Uplink polarization diversity measurement with human body effect at 900 MHz. *Electronics and Communications in Japan*, vol.85, No.7, (July 2002), pp. 32-44, ISSN 1520-6424.

Raleigh, G. & Cioffi, J. (1998). Spatio-temporal coding for wireless communication. *IEEE Transactions on Communications*, vol.46, No.3, (March 1998), pp. 357-366, ISSN 0090-6778.

Shin, J.; Chen, S. & Schaubert, D. (1999). A parameter study of stripline-fed Vivaldi notch-antenna arrays. *IEEE Transactions on Antenna and propagations*, vol.47, No.5, (May 1999), pp. 879-886, ISSN 0018-926X.

Sulonen, K.; Suvikmnas, P.; Vuokko, L.; Kivinen, J. and Vainikainen, P. (2003). Comparison of MIMO antenna configurations in Picocell and Microcell environments. *IEEE Journal on Selected Areas in Communications*, vol.21, No.5, (June 2003), pp. 703-712, ISSN 0733-8716.

Tarokh, V.; Seshadri, N.; & Calderbank, A. (1998). Space-time codes for high data rate wireless communication: performance criterion and code construction. *IEEE Transactions on Information Theory*, vol.44, No.2, (March 1998), pp. 744-765, ISSN 0018-9448.

Telatar, I. (1999). Capacity of multi-antenna Gaussian channels. *European Transactions on Telecommunications*, vol.10, No.6, (December 1999), pp. 585–595, ISSN 1541-8251.

Wallace, J.; Jensen, M. Swindlehurst, A. & Jeffs, B. (2003). Experimental characterization of the MIMO wireless channel: Data acquisition and analysis. *IEEE Transactions on Wireless Communication,* vol.2, No.2, (March 2003), pp. 335-343, ISSN 1536-1276.

Wallace, J. & Jensen, M. (2004). Mutual coupling in MIMO wireless systems: A rigorous network theory analysis. *IEEE Transactions on Wireless Communication,* vol.3, No.4, (July 2004), pp. 2437-2440, ISSN 1536-1276.

Winters, J. (1987). On the capacity of radio communication systems with diversity in a Rayleigh fading environment. *IEEE Journal on Selected Areas in Communications,* vol.5, No.5, (June 1987), pp. 871-878, ISSN 0733-8716.

Wu, S.; Chen, S. & Ma, T. (2008). A Wideband Slotted Bow-Tie Antenna with Reconfigurable CPW-to-Slotline Transition for Pattern Diversity, *IEEE Transactions on Antenna and propagations,* vol.56, No.2, (February 2008), pp. 327-334, ISSN 0018-926X.

3

Travelling Planar Wave Antenna for Wireless Communications

Onofrio Losito and Vincenzo Dimiccoli
Itel Telecomunicazioni srl, Ruvo di Puglia (BA)
Italy

1. Introduction

Microstrip antennas are one of the most widely used types of antennas in the microwave frequency range, and they are often used in the millimeter-wave frequency range. Actually as the demand for high data rates grows and microwave frequency bands become congested, the millimeter-wave spectrum is becoming increasingly attractive for emerging wireless applications. The abundance of bandwidth and large propagation losses at millimeter-wave frequencies makes these bands best-suited for short-range or localized systems that provide broad bandwidth. Automotive radar systems including cruise control, collision avoidance and radiolocation with operation up to 10 GHz have a large market potential in the near future of millimetre wave applications.

One advantage of the microstrip antenna is easy matching, fabrication simplicity and low profile, in the sense that the substrate is fairly thin. If the substrate is thin enough, the antenna actually becomes conformal, meaning that the substrate can be bent to conform to a curved surface. Disadvantages of the microstrip antenna include the fact that it is usually narrowband, with bandwidths of a few percent being typical. Also, the radiation efficiency of the microstrip antenna tends to be lower than some other types of antennas, with efficiencies between 70% and 90% being typical. A microstrip antenna operating in a travelling wave configuration could provide the bandwidth and the efficiencies needed.

Travelling-wave antennas are a class of antennas that use a travelling wave on a guiding structure as the main radiating mechanism. It is well known that antennas with open-ended wires where the current must go to zero (dipoles, monopoles, etc.) can be characterized as standing wave antennas or resonant antennas. The current on these antennas can be written as a sum of waves traveling in opposite directions (waves which travel toward the end of the wire and are reflected in the opposite direction). For example, the current on a dipole of length l is given by:

$$I(z) = I_o \sin\left[k\left(\frac{l}{2} - z'\right)\right]$$

$$= \frac{I_o}{2j}\left[e^{jk\left(\frac{l}{2}-z'\right)} - e^{-jk\left(\frac{l}{2}-z'\right)}\right] \tag{1}$$

$$= \frac{I_o}{2j} \left[e^{j\frac{kl}{2}} e^{-jkz'} - e^{-j\frac{kl}{2}} e^{jkz'} \right]$$

+z directed   z directed
wave   wave

Traveling wave antennas are characterized by matched terminations (not open circuits) so that the current is defined in terms of waves traveling in only one direction (a complex exponential as opposed to a sine or cosine). A traveling wave antenna can be formed by a single wire transmission line (single wire over ground) which is terminated with a matched load (no reflection). Typically, the length of the transmission line is several wavelengths.

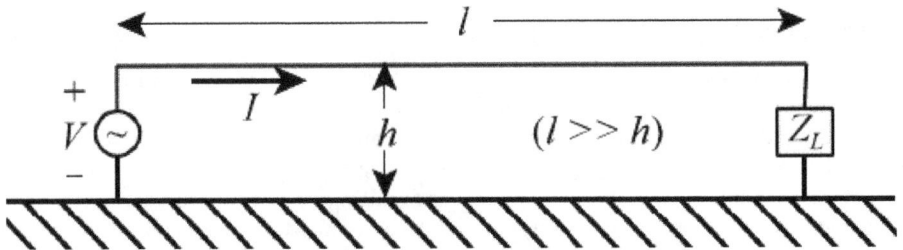

Fig. 1. Beverage or wave antenna.

The antenna shown in Fig. 1 above is commonly called a Beverage or wave antenna. This antenna can be analyzed as a rectangular loop, according to image theory. However, the effects of an imperfect ground may be significant and can be included using the reflection coefficient approach. The contribution to the far fields due to the vertical conductors is typically neglected since it is small if l >> h. Note that the antenna does not radiate efficiently if the height h is small relative to wavelength. In an alternative technique of analyzing this antenna, the far field produced by a long isolated wire of length l can be determined and the overall far field found using the 2 element array factor. Traveling wave antennas are commonly formed using wire segments with different geometries. Therefore, the antenna far field can be obtained by superposition using the far fields of the individual segments. Thus, the radiation characteristics of a long straight segment of wire carrying a traveling wave type of current are necessary to analyze the typical traveling wave antenna. Traveling-wave antennas are distinguished from other antennas by the presence of a traveling wave along the structure and by the propagation of power in a single direction. Linear wire antennas are the dominant type of traveling-wave antennas.

There are in general two types of traveling-wave antennas [1-2]. The first one is the surface-wave antenna, which is a slow-wave structure, where the phase velocity of the wave is smaller than the velocity of light in free space and the radiation occurs from discontinuities in the structure (typically the feed and the termination regions). The propagation wavenumber of the traveling wave is therefore a real number (ignoring conductors or other losses). Because the wave radiates only at the discontinuities, the radiation pattern physically arises from two equivalent sources, one at the beginning and one at the end of the structure. This makes it difficult to obtain highly-directive singlebeam radiation patterns. However, moderately directly patterns having a main beam near endfire can be achieved, although with a significant sidelobe level. For these antennas there is an optimum length depending on the desired location of the main beam. Examples include wires in free space or over a ground plane, helixes, dielectric slabs or rods, corrugated conductors, "beverage" antenna, or the V antenna. An independent control of the beam angle and the beam width is not possible.

The second type of the travelling wave antennas are a fast-wave structure as leaky-wave antenna (LWA) where the phase velocity of the wave is greater than the velocity of light in free space. The structure radiates all its power with the fields decaying in the direction of wave travel.

A popular and practical traveling-wave antenna is the Yagi–Uda antenna It uses an arrangement of parasitic elements around the feed element to act as reflectors and directors to produce an endfire beam. The elements are linear dipoles with a folded dipole used as the feed. The mutual coupling between the standing-wave current elements in the antenna is used to produce a traveling-wave unidirectional pattern. Recently has been developed a new simple analytical and technical design of meanderline antenna, taped leaky wave antenna (LWA) and taped composite right/left-handed transmission-line (CRLHTL) LWA. The meanderline antenna is a traveling-wave structure, which enables reduction of the antenna length. It has a periodical array structure of alternative square patterns. With this pattern, the extended wire can be made much longer than the initial antenna (dipole) length, so that the selfresonance can be attained. The resonance frequency is then lower and radiation resistance is higher than that of a dipole of the same length. This in turn implies that the antenna is effectively made small.

2. Leaky wave antennas

In detail this type of wave radiates continuously along its length, and hence the propagation wavenumber kz is complex, consisting of both a phase and an attenuation constant. Highly-directive beams at an arbitrary specified angle can be achieved with this type of antenna, with a low sidelobe level. The phase constant β of the wave controls the beam angle (and this can be varied changing the frequency), while the attenuation constant α controls the beamwidth. The aperture distribution can also be easily tapered to control the sidelobe level or beam shape.

All kinds of open planar transmission lines are predisposed to excite leaky waves. There are two kinds of leaky waves. Surface leaky waves radiate power into the substrate. These waves are in most cases undesirable as they increase losses, cause distortion of the transmitted signal and cross-talk to other parts of the circuit. Space leaky waves radiate power into a space and mostly also into the substrate. These waves can be utilized in leaky wave antennas. Leaky-wave antennas can be divided into two important categories, uniform and periodic, depending on the type of guiding structure. A uniform structure has a cross section that is uniform (constant) along the length of the structure, usually in the form of a waveguide that has been partially opened to allow radiation to occur. The guided wave on the uniform structure is a fast wave, and thus radiates as it propagates.

As said previously leaky-wave antennas form part of the general class of travelling-wave antennas which are a class of antennas that use a travelling wave on a guiding structure as the main radiating mechanism [3], as defined by standard IEEE 145-1993: "An antenna that couples power in small increments per unit length, either continuously or discretely, from a travelling wave structure to free space".

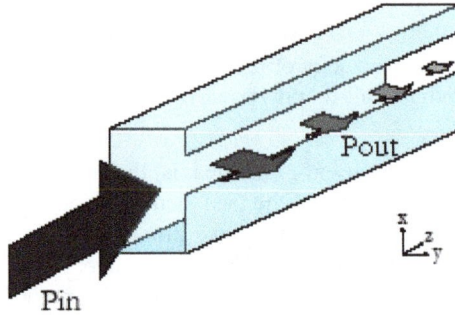

Fig. 2. Rectangular metal waveguide with a slit, aperture of the leaky wave antenna.

Leaky-wave antennas are a fast-wave travelling-wave antennas in wich the guided wave is a fast wave, meaning a wave that propagates with a phase velocity that is more than the speed of light in free space.

The slow wave travelling antenna does not fundamentally radiate by its nature, and radiation occurs only at discontinuities (typically the feed and the termination regions). The propagation wavenumber of the travelling wave is therefore a real number (ignoring conductors or other losses). Because the wave radiates only at the discontinuities, the radiation pattern physically arises from two equivalent sources, one at the beginning and one at the end of the structure. This makes it difficult to obtain highly-directive singlebeam radiation patterns. However, moderately directly patterns having a main beam near endfire can be achieved, although with a significant sidelobe level. For these antennas there is an optimum length depending on the desired location of the main beam. An independent control of the beam angle and the beam width is not possible. By contrast, the wave on a leaky-wave antenna (LWA) may be a fast wave, with a phase velocity greater than the speed of light. Leakage is caused by asymmetry, introduced in radiating structure transversal section (e.g.: aperture offset, waveguide shape, etc...), feeding modes or a combination of them. In this type of antennas, the power flux leaking from waveguide to free space (P_{out} in Fig. 2 and Fig. 3), introduces a loss inside structure, determining a complex propagation wavenumber k_z [4-5]:

$$(k_z = \beta - j\alpha)$$ (2)

Where α is the leakage constant and β is the propagation constant . The phase constant β of the wave controls the beam angle (and this can be varied changing the frequency), while the attenuation constant α controls the beamwidth. Highly-directive beams at an arbitrary specified angle can be achieved with this type of antenna, with a low sidelobe level.

Moreover the aperture distribution can also be easily tapered to control the sidelobe level or beam shape. Leaky-wave antennas can be divided into two important categories, uniform and periodic, depending on the type of guiding structure.

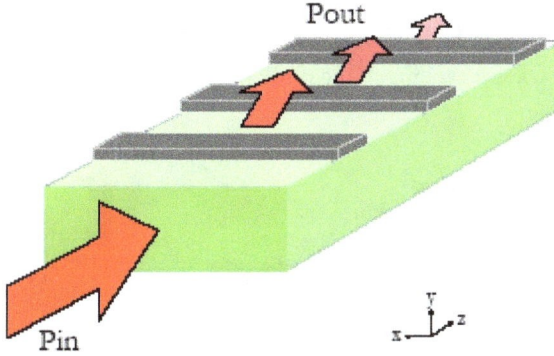

Fig. 3. Example of periodic leaky wave antenna, using a dielectric substrate upon which are placed rods of other material, even metal, in a periodic layout.

A uniform structure has a cross section that is uniform (constant) along the length of the structure, usually in the form of a waveguide that has been partially opened to allow radiation to occur [6]. The guided wave on the uniform structure is a fast wave, and thus radiates as it propagates. A periodic leaky-wave antenna structure is one that consists of a uniform structure that supports a slow (non radiating) wave that has been periodically modulated in some fashion. Since a slow wave radiates at discontinuities, the periodic modulations (discontinuities) cause the wave to radiate continuously along the length of the structure. From a more sophisticated point of view, the periodic modulation creates a guided wave that consists of an infinite number of space harmonics (Floquet modes) [7]. Although the main (n = 0) space harmonic is a slow wave, one of the space harmonics (usually the n = −1) is designed to be a fast wave, and hence a radiating wave.

3. LWA in waveguide

A typical example of a uniform leaky-wave antenna is a rectangular waveguide with a longitudinal slot. This simple structure illustrates the basic properties common to all uniform leaky-wave antennas. The fundamental TE_{10} waveguide mode is a fast wave, with

$\beta = \sqrt{k_0^2 + -(\frac{\pi}{a})^2}$ lower than k_0 . As mentioned, the radiation causes the wavenumber k_z of the propagating mode within the open waveguide structure to become complex. By means of an application of the stationary-phase principle, it can be found in fact that [5]:

$$\sin \vartheta_m \cong \frac{\beta}{k_0} = \frac{c}{v_{ph}}$$

(3)

where ϑ_m is the angle of maximum radiation taken from broadside. As is typical for a uniform LWA, the beam cannot be scanned too close to broadside ($\vartheta_m = 0$), since this corresponds to the cutoff frequency of the waveguide. In addition, the beam cannot be scanned too close to endfire ($\vartheta_m = 90$) since this requires operation at frequencies significantly above cutoff, where higher-order modes is in a bound condition or can propagate, at least for an air-filled waveguide. Scanning is limited to the forward quadrant only ($0 < \vartheta_m < \dfrac{\pi}{2}$) for a wave travelling in the positive z direction.

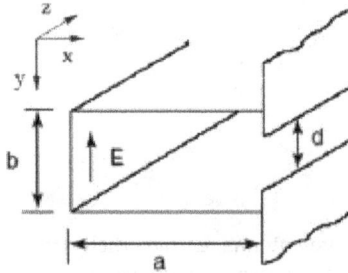

Fig. 4. Slotted guide (patented by W. W. Hansen in 1940).

This one-dimensional (1D) leaky-wave aperture distribution (see Fig. 4), results in a "fan beam" having a narrow beam in the x-z plane (H plane), and a broad beam in the cross-plane. Unlike the slow-wave structure, a very narrow beam can be created at any angle by choosing a sufficiently small value of α. From diffraction theory, a simple formula for the beam width, measured between half power points (3dB), is:

$$\Delta\vartheta \cong \frac{const}{\dfrac{L}{\lambda_0}\cos\vartheta_m} \qquad (4)$$

"const" is a parameter which is influenced by the type of aperture and illumination; for example, if at the aperture there's a constant field, const = 0.88 and, if the structure is uniform, const =0.91. As a rule of thumb, supposing:

$$\Delta\vartheta \cong \frac{1}{\dfrac{L}{\lambda_0}\cos\vartheta_m} \qquad (5)$$

a good approximation of beamwidth is yielded, where L is the length of the leaky-wave antenna, and $\Delta\vartheta$ is expressed in radians. For 90% of the power radiated it can be assumed:

$$\frac{L}{\lambda_0} \cong \frac{0.18}{\dfrac{\alpha}{k_0}} \Rightarrow$$

$$\Delta \vartheta \cong \frac{\alpha}{k_0}$$

If the antenna has a constant attenuation throughout its length $\alpha_z(z) = \alpha_z$ results:

$$P(z) = P(0)e^{-2\alpha_z}$$

Therefore, being L the length of antenna, if a perfectly matched load is connected at the end of it, it's possible to express antenna efficiency as:

$$\eta_{rad} = \frac{P(0) - P(L)}{P(0)} = 1 - \frac{P(L)}{P(0)} = 1 - e^{-2\alpha_z L} \qquad (6)$$

Rearranging:

$$L = -\frac{\ln(1 - \eta_{rad})}{2\alpha_z} \qquad (7)$$

For most application, to gain a 90% efficiency, means that the antenna length is within $10\lambda_0 \div 100\lambda_0$ interval.

Fixing the antenna efficiency, using (7), makes possible to express attenuation constant in terms of antenna length, and vice versa. Using antenna efficiencies grater than 90%-95% is not advisable; in fact, supposing constant antenna cross section and, as a consequence, fixed leakage constant, the necessary length L grows exactly as $\alpha_z L$, which increases asymptotically, as shown in Fig 5. If we want a 100% efficiency ($\eta_{rad} = 1$) from (6):

$$P(L) = 0 \Rightarrow e^{-2\alpha_z L} = 0 \Rightarrow L = \infty$$

we note that is necessary an infinite antenna length.

Substuting (7) in (5), being $\lambda_0 = 2\pi / k_0$:

$$\Delta \vartheta \approx \left(\frac{-4\pi}{\ln(1 - \eta_{rad})\cos \vartheta_m} \right) \frac{\alpha_z}{k_0}$$

Because $\cos \vartheta_m = \sqrt{1 - \sin^2 \theta_m}$, considering (7)

$$\Delta \vartheta \approx \left(\frac{-4\pi}{\ln(1 - \eta_{rad})\sqrt{1 - \left(\frac{\beta_z}{k_0}\right)^2}} \right) \frac{\alpha_z}{k_0} \qquad (8)$$

Since $k_0^2 = k_c^2 + k_z^2$ and having supposed the attenuation constant much smaller than the phase constant, $k_z \approx \beta_z$, getting:

$$\Delta \vartheta \approx \left(\frac{-4\pi}{\ln(1-\eta_{rad})\dfrac{k_c}{k_0}} \right) \frac{\alpha_z}{k_0} \tag{9}$$

where k_c is the transverse propagation constant. Alternatively, considering (7):

$$\Delta \vartheta \approx \frac{2\pi}{L \cdot k_c}$$

Using waveguide theory notation, supposing λ_c the cut-off wavelength:

$$\Delta \vartheta \approx \frac{\lambda_c}{L} \tag{10}$$

(3) and (10), provided the approximations used to be valid, are a valid tool for describing the main parameters of radiated beam.

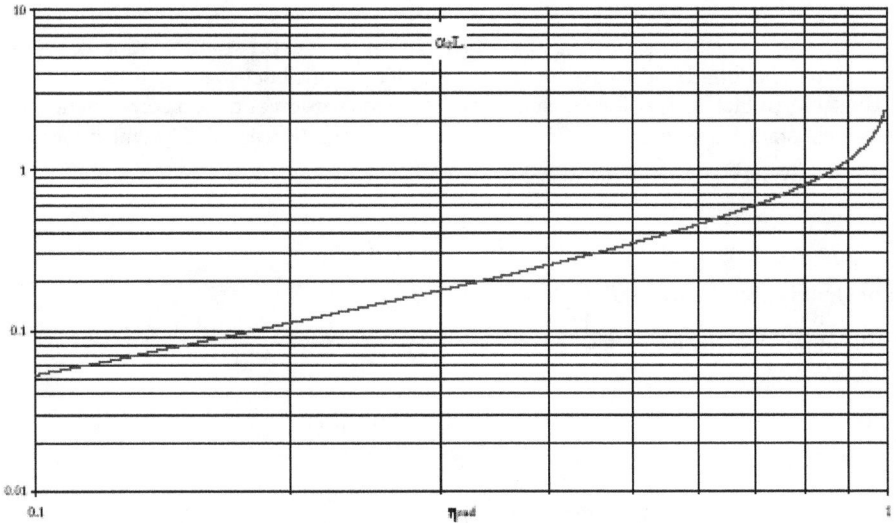

Fig. 5. Variation of $\alpha_z L$ versus antenna efficiency.

Radiation properties of leaky wave antennas are well described by dispersion diagrams. In fact since leakage occurs over the length of the slit in the waveguiding structure, the whole length constitutes the antenna's effective aperture unless the leakage rate is so great that the power has effectively leaked away before reaching the end of the slit. A large attenuation constant implies a short effective aperture, so that the radiated beam has a large beamwidth. Conversely, a low value of α results in a long effective aperture and a narrow beam, provided the physical aperture is sufficiently long.

Moreover since power is radiated continuously along the length, the aperture field of a leakywave antenna with strictly uniform geometry has an exponential decay (usually slow), so that the sidelobe behaviour is poor. The presence of the sidelobes is essentially due to the fact that the structure is finite along z.

When we change the cross-sectional geometry of the guiding structure to modify the value of α at some point z, however, it is likely that the value of β at that point is also modified slightly. However, since β must not be changed, the geometry must be further altered to restore the value of β, thereby changing α somewhat as well.

In practice, this difficulty may require a two-step process. The practice is then to vary the value of α slowly along the length in a specified way while maintaining β constant (that is the angle of maximum radiation), so as to adjust the amplitude of the aperture distribution to yield the desired sidelobe performance.

Radiation modes

Let us consider a generic plane wave, whose propagation vector belongs to plane (y-z), directed towards a dielectric film grounded on a perfect electric conductor (PEC) parallel to plane (x-z), as shown in Fig. 6 [8-9].

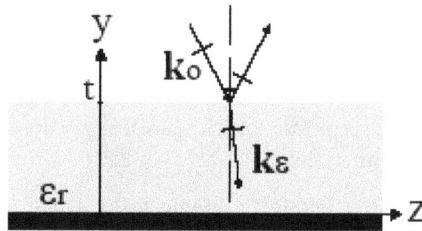

Fig. 6. Incident wave on a grounded dielectric film, whose thickness is t.

If the incident wave polarization is linear and parallel to the x axis, since both reflection and refraction occur:

$$\begin{cases} E_{x_0} = Ae^{-jk_{y0}(y-t)} + Ce^{jk_{y0}(y-t)} & y \geq t \\ E_{x_\varepsilon} = B\cos(k_{y_\varepsilon}y) + D\sin(k_{y_\varepsilon}y) & t \geq y \geq 0 \end{cases}$$

Being the tangent components of electric field null on a PEC surface, B = 0:

$$\begin{cases} E_{x_0} = Ae^{-jk_{y0}(y-t)} + Ce^{jk_{y0}(y-t)} & y \geq t \\ E_{x_\varepsilon} = D\sin(k_{y_\varepsilon}y) & t \geq y \geq 0 \end{cases} \tag{11}$$

One constant can be expressed by the remaining two, as soon as continuity of tangent components of electric field is considered in y = t. First equation of (11) contains an exponential term which, diverging for $y \rightarrow \infty$, violates the radiation condition at infinite distance. Therefore, $C \neq 0$ only near plane y = t, at the incidence point. k_z can assume any value from 0 to k_0 (i.e.: radiating modes); above it, only discrete values of kz exist, identifying the associated guided modes. Since separability condition must be satisfied, in air:

$$k_0^2 = \omega^2 \mu_0 \varepsilon_0 = k_{y_0}^2 + k_z^2 \text{ where } k_0 \in \Re \tag{12}$$

For every k_z, it's now possible calculate k_{y_0}. In fact, considering only positive solutions:

$$k_{y_0} = \sqrt{k_0^2 - k_z^2}$$

Obtaining:

Mode	Wave numbers	
Guided	$k_z > k_0$	$k_{y_0} \in \Im$
Radiating	$k_0 > k_z > 0$	$k_0 > k_{y_0} > 0$
Evanescent	$0 > k_z > -j\infty$	$\infty > k_{y_0} > k_0$

Table 1. Wave modes identified by k_z.

Thus, a spectral representation of electromagnetic field near the air-dielectric interface, must contain all values of k_{y_0}, from 0 to ∞: the associated integral is complex and slowly convergent. Alternatively, a description, which uses leaky waves and guided modes, both discrete, can well approximate such field.

It's been observed that it's often enough a single leaky wave to obtain a good far field description.

Letting $k_y = k_{y_0}$, from (2) and (12), in general:

$$\begin{cases} k_0 = \beta_y^2 + \beta_z^2 - \alpha_y^2 - \alpha_z^2 \\ 0 = \beta_y \alpha_y + \beta_z \alpha_z \end{cases}$$

alternatively

$$\begin{cases} k_0^2 = |\overline{\beta}|^2 \cdot |\overline{\alpha}|^2 \\ 0 = \overline{\beta} \cdot \overline{\alpha} \end{cases} \tag{13}$$

Having defined the attenuation and the phase vectors, respectively, as:

$$\alpha = \alpha_y \overline{y}_0 + \alpha_z z_0$$

$$\beta = \beta_y \overline{y}_0 + \beta_z z_0$$

Being $k_0 \in \Re$, from (9) $|\vec{\beta}| \neq 0$, and $|\vec{\beta}| > |\vec{\alpha}|$

Considering waves propagating in the positive direction of z axis, $\beta_z > 0$ and supposing no losses in z direction, $\alpha_z = 0$, from (13):

$$0 = \beta \cdot \alpha$$

Leaving out $\beta_y, \alpha_y = 0$, two situations can occur: $\alpha_y = 0$ and $\beta_y = 0$. If $\alpha_y = 0$, equations describe a uniform plane wave passing the air-dielectric interface. On the other hand, if $\beta_y = 0$, two types of superficial waves exist, depending on $\alpha_y = 0$ sign:

$\alpha_y > 0$	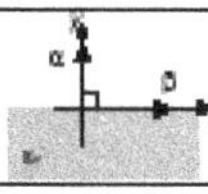	Confined superficial wave
$\alpha_y < 0$	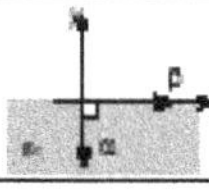	Improper superficial wave

Fig. 7. Superficial waves at air-dielectric interface when $\beta_y = 0$.

Because confined superficial waves amplitude decreases exponentially as distance from interface increases, when y is greater than 10 times radiation wavelength, electromagnetic field practically ceases to exist. Improper superficial wave, whose amplitude increases exponentially as distance from interface increases, are not physically possible because they violate the infinite radiation condition.

Removing the hypothesis $\alpha_z = 0$, both $\alpha_y \neq 0$, and $\beta_y \neq 0$.

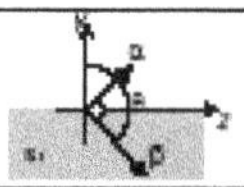	Losses in Dielectric
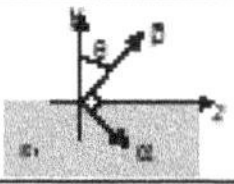	Leaky Wave

Fig. 8. General mutual β and α configurations depicting condition $0 = \beta \cdot \alpha$.

When losses in dielectric occur, β must point towards the inner part of dielectric to compensate such losses (see Fig.8). In the other configuration, when β points upwards, even though a non-physical solution is described, the associated wave is useful to describe electromagnetic field near air-dielectric interface.

4. LWA in microstrip

Microstrip antenna technology has been the most rapidly developing topic in antennas during the last twenty years [10]. Microstrip is an open structure that consists of a very thin metallic strip or patch of a width, w, separated from a ground plate by a dielectric sheet called substrate (Fig. 9). The thickness of the conductor, t, is much less than a wavelength, and may be of various shapes. The height of the substrate, h, is usually very thin compared to the wavelength (.0003 λ ≤ h ≤ 0.05 λ) [11]. The substrate is designed to have a known relative permittivity, ε_r, that is homogeneous within specified temperature limits.

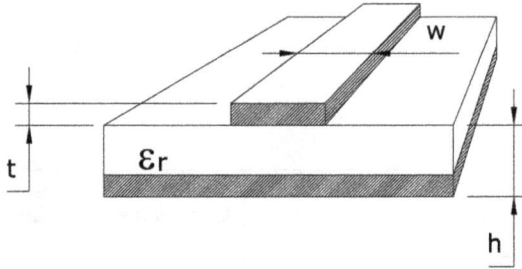

Fig. 9. Geometry of a microstrip transmission line.

The antenna can be excited directly by a microstrip line, by a coaxial cable, or a combination of the two. The antenna can also be fed from a microstrip line without direct contact through electromagnetic coupling. Feeding by electromagnetic coupling through an aperture in the ground plane tends to improve bandwidth. To maximize efficiency, the impedance of the feed must be matched to the input impedance of the antenna. There are a variety of stubs, shunts, and other devices used for matching. The major disadvantages of microstrip are lower gain, very narrow bandwidth, low efficiency and low power handling ability. In addition, antennas made with microstrip typically have poor polarization purity and poor scan performance [12].

Operating above the cutoff frequency, the field lines of microstrip extend throughout the substrate as well as into the free space region above the substrate, as seen in Fig. 10. The phase velocity of the field in the free space surrounding the structure is the speed of light, c, and the phase velocity of the field in the substrate is given by Equation (14)

$$v_p = \frac{c}{\sqrt{\varepsilon_r}} \tag{14}$$

This difference in phase velocity at the interface between the substrate and free space makes the TEM mode impossible. Instead, the fundamental mode for microstrip is a quasi-TEM mode, in which both the electric and magnetic fields have a component in the direction of propagation. Likewise, a higher order mode in microstrip is not purely TE or TM, but a hybrid combination of the two. The nth higher order mode is termed the TE_n mode. The fundamental mode of microstrip, as seen in Fig. 10, does not radiate since the fields produced do not decouple from the structure. If the fundamental mode is not allowed to propagate, the next higher order mode will dominate. Fig. 11 shows the fields due to the first higher order mode, TE_{10}. A phase reversal, or null, appears along the centerline, allowing the fields to decouple and radiate.

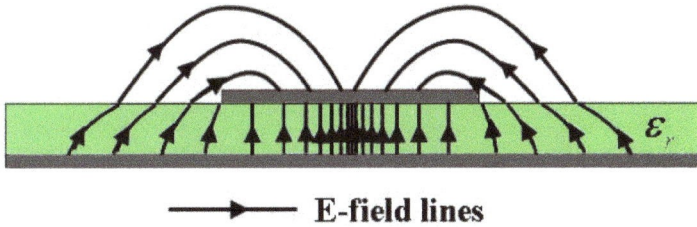

E-field lines

Fig. 10. Field pattern associated with the fundamental mode of microstrip.

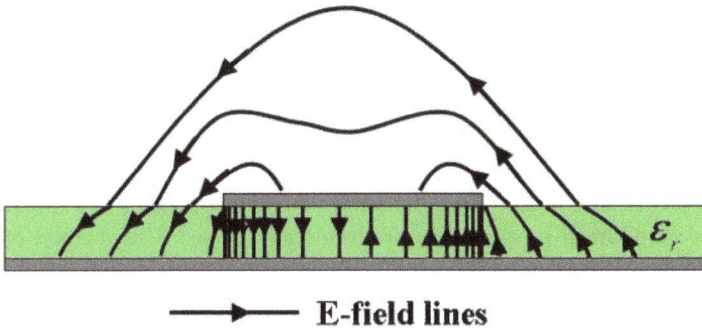

E-field lines

Fig. 11. Field pattern associated with the first higher order mode of microstrip.

Recently, there has been significant interest in the microstrip leaky-wave antenna which utilizes a higher order radiative microstrip mode. Since Menzel in 1979, published the first account of a travelling wave microstrip antenna that used a higher order mode to produce leaky waves [13], many microstrip leaky-wave antenna designs incorporating various modifications have been investigated. The design of Menzel antenna [13], can be seen in Fig. 12. Menzel's antenna uses seven slots cut from the conductor along the centerline to suppress the fundamental mode allowing leaky wave radiation via the first higher order mode. Menzel's antenna has been analyzed by a host of researchers over the past 25 years [14] and its performance is known and reproducible. Instead of transverse slots, we can uses a metal wall down the centerline of the antenna to block the fundamental mode. Symmetry along this metal wall invites the application of image theory. One entire side of the antenna

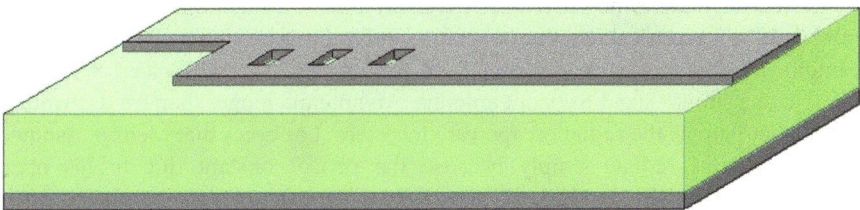

Fig. 12. Menzel's original antenna [13].

is now an image of the other side, making it redundant and unneeded. This property allows to design the resulting antenna half of the width of Menzel's antenna, as shown in Fig. 13.

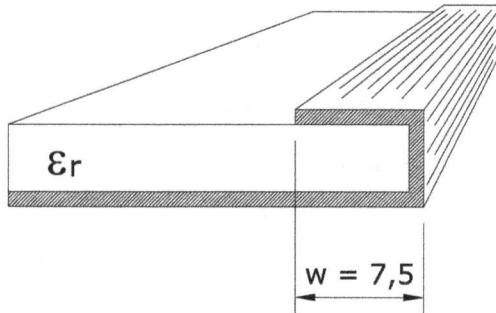

Fig. 13. Half Width Leaky Wave Antenna.

As mentioned the microstrip structures do not radiate for the fundamental mode, therefore, a higher order mode must be excited to produce leaky waves. This method of producing radiation by exciting higher order modes in a transmission line has been documented since the 1950's [6]. By the 1970's, rectangular waveguides, circular waveguides, and coaxial cables were in use as leaky traveling wave antennas. However, until Menzel, the jump to microstrip had not been made. By looking at a cross section of microstrip excited in the fundamental mode, the E field is strongest in the center and tapers off to zero at the sides, as depicted in Fig 10. If the electric field down the centerline is suppressed, the fundamental mode will be prohibited, forcing the energy to propagate at the next higher mode, TE_{10} . As seen in Fig. 11, TE_{10} mode causes E to be strongest at the edges. Menzel attempted to force the TE_{10} mode using several means. Feeding two equal magnitude waves 180° out of phase with a "T" or "Y" feed produced TE_{10} as desired, but did not fully eliminate the foundamental mode. Easier to produce and providing an even better response was given using transverse slots down the centerline (Fig. 12). The multiple feeds were not necessary to produce the TE_{10} mode when the fundamental mode was suppressed. Menzel demonstrated that the beam angle can be predictively steered by input frequency if the electrical length of the antenna is at least 3λ . If the length is less than 3λ , too little of the incident wave is being radiated and a resonance standing wave pattern is forcing the beam toward broadside. Qualitative analysis shows that the beamwidth of Menzel's antenna is not frequency dependent, however, it is inversely related to length. The 3 dB beamwidth approaches 10° for electrical length of over 6λ and approaches nearly 90° for fractions of a wavelength. Menzel's gain varied from 7 dB for l = 0.2λ to 14 dB for l = 4λ. 7 dB is comparable to a similar sized resonant antenna. An antenna longer than l = 4λ would have an even higher gain as the radiation aperture increases. Lee notes that Menzel assumed that his antenna should radiate simply because the phase constant due to his operating frequency was less than k_0 [15]. If Menzel had considered the complex propagation constant, he would have realized that his antenna was operating in a leaky regime. The length would need to be roughly 220 mm, or more than twice as long as his design, to radiate at 90% efficiency. Radiation patterns in Menzel's paper clearly show the presence of a large backlobe due to the reflected traveling wave.

Now this class of printed antennas that is particularly well suited for operation at mm-wave frequencies, alleviate some of the problems associated with resonant antennas since they provide higher gain, broader bandwith performance, and frequency scanning capabilities. These microwave and millimeter leaky wave antennas, have the same properties of the waveguide leaky wave antennas described previously. In addition, when opening a waveguide to free space, a discrete spectrum is not enough to express an arbitrary solution [16].

In fact, when considering a closed region, all characteristic solutions, individuated by the associated eigenvalues, constitute a complete and orthogonal set of modes, whose linear combination can express any field satisfying boundary conditions. As soon as the region is not perfectly bounded, an arbitrary field solution cannot be expressed only using discrete eigenmodes but, generally, a continuous spectrum of modes, which don't necessarily have finite energy (e.g.: plane waves), must be considered, too.

Fortunately, for leaky wave antennas, an approximation that uses particular waves, called leaky, can be used instead of the continuous spectrum. Moreover, leaky waves are well described by dispersion constants (i.e.: leakage and phase constants) that strongly affect the radiated beam width and elevation.

5. Dispersion curves, spectral-gap

Dispersion curves, describing how attenuation and phase vectors, solutions of dispersion equation (12), evolve, are a valid tool to study leaky waves.

As discussed previously, the radiation mechanism of higher order modes on microstrip LWA is attributed to a traveling wave instead of the standing wave as in patch antennas and above cutoff frequency, where the phase constant equals the attenuation constant ($\alpha_c = \beta_c$), it is possible to observe three different range of propagation: bound wave, surface wave and leaky wave [15]. At low frequency, below the cutoff frequency, we have the reactive region due to evanescent property of LWA.

From (3) we can observe that the leaky mode leaks away in the form of space wave when $\beta < K_0$, therefore we can define the radiation leaky region from the cutoff frequency to the frequency at which the phase constant equals the free-space wavenumber $(\beta = K_0)$. For $(\beta > K_s)$ we have the bound mode region and for $K_0 < \beta < K_s$, exists a narrow frequency range ($K_s < \dfrac{1}{\sqrt{\varepsilon_r}}$), in which we can have surface-wave leakage, where K_s is the surface wavenumber.

Moreover the transition region between surface wave leakage and space wave leakage including a small range in frequency for which the solution is non-physical, and it therefore cannot be seen. For this reason, the transition region is called a spectral gap. Such a spectral gap occurs commonly (but not always) at such transitions in printed circuits, but it also occurs in almost all situations for which there is a change from a bound mode to a leaky

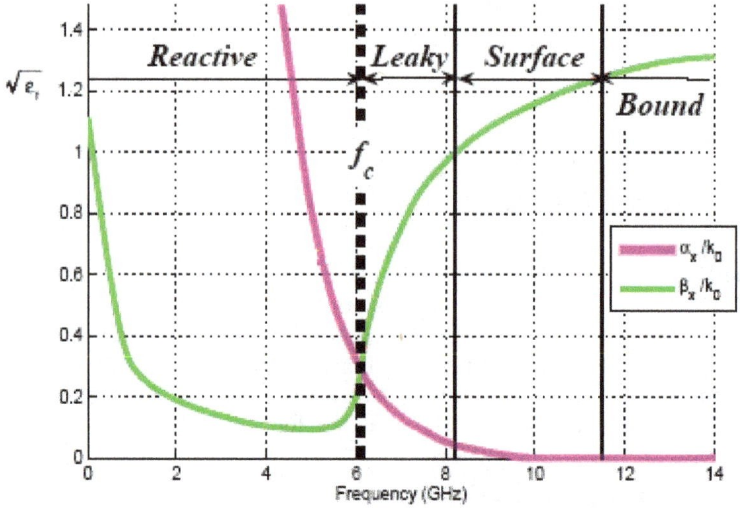

Fig. 14. The typical normalized attenuation constant, α / k_0, and phase constant β / k_0, in the direction of propagation of the first higher order mode, TE_{10}. There are four frequency regions associated with propagation regimes: Reactive, Leaky, Surface, and Bound.

mode, or vice versa. For example, a spectral gap will appear when the beam approaches endfire in all leaky-wave antennas whose cross section is partly loaded with dielectric material. It is necessary to employ a greatly enlarged scale, on which the dispersion plot is sketched qualitatively. The transition region itself is divided into two distinct frequency ranges, one from point A to point B and the second from point B to point C. Before point B, a leaky wave occurs. As soon as frequency reaches f_1 (point B), an improper superficial wave is solution of dispersion equation [9]. Because, both α_z and β_z cannot increase with frequency between f_1 and f_2, their trend will change until point C, from which a confined superficial wave is an acceptable solution for increasing values of frequency.

To depict normalized constants behaviour around the spectral-gap, it's necessary a very precise numerical method since, leaving out particular structures, its width (Δf) is very small compared to working frequencies.

The dispersion characteristics for microstrip has been investigated by a number of authors using different full wave methods and evaluating different regimes of the dispersion characteristic.

The spectral domain analysis has proven to be one of the most efficient and fruitful techniques to study the dispersion characteristics of printed circuit lines [17]. As is explained in literature, the Galerkin method in conjunction with Parseval theorem can be used to pose the dispersion relation of an infinite printed circuit line as the zeros of the following equation:

$$F(k_z) = \int_{C_x} \tilde{G}_{zz}(k_x; k_z)\tilde{T}^2(k_x)dk_x = 0 \tag{15}$$

where $\tilde{T}\,(k_x)$, is the Fourier transform of the basis function $T(k_x)$ used to expand the longitudinal current density on the strip conductor as

$$J_{sz}(x,z) = T(x)e^{-jk_z z}$$

The term $\tilde{G}_{zz}(k_x, k_{z,\omega})$ is the zz component of the spectral dyadic Green's function, and C_x is an appropriate integration path in the complex k_x plane to allow for an inverse Fourier transform non uniformly convergent function. The spectral dyadic Green's function has the following singularities in the complex k_x plane: branch point, a finite set of poles on the proper sheet and a infinite set of poles on the improper sheet. For a fixed frequency, the function $F(k_z)$ is not uniquely defined because of the many possible different C_x integration paths that can be used to carry out the integral (15) [18].

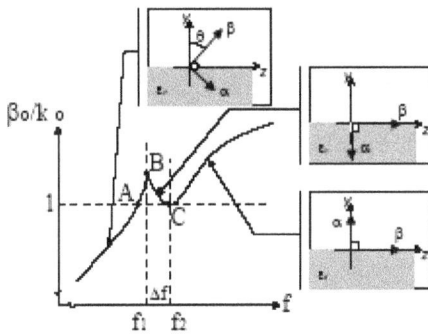

Fig. 15. Transition region between leaky wave and confined superficial wave showing the spectral-gap occurring f_1 and f_2.

The different C_x paths come from the different singularities of the spectral dyadic Green's function, that can be detoured around. For complex leaky mode solution, an integration path detouring around only the proper poles of the spectral dyadic Green's function is associated with an surface-wave leaky mode solution. If the path also detours around the branch points, passing trough the branch cuts and, therefore, lying partly on the lower Riemann sheet, the path will be associated with an space-wave leaky mode solution (see Fig. 16). This procedure, is not trivial. We shown in the next chapters how it is possible to extract the propagation constant of a microstrip LWA more simply using an FDTD code with UPML boundary condition, who directly solves the _elds in the time domain using Maxwell's equations and with which the analysis is easy modifying the geometries of the LWAs. The results are in a good agreement with transverse resonance approximation (a full wave method) derived by Kuestner [19].

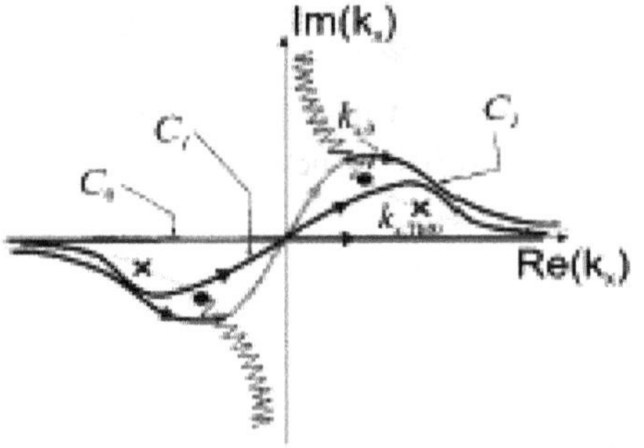

Fig. 16. Possible integration paths C . The three different are denoted as C_0 for the real-axis path (bound mode solution), C_1 for the path that detours around only the spectral dyadic Green's function poles (surface-wave leaky modes solution), and C_2 for the path that also passes around the branch points (space-wave leaky modes solution).

6. Tapered leaky wave antennas

Nowdays, some applications especially with regard to communication applications like the indoor wireless LAN(WLAN) actually are increasing the use of millimeterwave antennas like leaky-wave antennas (LWA), suited for more purpose. In detail, the transmitting/ receiving antennas with relatively broadbeam and broadband can be obtained from the curved and tapered leakywave structures. In fact, the microstrips (LWA), are very popular and widely used in applications thanks to their advantages of low-profile, easy matching, narrow beamwidth, fabrication simplicity, and frequency/electrical scanning capability. Is well know that the radiation mechanism of the higher order mode on microstrip LWAs is attributed to a traveling wave instead of the standing wave as in patch antennas. Moreover the symmetry of the structure along this physical grounding structure, thanks to the image theory, allows to design only half of an antenna with the same property of one in its entirety, and reducing up to 60% the antenna's dimensions. Using this tapered antenna we can obtained a quasi linear variations of the phase normalized constant and than a quasi linear variations of the its radiation angle. Moreover the profile of the longitudinal edges of the LWA, was designed, by means of the reciprocal slope of the cutoff curve, symmetrically to the centerline of the antenna, allows a liner started of leaky region.

Nevertheless the variation of the cross section of the antenna, allowing a non-parallel emitted rays, such as happens in a non-tapered LWA. In fact, using the alternative geometrical optics approach proposed in the tapering of the LWA, for a fixed frequency, involves the variation of the phase constant β and the attenuation constant a, obtained as a cut plane of 3D dispersion surface plot varying width and frequency. We can be determined a corresponding beam radiation interval with respect to endfire direction. As mentioned previously, for a tapered antenna with a curve profile (square root law profile) the radiation angle in the leaky regions, vary quasi linearly whit the longitudinal dimension, so it is possible to calculate the radiation angle of the antenna as a average of the phase constant using the simple formula.

Alternatively using the geometrical optics approach it is easy to determine the closed formula to predict the angle of main beam of a tapered LWA.

7. Design of tapered LWA

The radiation mechanism of the higher order mode on microstrip LWAs is attributed to a traveling wave instead of the standing wave as in patch antennas [13,20].

We can explain the character of microstrip LWAs trough the complex propagation constant $k = \beta - j\alpha$, where β is the phase constant of the first higher mode, and α is the leakage constant. Above the cutoff frequency, where the phase constant equals the attenuation constant ($\alpha_c = \beta_c$), it is possible to observe three different propagation regions: bound wave, surface wave and leaky wave.

The main-beam radiation angle of LWA can be approximated by:

$$\theta = \cos^{-1}\left(\frac{\beta}{K_0}\right)$$
(16)

where θ is the angle measured from the endfire direction and K_0 is the free space wavenumber. According to (16) we can observe that the leaky mode leaks away in the form of space wave when $\beta < K_0$, therefore we can define the radiation leaky region, from the cutoff frequency to the frequency at which the phase constant equals the free-space wavenumber $(\beta = K_0)$. An example of tapered LWA was proposed in [21-22], using an appropriate curve design to taper LWA.

In fact through the dispersion characteristic equation, evaluated with FDTD code, we can obtain the radiation region of the leaky waves indicated in the more useful way for the design of our antenna:

$$\frac{c}{2w_{eff}\sqrt{\varepsilon_r}} = f_c < f < \frac{f_c\sqrt{\varepsilon_r}}{\sqrt{\varepsilon_r - 1}} \; .$$
(17)

From equation (17) we can observe that the cutoff frequency increases when the width of the antenna decrease, shift toward high frequencies, the beginning of the radiation region as shown in Fig. 17. 1.

Therefore it is possible to design a multisection microstrip antenna [as Type I antenna in Fig. 17.a], in which each section able to radiate at a desired frequency range, can be superimposed, obtaining an antenna with the bandwidth more than an uniform microstrip antenna. In this way every infinitesimal section of the multisection LWA obtained overlapping different section should be into bound region, radiation region or reactive region, permitting the power, to uniformly radiated at different frequencies.

Using the same start width and substrate of Menzel travelling microstrip antenna (TMA) [13], and total length of 120 mm., we have started the iterative procedure mentioned in [23] to obtain the number, the width and the length of each microstrip section. From Menzel

TMA width, we have calculated the f_{START} (onset cutoff frequency) of the curve tapered LWA, than, choosing the survival power ratio ($\tau = e^{-2\alpha_i L_i}$) opportunely, at the end of the first section, we have obtained the length of this section. The cutoff frequency of subsequent section (f_i), was determined by FDTD code, while the length of this section was determined, repeating the process described previously. This iterative procedure was repeated, until the upper cutoff frequency of the last microstrip section.

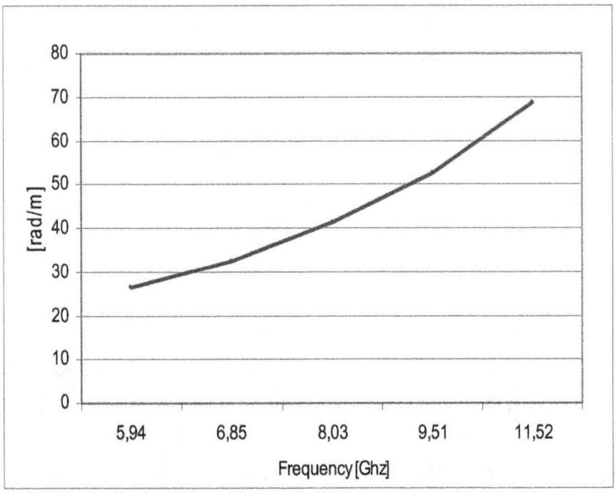

Fig. 17.1 Cutoff frequency of multisection microstrip LWA.

The presence of ripples in return loss curve and the presence of spurious sidelobes shows the impedence mismatch and discontinuity effect of this multisection LWA that reduce the bandwidth. A simple way to reducing these effects is to design a tapered antenna in which the begin and the end respectively of the first and the last sections are linearly connected together (as the Type II antenna in Fig. 17.a).

Alternatively the ours idea was to design a LWA using a physical grounding structure along the length of the antenna, with the same contour of the cutoff phase constant or attenuation constant curve ($\alpha_c = \beta_c$), obtained varying the frequency (the cutoff frequency fc is the frequency at which $\alpha_c = \beta_c$) , for different width and length of each microstrip section as shown in Fig. 17.1, employing the following simple equation (18):

$$\beta_c = c_1 f^2 + c_2 f + c_3 \tag{18}$$

obtained from linear polynomials interpolation, where $c_1 = 0.0016$, $c_2 = 0.03$, $c_3 = -15.56$.

The antenna layout (as the Type III antenna in Fig. 17.a), was optimized through an 3D electromagnetic simulator, and the return loss and the radiation pattern was compared with Type I antenna and Type II antenna.

8. Simulation results

An asymmetrical planar 50 Ω feeding line was used to excite the first higher-order mode while a metal wall down the centerline connecting the conductor strip and the ground plate was used to suppress the dominant mode for Type I - III. The chosen substrate had a dielectric constant of 2.32 and a thickness of 0.787 mm, while the total length of the leaky wave antenna was chosen to be 120 mm.

The leaky multisection tapered antenna Type I was open-circuited, with a 15 mm start width, and 8.9 mm of final width obtained according to [23]. For LWA layout Type I, we used four microstrip steps, for layout Type II we tapered the steps linearly, while the curve contour of the LWA layout Type III, was designed through equation (18).

Fig. 17.b shows the simulated return loss of three layouts. We can see that the return loss (S11) of Type I is below -5 dB from 6 to 10.3 GHz, but only three short-range frequencies are below -10 dB. S11 of Type II is below -5 dB from 6.1 to 9.1 GHz, and below -10 dB from 6.8 to 8.6 GHz. At last, S11 of Type III is below -5 dB from 6.8 to 11.8 GHz, and below -10 dB from 8.0 to 11.2 GHz. In Fig.17.c are shows the mainlobe direction at 9.5 GHz for the different Type I to Type III. We can see a reduction of sidelobe and only few degrees of mainlobe variation between Type I to Type III. Moreover, in Fig. 18 is shown the variation of mainlobe of antenna Type III, for different frequency, while in Fig. 19 is shown the trend of gain versus frequency of the same antenna. It is clear that, the peak power gain is more than 12 dBi, which is almost 3 dBi higher than uniform LWAs.

Finally the simulated VSWR is less than 2 between 8.01 and 11.17 GHz (33%), yielding an interesting relative bandwidth of 1.39:1, as shown in Fig. 20, compared with uniform microstrip LWAs (20% for VSWR < 2) as mentioned in [24].

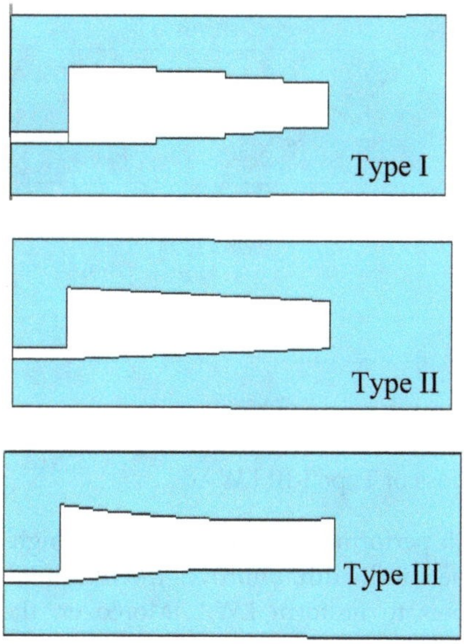

Fig. 17a. Layout of leaky wave antennas Type I-III. A physical grounding structure was used to connecting the conductor strip and the ground plane.

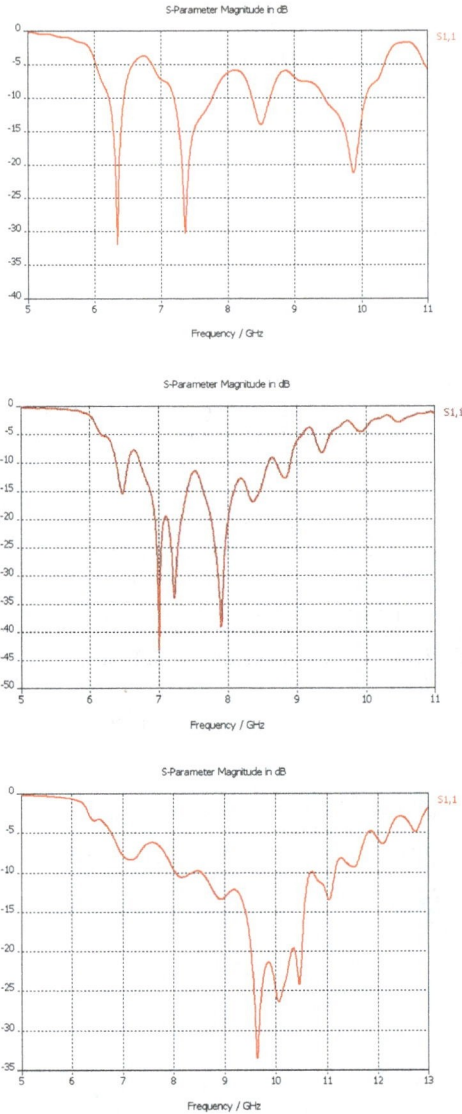

Fig. 17b. Simulated Return loss of Type I-III LWA.

These results indicate a high performance of Type III LWA: high efficiency excitation of the leaky mode, increases of the bandwidth, improves the return loss and reduction of 19% of metallic surface with respect to uniform LWA. Moreover, these results are in a good agreement whit the experimental results of return loss and radiation pattern of a prototype made using a RT/Duroid 5880 substrate with thickness of 0.787 mm and relative dielectric constant of 2.32, as shown in Fig.21 and Fig.22.

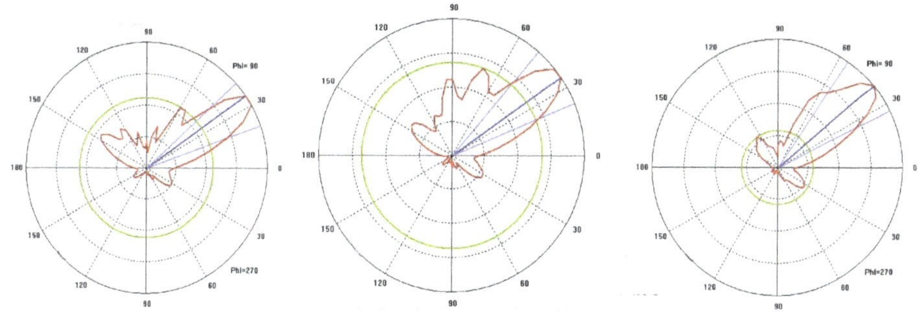

Fig. 17c. Radiation patterns of Electric field (H plane) of Type I-III, LWA at 9.5 GHZ.

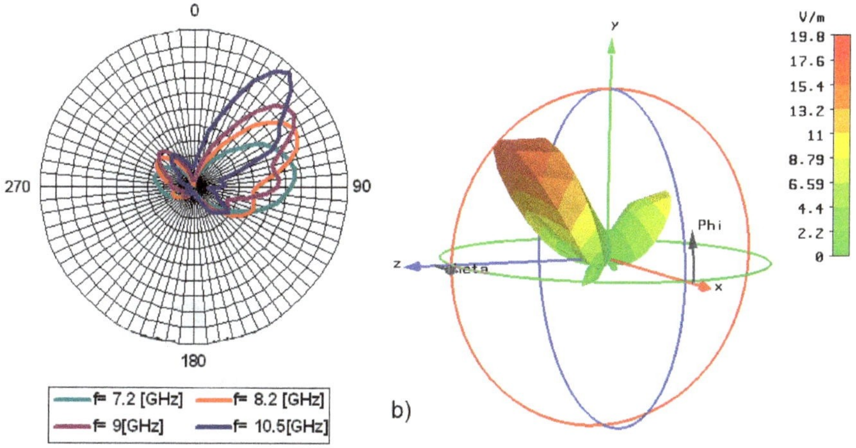

a)

| f= 7.2 [GHz] | f= 8.2 [GHz] |
| f= 9[GHz] | f= 10.5[GHz] |

b)

Fig. 18. a) Simulated radiation patterns of E field of LWA Type III for different frequency. b) 3-D radiation pattern of Electric field.

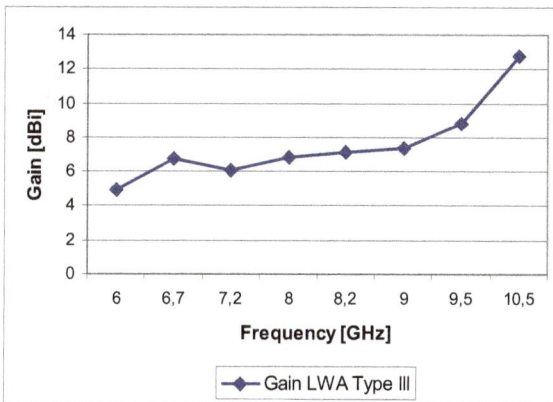

Fig. 19. The gain versus frequency of the LWA Type III.

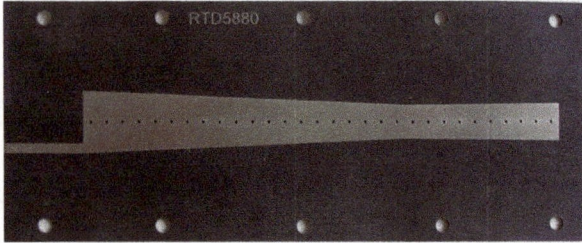

Fig. 20. Simulated VSWR of LWA Type III.

Fig. 21. A prototype of tapered LWA Type III with holes made in the centerline of the antenna.

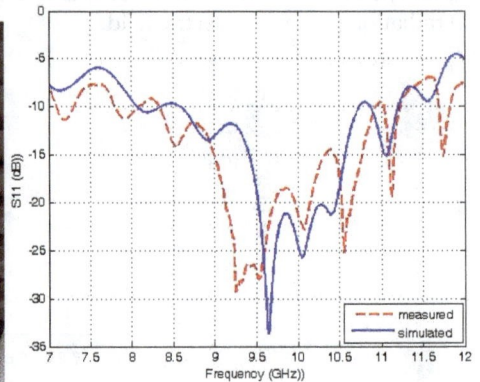

a)   b)

Fig. 22. a) Measurement set-up of LWA Type III. b) Experimental and simulated return loss of LWA Type III.

Fig. 23. A prototype of half tapered LWA.

Moreover the use of a physical grounding structure along the length of the antenna, as suggested in [21-22], allows the suppression of the dominant mode (the bound mode), the adoption of a simple feeding, and due to the image theory, it is also possible to design only half LWA (see Fig. 23) with the same property of one entire, as shown in Fig. 24 and in Fig. 25, reducing up to 60% the antenna's dimensions compared to uniform LWAs [24].

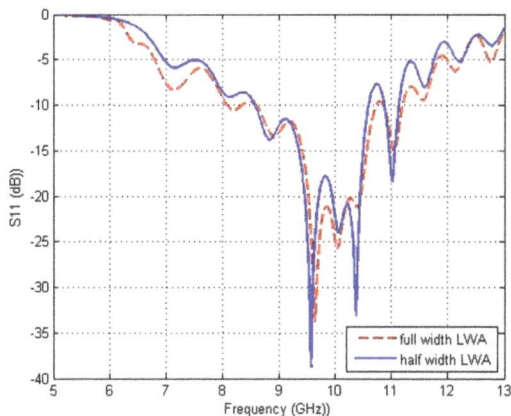

Fig. 24. Measured return loss of full and half Leaky Wave Antennas

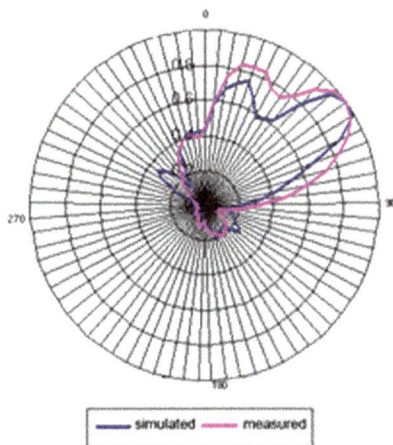

Fig. 25. The measured and simulated radiation patterns of E field of half tapered LWA at 8 GHz.

9. Focusing-diverging property

As described in [21-22], the profile of the longitudinal edges of the LWA, was designed, by means of the reciprocal slope of the cutoff curve, symmetrically to the centerline of the antenna, allows a liner started of leaky region. Using this tapered antenna we can obtained a quasi linear variations of the phase normalized constant and than a quasi linear variations of the its radiation angle as we can see in Fig. 26 and in Fig. 27. Nevertheless the variation of the cross section of the antenna, allowing a non-parallel emitted rays, such as happens in a non-tapered LWA (see Fig. 28). In fact, as was described in the alternative geometrical optics approach proposed in [24] the tapering of the LWA, for a fixed frequency, involves the variation of the phase constant β and the attenuation constant α, as shown in Fig. 29, obtained as a cut plane of 3D dispersion surface plot varying width and frequency (see Fig. 30).

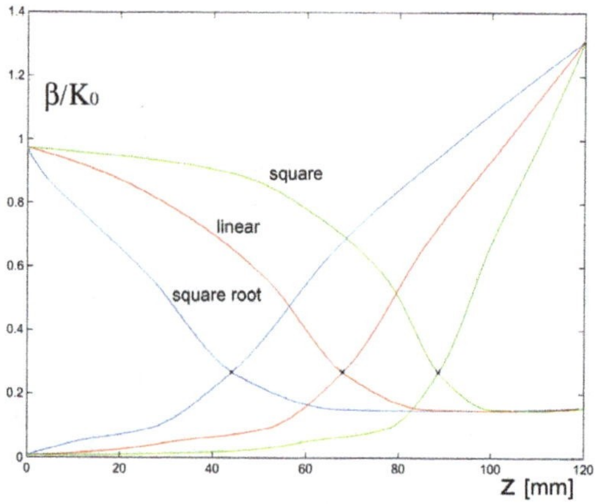

Fig. 26. The variation of the main beam radiation angle versus length of the antenna, at f= 8 GHz, for linear, square and square root profile of the LWA.

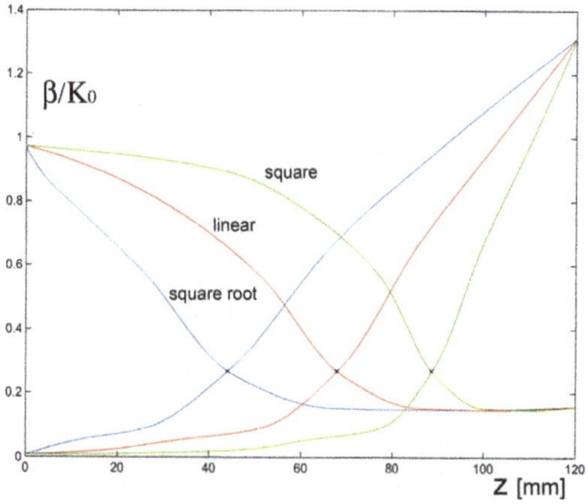

Fig. 27. The variation of the phase constant versus length of the antenna, at f= 8 GHz, for linear, square and square root profile of the LWA.

From (16) can be determined in the leaky regions of the antenna, a corresponding beam radiation interval [ϑ_{min} , ϑ_{max}] , , with respect to endfire direction.

As mentioned previously, for a tapered antenna with a curve profile (square root law profile) the radiation angle in the leaky regions, vary quasi linearly whit the longitudinal dimension, so it is possible to calculate the radiation angle of the antenna as a average of the phase constant using the simple equation (19).

$$\vartheta_m = sen^{-1}\left(-\frac{1}{K_0 L} \int_0^L \beta(z)dz \right) \tag{19}$$

Alternatively using the geometrical optics it is easy to determine the closed formula to predict the angle of main beam of a tapered LWA. Through simple mathematical passages, the main beam angle ϑ_m can be obtained by the equation (20).

$$\vartheta_m = sen^{-1}\left(\frac{A sen\vartheta_{min}}{\frac{1}{2}\sqrt{(2A sen\vartheta_{min})^2 + (L + 2C \cos\vartheta_{max})^2}} \right) \tag{20}$$

Where A and C are respectively the distance between real focus F and the beginning and the end of the length of the antenna L . Therefore, if we know the begin width and the end width of the antenna, from the curves of normalized phase and attenuation constant at fixed frequency, we can determine the beam radiation range from (16), and the main beam angle through (20).

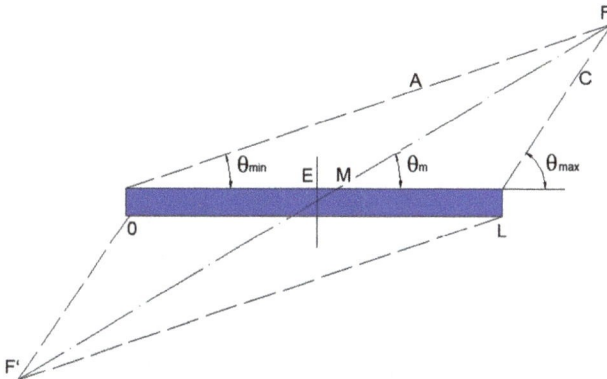

Fig. 28. The ray optical model for a tapered Leaky Wave antenna.

Furthermore this focusing phenomena of a tapered LWA can determine a wide-beam pattern in a beam radiation range which is evident when the antenna length is increased ($L \cong 50\lambda_0$) [25].

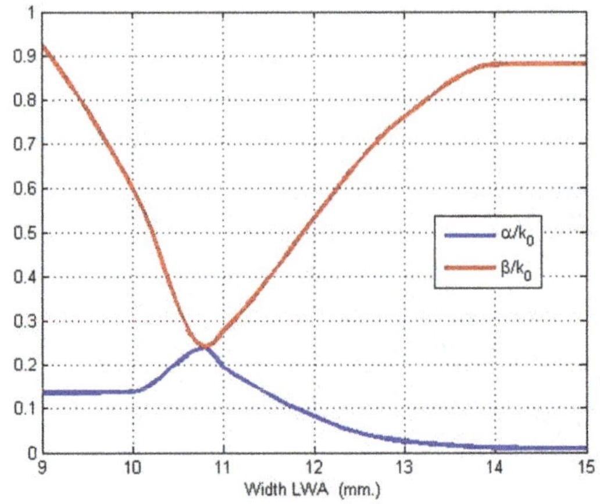

Fig. 29. The curve of normalized phase and attenuation constants versus the width, at 8 GHz for the LWA with the angular range [28°, 76°]. The leaky region start from 10.8 mm. (cutoff frequency).

To obtain a broad beam pattern without the use of a longer LWA, we can bend a tapered LWA (see Fig. 31), leading the electromagnetic waves to diverge. This, increases the beam of the radiation pattern and reduce furthermore the back lobes as we can see compared the curves of Fig. 32. Finally in Fig. 33 is shown the measured return loss of half bend LWA Type III.

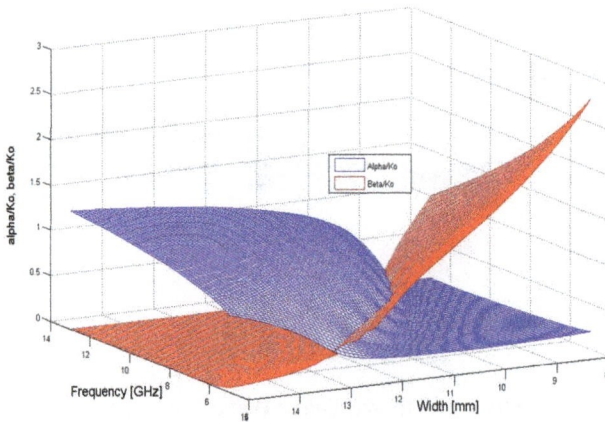

Fig. 30. The 3D normalized phase constant and attenuation constant of tapered LWA versus frequency and width.

a)

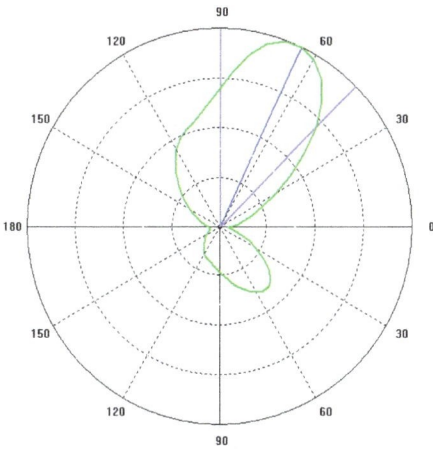

b)

Fig. 31. a)Layout of a bend tapered LWA. b) A prototype of bend half LWA Type III made using Roger 5880 RT/Duroid.

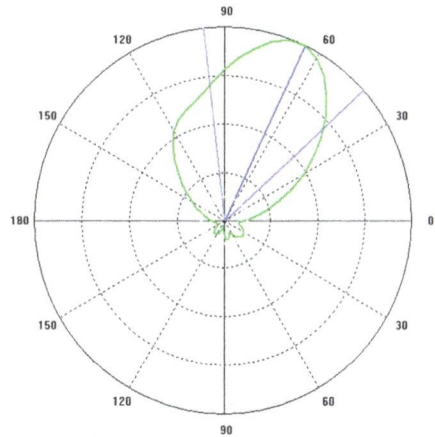

a)   b)

Fig. 32. a)The radiation patterns of E field of tapered LWA at f= 8 GHz. b) The radiation patterns of E field of bend tapered LWA at f= 8 GHz.

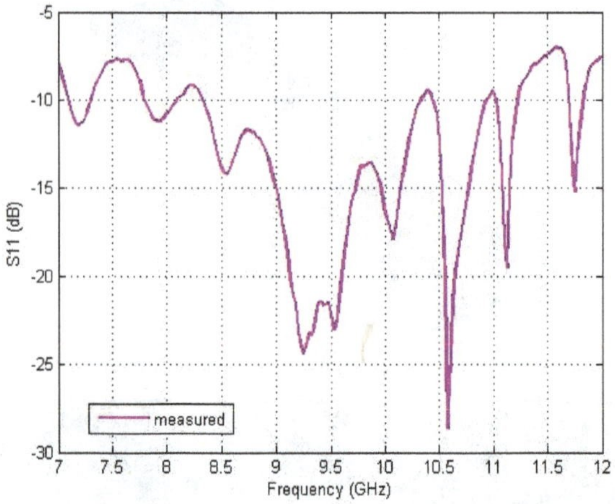

Fig. 33. Experimental return loss of half bend LWA Type III.

10. Tapered composite right/left-handed transmission-line (CRLH-TL) leaky-wave antennas (LWAs)

Recently composite right/left-handed (CRLH) leaky-wave antennas (LWAs) have been shown as one of the applications of the CRLH transmission line (TL) metamaterials thanks to their advantages of fabrication simplicity and frequency/electrically scanning capability without any complex feeding network. Neverthless the fixed geometrical size of a unit cell of the CRLH-TL Leaky-wave antennas, prevents the possibility to improve the antenna bandwidth "tapering" the geometrical size of unit cell.

It is well known as a composite right/left-handed transmission-line (CRLH-TL) metamaterials, used for the leaky-wave antennas (LWAs) allow to obtain a superior frequency scanning ability than its conventional counterpart [26-27]. The leaky-wave antennas possess the advantages of low-profile, easy matching, fabrication simplicity, and frequency/electrically scanning capability without any complex feeding network.

However, the conventional leaky-wave antennas suffer from major limitations in their scanning capabilities. In fact the radiation pattern is restricted to strictly positive θ for uniform configurations, or to a discontinuous range of negative or positive θ excluding broadside direction, for periodic configurations. The CRLH LWAs have essentially suppressed these limitations, being able to scans the entire space from θ = -90° to θ = +90° and thereby paved the way for novel perspectives for leaky-wave antennas.

Although actually the designs of CRLH-TL for LWAs available in the literature, are developed as a different number of unit cell with a fixed geometrical size for all the unit cells of the entire antenna.

These design prevents the possibility to improve the antenna bandwidth "tapering" the geometrical size of unit cell. In order to obtain an improvement of the antenna bandwidth a novel design of CRLH LWAs was used in our work. The simulation results of the the CRLH unit-cell with different size, obtained by a commercial 3D EM simulator has shown the good performance of this antenna compared with the performance of the uniform CRLH TL LWA antenna.

The good performance of this composite right/left-handed LWA are also demonstrated by measured results, which shown a good agreement with simulation results paving the way for the future applications of the antenna.

11. Antenna design

It has been shown that the leakage rate of the CRLH-TL LWA can be altered by using different sizes of the unit-cell [27] as shown in Fig.34.

Fig. 34. Different size and number of fingers of CRLH-TL unit cell.

In detail the radiation resistance, of the unit-cell having four fingers and the unit-cell of six fingers, both the unit-cells designed to have the phase origin at the same frequency, shows two different bandwidth as mentioned in [27-28].

The radiation resistance of the four finger antenna is always higher than that of the six-finger one, which implies faster decay of power (more leakage) along the structure for the former.

Moreover it should be noted that increasing the number of fingers the size of the unit-cell has to be reduced in order to have the same centre-frequency for the antenna antennas, otherwise, the centre-frequency for the antenna with unit-cell which have the larger number of fingers will shift down to a lower frequency [29-30].

As shown in [21-22] for a simple microstrip leaky wave antenna the radiation bandwidth is governed by the line width once the substrate is fixed. The bandwidth can be improved by adopting a tapered line structure (Fig.35), where, the radiation of different frequency regions leaks from different parts of the antenna.

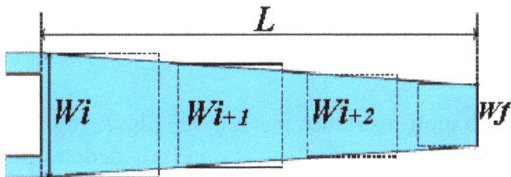

Fig. 35. A taper layout of LWA with a different frequency regions leaks from different parts of the antenna.

In fact from the propagation characteristics of the leaky wave antenna, we known that the leakage radiate phenomena, can only be noted above the cutoff frequency of higher order mode, and below the frequency such that, the phase constant is equal at the free space wave number. Decreasing the width of the antenna for a microstrip leaky-wave antenna the cutoff frequency increases shift toward high frequency. This behaviour allows to design a multisection microstrip LWA according [21-22] superimposing different section, in which each section can radiate in a different and subsequence frequency range, obtaining a broadband antennas. In this way each section should be into bound region, radiation region or reactive region, permitting the power, to uniformly radiated at different frequencies.

Following these idea in the our developed procedure we have applied a process to get the dimension of the physical parameters of the unit cell shows in Fig. 36 whit different optimized number of fingers (see Fig. 37). Naturally we have calculated the extraction parameters of every cell of CRLH implementation: LR, CR, LL, and CL using the equation mentioned in [26].

In the case of CRLH transmission line based LWA the amount of radiation by the unit cell can be related to the beam shape required and thus can be used to determine the total size of the structure as mentioned in [31].

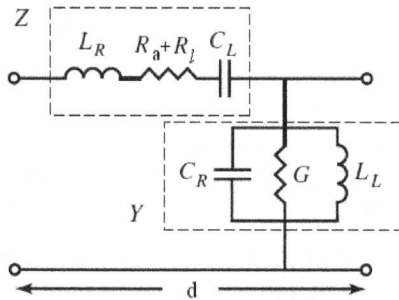

Fig. 36. Unit cell equivalent circuit with radiation resistance.

Given the unit cell equivalent circuit in Fig. 36, we have the per unit length series impedance and per unit length shunt admittance as follows [31]:

$$Z' = R'_a + R'_l + j\omega L'_R - \frac{j}{\omega C'_L} \tag{21}$$

$$Y' = G' + R'_l + j\omega C'_R - \frac{j}{\omega L'_L} \tag{22}$$

Where R'_a represents radiation resistance per unit length, R'_l represents the per unit length resistance associated with transmission loss, L'_R and C'_R denotes per unit length parasitic inductance and capacitance respectively, C'_L and L'_L denotes times unit cell length left-hand capacitance and inductance respectively.

Propagation constant and characteristics impedance are given by the following relations:

$$\gamma = \alpha + j\beta = \sqrt{Z'Y'} \tag{23}$$

$$Z_c = \sqrt{\frac{Z'}{Y'}} \tag{24}$$

From the above expression of propagation constant and line impedance we can find the centre frequency of the CRLH LWA [31].

These procedure was applied for subsequency frequency range of interest able to obtain a broadband antenna and a narrow-beam radiation pattern more than the uniform CRLH-TL LWA.

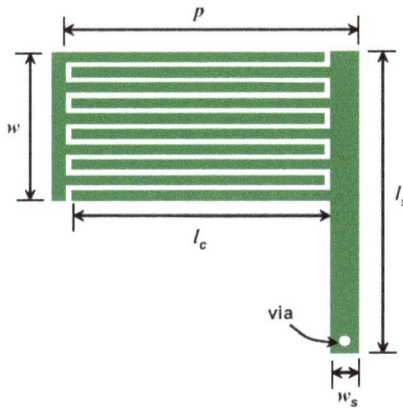

Fig. 37. Layout of a single unit cell of CRLH-TL LWA were p is the length of the unit cell period, lc, is the length of the capacitor w and ws represent, the overall width of its finger and the width of the stub respectively.

The optimized antenna design as we can see in Fig. 38 was obtained considering the sequence of 16 cells composed respectively by 4 cell with 12 fingers, 4 cells with 10 fingers,

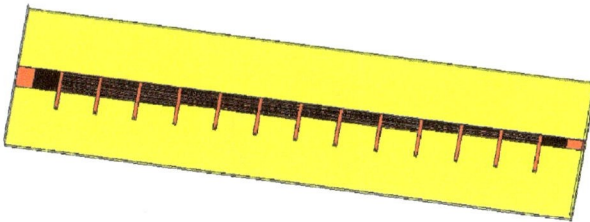

Fig. 38. Layout of the 16 unit-cell CRLH-TL LWA with cells of 10, 8 and 6 fingers.

4 cell with 8 fingers and 4 cell with 6 fingers for the entire length of the antenna of 207.55 mm and width between 2.9 mm (cells with 12 fingers) and 5.9 mm (cells with 6 fingers).

12. Simulation and experimental results

In the following Fig. 39 and Fig. 40 are showing the simulation data of the return loss obtained with a 3D EM commercial software, of the uniform 16 unit-cell CRLH-TL LWA of 10 fingers compared with the results of tapered 16 unit-cell CRLH-TL LWA. Instead in Fig. 41 and in Fig. 42, are shown the results of the radiation pattern of the uniform CRLH-TL LWA compared with 16 unit-cell of 10 fingers and tapered 16 unit-cell CRLH-TL LWA. It is evident the good performance of the tapered 16 unit-cell CRLH-TL LWA compared with uniform CRLH-TL LWA in term of broadband and narrowbeam.

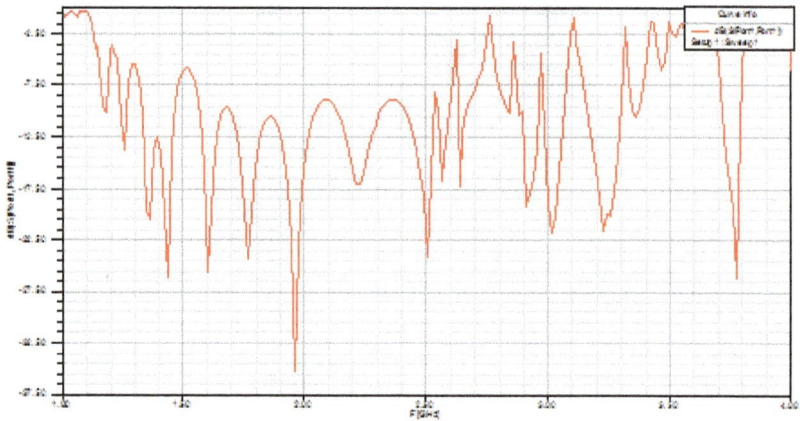

Fig. 39. Return loss (S11) of the uniform 16 unit-cell CRLH-TL LWA with 10 fingers.

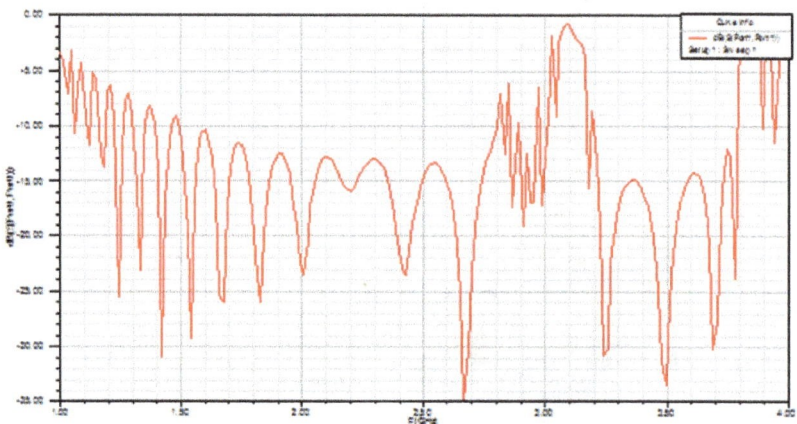

Fig. 40. Return loss (S11) of the 16 unit-cell CRLH-TL LWA with cells of 12, 10, 8 and 6 fingers.

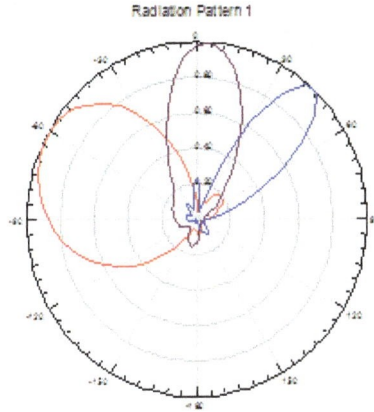

Fig. 41. Radiation pattern of the E field of the uniform CRLH-TL LWA for f=1.12 GHz (red line), f=2.30 GHz (brown line), f= 3.20 GHz (blu line).

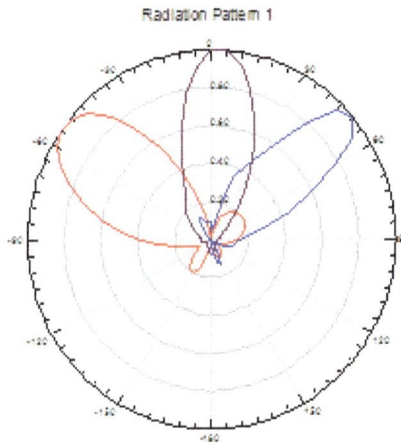

Fig. 42. Radiation pattern of the E field of the tapered CRLH-TL LWA for f=1.12 GHz (red line), f=2.30 GHz (brown line), f= 3.20 GHz (blu line).

The simulation results were compared with experimental results made on a prototype of CRLH-TL LWA (see Fig. 43) designed with 16 unit-cell on Rogers RT/duroid 5880 substrate with dielectric constant εr = 2.2 and thickness h = 62 mil (loss tangent = 0.0009) showing a quite good agreement with simulated results of tapered 16 unit-cell CRLH-TL LWA as we can see in Fig. 44 and Fig. 45.

Fig. 43. A prototype and its detail of Radiation patter of tapered 16 unit-cell CRLH-TL LWA made on Rogers RT/duroid 5880.

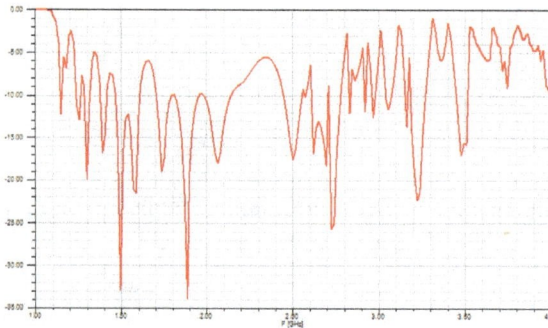

Fig. 44. Experimental return loss (S11) of the 16 unit-cell prototype CRLH-TL LWA with cells of 12, 10, 8 and 6 fingers.

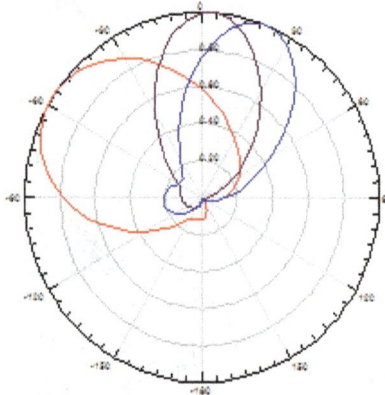

Fig. 45. Experimental E field radiation pattern of tapered prototype CRLH-TL LWA for f=1.12 GHz (red line), f=2.30 GHz (brown line), f= 3.20 GHz (blu line).

13. Meander antenna

Nowadays, miniaturization of electronic devices is the main request that productions have to fulfil. In this process, the reduction of the antenna size is the crucial challenge to face, being its dimensions related to the working frequency.

The rapid developing of the modern society has become to a crescent interest for the wireless communications. Nowadays, everybody wants to be connected everywhere without the use of cumbersome devices. The current tendency goes towards portable terminals that have to be light and hand pocket. A suitable antenna for the portable terminals should be low cost, low profile, light weight and especially small size.

Printed antennas are commonly used for their simple structure and easy fabrication. As applications are space limited, it is challenging to design an antenna of small size but with simple tunable feature. For microstrip antennas, some techniques, such as making slots in their structure or using high dielectric constant substrates, can be used to reduce the antenna size. However, it results in the narrow bandwidths for its high Q factor and the low radiation efficiency.

In the following sentences is described an antenna printed on a substrate with low dielectric constant in order to get a reliable bandwidth. Moreover, the meander configuration allows us to reduce the antenna size keeping good radiation performance.

Meander dipole antennas have been already designed through numerical techniques that apply either time- or frequency-domain algorithms demanding high computational efforts and long-time processing. The common approach in the design of meander antenna is to draw a meander path with a suited length to the working frequency in commercial software and run simulations. Nevertheless, This empirical approach may lead to several consecutive trials and verifications. In order to decrease the long-time processing and avoid these cut and try methods, it would be convenient to start simulations with a commercial software having an antenna size close to its optimized dimensions. Thus, a good initial configuration can strongly affect the numerical convergence efficiency and the design process would be quicker.

This paper presents a transmission line model that provides an initial geometrical configuration of the antenna that allows us a computational improvement in the design of meander antennas. The dimensions obtained from the model have been used to run a simulation with a commercial software and an antenna resonating very close to that working frequency has been achieved. Finally, a quick optimization has been performed to definitely tune the antenna according to the ISM band.

14. TL model for meander antennas

Commercial and military mobile wireless systems demand for high compactness devices. An important component of any wireless system is its antenna. Whereas significant efforts have been devoted towards achieving low power and miniaturized electronic and RF components, issues related to design and fabrication of efficient, miniaturized, and easily integrable antennas have been overlooked. In this paper a novel approach for antenna miniaturization is presented. The meander topology is proposed as a good approach to achieve miniaturization and a transmission line model for the analysis and synthesis of meander antennas is developed.

Indeed, the miniaturization of an antenna can be accomplished through loads placed on the radiating structure. [32-33]. For example, monopoles were made shorter through center loaded (inductive) or top loaded (capacitive). Hence appropriate loading of a radiating element can drastically reduce the size, however, antenna efficiency may be reduced as well. To overcome this drawback, lumped elements of large dimensions can be created using distributed reactive elements. For this reason, we propose a meander topology that allows us to distribute loading through short-circuited transmission lines.

In the past, meander structures were suitably introduced to reduce the resonant length of an antenna without great deterioration of its performances [34-36].

To exploit the meander topology to miniaturize printed antennas and develop a transmission line model for the analysis and the synthesis of this kind of antennas is proposed an antenna shown in Fig 46. It is a meander printed on the same side of the chassis of a circuit board on a FR4 substrate with εr =3.38 and thickness 0.787 mm. The feeding is between the meander structure and the ground plane. Even if the antenna is a printed monopole, it can be studied as an asymmetric dipole and its input reactance has been studied through a transmission line model. It is well known that the resonance condition is obtained when the input impedance is purely resistive [37-39]. The antenna has been modelled as a transmission line periodically loaded from inductive reactances Xm represented by the half meanders shown in Fig 46. We have named half-meander the shorted transmission line of length $w/2$.

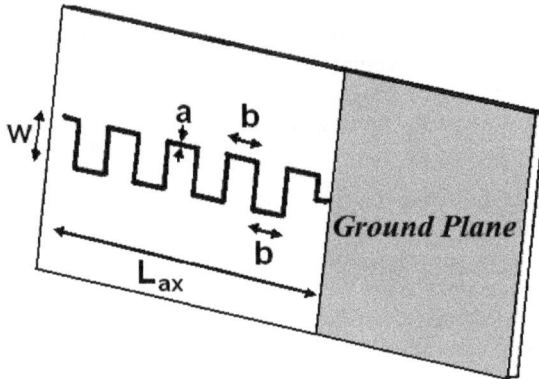

Fig. 46. Meander antenna monopole printed on the same side of the ground plane of the substrate.

The height b of each half meander and their total number $2n$ are related to the total length of the monopole Lax with the following formula:

$$L_{ax} = (2n + 1)b \tag{25}$$

Each half meander was studied as a short transmission line $w/2$ long with a characteristic impedance Zcm obtained as:

$$Z_{cm} = 120 \ln(b / a) \tag{26}$$

and terminating with a metallic strip having an inductance Lsc:

$$L_{sc} = 2 * 10^7 b[\ln(8b / a) - 1]. \tag{27}$$

The inductance Lsc of the strip with the length b and width a was substituted by a line $Lall$ long terminating with a short circuit. The length $Lall$ was properly chosen because this line had the same inductance of the strip (Lsc).

At the end, the inductance of each meander Xm was obtained by the formula (28):

$$X_m = Z_{cm}tg[2\pi\sqrt{\varepsilon_{e_{eff}}}\,(w/2 + L_{all} - a/2)] \tag{28}$$

The total characteristic impedance of the transmission line with a length Lax and loaded by $2n$ half meanders was:

$$Z_c = 120[\ln(8L_{ax}/a) - 1] \tag{29}$$

In Fig. 47 the normalized length Lax of the printed monopole versus the normalized thickness a according to the transmission line model for $w/b = 1$ is shown. It is pointed out from the figure that, when $b=w$, the meander length is smaller than the conventional monopole at its resonant frequency.

In Fig 48, for several values of the parameter $x=w/\lambda$, the resonant length Lax/λ versus the ratio w/b is plotted.

It can be seen that, for each value of w/b, a remarkable reduction of the antenna length is obtained by increasing the values of w at the resonant frequency. Therefore, by choosing a value of w/b, the model allows to detect the correspondent resonant length of the antenna.

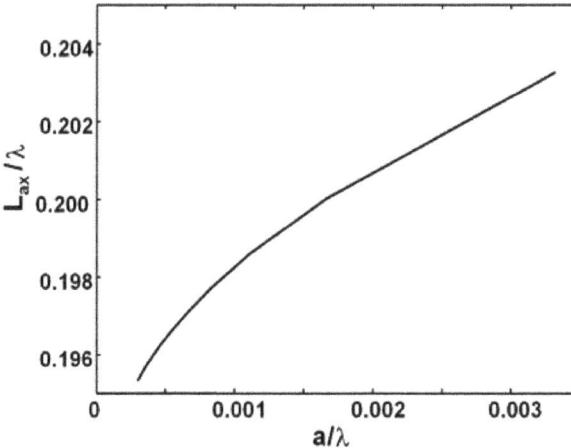

Fig. 47. The normalized length Lax of the printed monopole versus the normalized thickness a according to the transmission line model for $w/b = 1$.

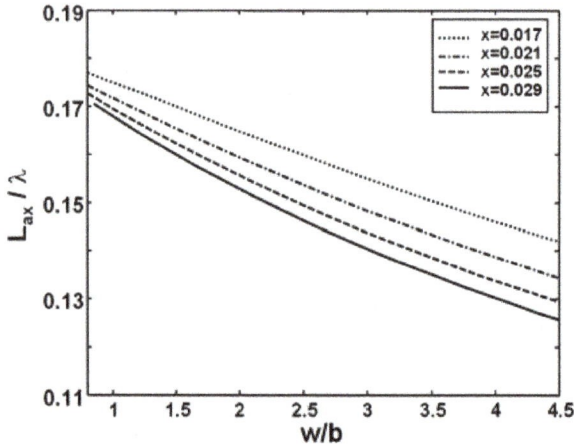

Fig. 48. Transmission line model for meander antenna printed on substrate with εr =3.38 and s = 0.81 mm for different $x=w/\lambda$.

15. Simulated and experimental results

To test the validity of the model, simulations were run with full wave commercial software CST Microwave Studio © at a frequency of 2.45 GHz with appropriate model as shows in Fig 49.

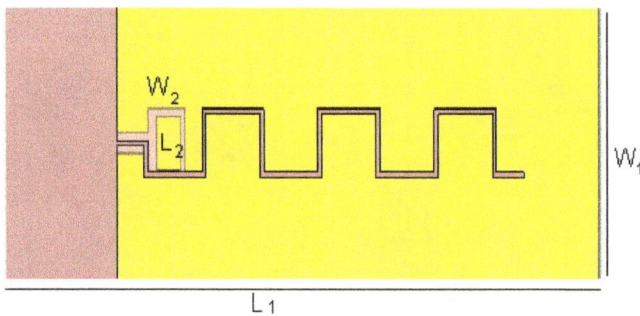

Fig. 49. Top and bottom model of meander antenna monopole designed with CST Microwave Studio ©.

The transmission line model (TLM) and the full wave simulation were in a good agreement and the difference between the resonant length obtained by TLM and the full wave simulations was within 250 MHz as it has been summarised in Table 2. Figs 47 and Fig. 48 show, respectively, the normalized length Lax versus antenna thickness a for different values of w/b and the normalized length Lax versus w/b for different values of $x=w/\lambda$.

			w/b=1.4				
		model		1° run		2° run	
w/λ	n	L_{ax}/λ	fr sim.	L_{ax}/λ	fr sim.	L_{ax}/λ	fr sim.
			[GHz]		[GHz]		[GHz]
0.029	3	0.158	2.26	0.137	2.54	0.143	2.448
0.025	4	0.177	2.13	0.154	2.40	-	-
0.021	5	0.185	2.12	0.162	2.37	0.157	2.430
0.017	6	0.180	2.23	0.159	2.49	-	-

			w/b=1				
		model		1° run		2° run	
w/λ	n	L_{ax}/λ	fr sim.	L_{ax}/λ	fr sim.	L_{ax}/λ	fr sim.
			[GHz]		[GHz]		[GHz]
0.029	2	0.155	2.43	-	-	-	-
0.025	3	0.189	2.14	0.160	2.446	-	-
0.021	3	0.160	2.45	-	-	-	-
0.017	4	0.169	2.41	-	-	-	-

Table 2. Resonant frequencies calculated with FIT method.

The antenna sizes derived from the model allow us to obtain a design very close to the final project which can be quickly optimized by avoiding long simulations with commercial software.

Table 2 shows that the antenna sizes derived from the model allow us to get antenna sizes close to the final structure as the antenna resonates almost at 2.45 GHz. Moreover, in order to get exactly 2.45GHz, a quick optimization has to be carried out by running few simulations with a commercial software.

To validate the proposed TLM method, simulations and measurements have been performed. The antenna has been printed on a Rogers R04003C with εr =3.38 and thickness 0.81 mm. A prototype is presented in Fig 50. The geometrical sizes chosen were a=0.5 mm, b=8 mm and w=b that has led to a length Lax =56mm by considering 6 half meanders (Fig 50). The board total size is $L1$=72mm and $W1$=32mm, by considering also the chassis. The antenna is fed by a microstrip printed on the back of the chassis by terminating with a stub for achieving good matching. The microstrip is 20mm length and the stub is L2=8mm and W2= 4mm. The dimensions of the microstrip line has been optimized using full wave software to provide better impedance matching for the frequency antenna-resonance.

The simulated and measured return loss is shown in Fig 51. Simulation has been performed by using CST Microwave Studio © 2009 and it has shown a value of -44dB at 2.45 GHz.

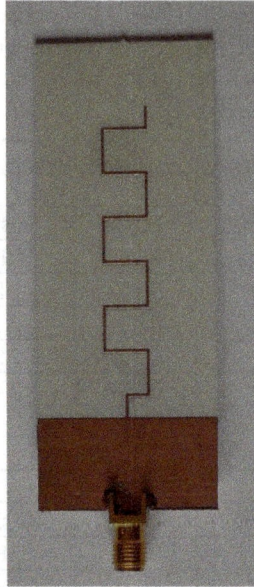

Fig. 50. Meander antenna monopole printed on the Rogers R04003C substrate.

The measurements were carried out in an anechoic chamber by connecting the antenna at a network analyser through coaxial cables.

The measured return loss in Fig 51 shows a slight shift of the antenna resonant frequency towards lower frequencies from 2.45 GHz to 2.42 GHz. Nevertheless, a good matching is still observed because the reflection coefficient assumes the value -26 dB instead of -44 dB.

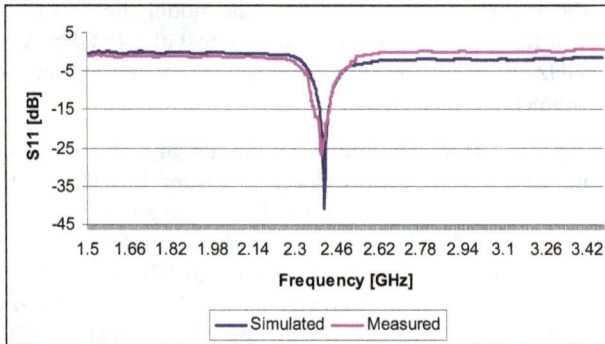

Fig. 51. Comparison of S11 simulation results with measured results.

Fig. 52 shows the E field radiation patterns of the antenna at 2.45 GHz on two principal planes, xz plane ($\Phi=0°$) and yz plane ($\Phi=90°$). The comparison of the radiation patterns shows that simulations and measurements are in a good agreement.

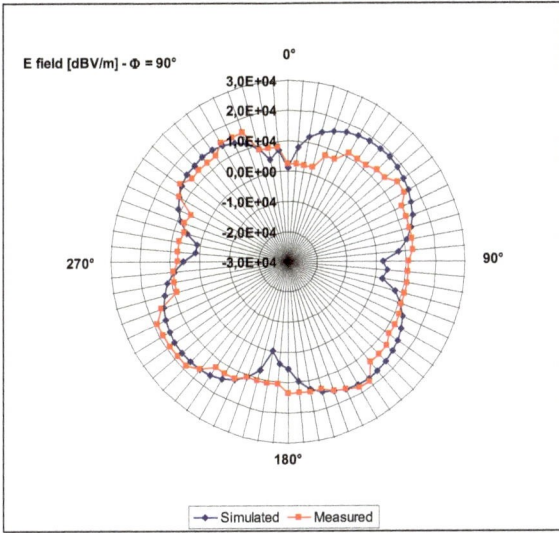

a)

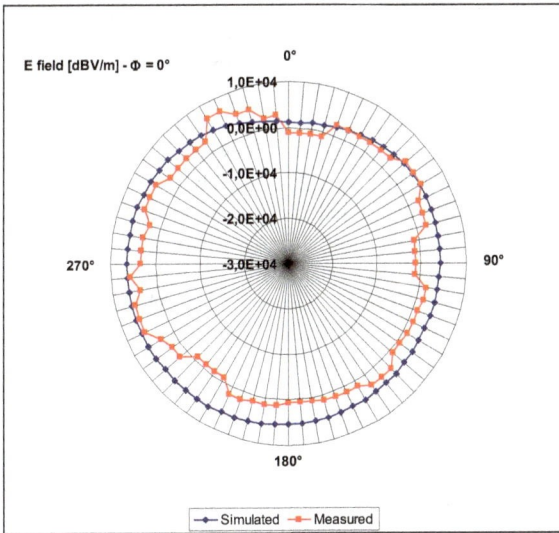

b)

Fig. 52. Comparison of measured and simulated E-field at 2.45 GHz for a) Φ= 0° and b) Φ= 90°.

Fig 53 shows the current distribution on the antenna. It can be observed that the current is particularly intense at the end of each half meander. Full wave simulations confirm that each half meander can be studied as a transmission line terminating in a short circuit.

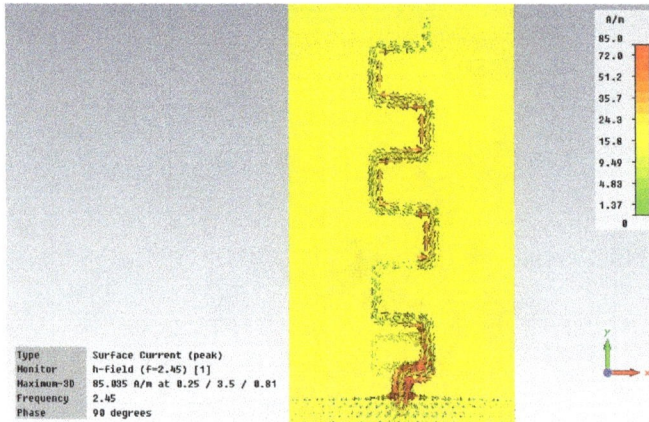

Fig. 53. Current distribution on the meander antenna.

16. Reference

[1] T. Tamir, \Leaky-wave antennas", ch. 20 in Antenna Theory, Part 2, R. E. Colin and F. J. Zucher, Eds., McGraw-Hill, New York, 1969.

[2] A. A. Oliner, \Leaky-wave antennas", ch. 10 in Antenna Engineering Handbook, 3rd ed., R. C. Hansen, Ed., McGraw-Hill, New York, 1993.

[3] C. H. Walter, \Traveling Wave Antennas", McGraw-Hill, New York, 1965.

[4] T. Tamir, A. A. Oliner, \Guided complex waves, part I: _elds at an interface", Proc. Inst. Elec. Eng., vol. 110, pp. 310-324, Feb. 1963.

[5] T. Tamir, A. A. Oliner, \Guided complex waves, part II: relation to radiation patterns", Proc. Inst. Elec. Eng., vol. 110, pp. 325-334, Feb. 1963.

[6] L. O. Goldstone and A. a. Oliner, \Leaky-wave antennas I: rectangular waveguide", IRE Trans. Antennas and Propagation, vol. AP-7, pp. 307-319, Oct. 1959.

[7] A. Hessel, \General characteristics of traveling -wave antennas", ch. 19 in Antenna Theory, Part 2, R. E. Colin and F. J. Zucher, Eds., McGraw-Hill, New York, 1969.

[8] G. Gerosa and P. Lampariello, \Lezioni di Campi Elettromagnetici I", Edizioni Ingegneria 2000, 1995.

[9] F. Frezza, Lezioni di Campi Elettromagnetici II, March 2004.

[10] Pozar, David M. and David H. Schaubert, \Microstrip Antennas: The Analysis and Design of Microstrip Antennas and Arrays", John Wiley, New York, NY, 1995.

[11] Pozar, David M., \Microwave Engineering", John Wiley, New York, NY, second edition, 1998.

[12] Kumar, Girish and K. P. Ray, \Broadband Microstrip Antennas", Artech House, Boston, MA, 2003.

[13] Menzel Wolfgang, \A New Travelling-Wave Antenna in Microstrip", Archiv fur Elektronik und Ubertragungstechnik (AEU), Band 33, Heft 4, pp. 137-140, April 1979.

[14] Yau, D., N. V. Shuley, and L. O. McMillan, \Characteristics of Microstrip Leaky-Wave Antenna Using the Method of Moments", IEE Proc. Microwave Antennas and Propag, Vol. 146, No. 5, pp. 324-328, October 1999.

[15] Lee, Kun Sam, \Microstrip Line Leaky Wave Antennas", Ph.D. thesis, Polytechnic Institute of New York, 1986.

[16] N. Marcuvitz, \On _eld representations in terms of leaky waves or Eigenmodes", IRE Transactions on Antenna and Propagation, Vol. AP-4, pp. 192-194, July 1956.

[17] T. Itoh, R. Mittra, \Spectral domain approach for calculating the dispersion characteristics of microstrip lines", IEEE Trans. Microwave Theory Techn., Vol. MTT-21, pp. 496-499, 1973.

[18] Mesa, Francisco and David R. Jackson, \Investigation of Integration Paths in the Spectral Domain Analysis of Leaky Modes on Printed Circuit Lines", IEEE Trans. Microwave Theory Techn., Vol.50, No. 10, pp. 2267-2275, October 2002.

[19] Kuester, Edward F., Robert T. Johnk, and David C. Chang, \The Thin-Substrate Approximation for Reection from the End of a Slab-loaded Parallel-Plate Waveguide with Applications to Microstrip Patch Antennas", IEEE Transactions on Antennas and Propagation, Vol. AP-30, No.5, pp. 910-917, September 1982.

[20] A. Oliner, "Leakage from higher modes on microstrip line with application to antennas," Radio Scienze, Vol. 22, pp. 907-912, 1987.

[21] O. Losito, "A New Broadband Microstrip Leaky-Wave Antenna" *Applied Computational Electromagnetics Society Journal*, Vol.23, n.3, Pg. 243-248 September 2008.

[22] O. Losito, "A Simple Design of Broadband Tapered Leaky-Wave Antenna" *Microwave and Optical Technology Letters*, vol. 49, pp.2833-2838, 2007.

[23] W. Hong, T. L. Chen, C. Y. Chang, J. W. Sheen, Y. D. Lin "Broadband Tapered Microstrip Leaky-Wave Antenna", IEEE Trans. Antennas and Propagation, Vol. 51, pp. 1922-1928, 2003.

[24] Y. Qian, B. C. C. Chang, T. Itoh, K. C. Chen and C. K. C. Tzuang, "High Efficiency and Broadband Excitation of Leaky mode in Microstrip Structures", *IEEE MTT-S Microwave Symposium Digest*, vol. 4, pp. 1419-1422, 1999.

[25] P. Burghignoli, F. Frezza, A. Galli, G. Schettini "Synthesis of broad-beam patterns through leaky-wave antennas with rectilinear geometry" *IEEE Antennas and Wireless Prop. Letters.*, vol. 2, pp. 136-139, 2003.

[26] Christophe Caloz, Tatsuo Itoh, 'Electromagnetic Metamaterials: transmission line theory and microwave applications', Wiley-IEEE Press December 2005, Chapter 3, pp. 122-124.

[27] A. Rahman, Y. Hao, Y. Lee and C.G. Parini, "Effect of unit-cell size on performance of composite right/left-handed transmission line based leaky-wave antenna", Electronics Letters, 19th June 2008 Vol. 44 No. 13.

[28] C. Caloz and T. Itoh, "Novel microwave devices and structures based o n the transmission line approach of meta-materials", in IEEE-MTT Int. Symp. Dig., June 2003, pp. 195 – 198.

[29] C. Caloz, I. Lin, and T. Itoh, "Orthogonal anisotropy in 2-D PBG structures and metamaterials," in IEEE-APS Int. Symp. Dig., vol. 1, June 2003, p. 199.11.

[30] C. Caloz and T. Itoh, "Application of the transmission line theory of lefthanded (LH) materials to the realization of a microstrip LH transmission line", in IEEE-APS Int. Symp. Dig., vol. 2, June 2002, pp. 412 – 415.

[31] A. Rahman, Y. Lee, Y.Hao and C. G. Parini "Limitations in bandwidth and unit cell size of composite right-left handed transmission line based leaky-wave antenna". Proceeding of EuCAP 2007 11 - 16 November 2007, EICC, Edinburgh, UK.

[32] L.C. Godara, Handbook on antennas in wireless communications, CRC Press, Boca Raton, FL, 2002, Ch. 12.

[33] C.W.Harrison, "Monopole with inductive loading", IEEE Trans. *Antennas Propagation* Vol AP-11,pp 394-400, 1963.

[34] R.C. Hansen, "Efficiency and matching tradeoffs for inductively loaded short antennas", *IEEE Trans. Commun.*, vol. COM-23, pp 430-435, 1975.

[35] H.Nakano, H. Tagami, A.Yoshizawa, and J. Yamauchi, "Shortening ratio of modified dipole antennas", *IEEE Trans . Antennas Propagation,* Vol. AP-32, pp. 385-386, 1984.

[36] J. Rashed and C.Tai, "A new class of resonant antennas", *IEEE Trans. Antennas Propagation,* Vol 39, Sett. 1991.

[37] C. T. P. Song, Peter S. Hall, and H. Ghafouri-Shiraz, "Perturbed Sierpinski Multiband Fractal Antenna With Improved Feeding Technique", *IEEE Transactions On Antennas And Propagation,* Vol. 51, No. 5, May 2003.

[38] R.P. Clayton, Compatibilità Elettromagnetica, *Ulrico Hoepli* Milano, 1992.

[39] R.K. Hoffmann, "Handbook of Microwave Integrated Circuits", Artech House, Norwood, MA, 1987.

4

Superstrate Antennas for Wide Bandwidth and High Efficiency for 60 GHz Indoor Communications

Hamsakutty Vettikalladi, Olivier Lafond and Mohamed Himdi
Institute of Electronics and Telecommunication of Rennes (IETR)
University of Rennes 1
France

1. Introduction

Modern multimedia applications demand higher data rates and the trend towards wireless is evident, not only in telephony but also in home and office networking and customer electronics. This has been recently proven by the accelerating sales of IEEE 802.11 family WLAN hardware. Current WLANs are, however, capable of delivering only 30-100 Mb/s connection speeds, which is insufficient for future applications like wireless high-quality video conferencing, multiple simultaneous wireless IEEE 1394 (Firewire) connections or wireless LAN bridges across network segments. For these and many other purposes, more capacity — wirelessly — is needed. Service provided by IEEE 802.11 WLANs fulfills casual internet users and office workers actual needs. But, bandwidth demands are still rising. Millimetre-wave technology is one solution to provide up to multi-Gbps wireless connectivity for short distances between electronic devices. The data rate is expected to be 40-100 times faster than today's wireless LAN systems, transmitting an entire DVD's data in roughly 15 seconds. 60 GHz is ideally suited for personal area network (PAN) applications. A 60 GHz link can replace various cables used today in the office or in home by wireless link as shown in Fig.1, including gigabit Ethernet (1000Mbps), USB 2.0 (480Mbps), or IEEE 1394 (~800Mbps). Currently, the data rates of these connections have precluded wireless links, since they require so much bandwidth. While other standards are evolving to address this market (802.11n and UWB), 60 GHz is another viable candidate. In such a context, 60 GHz millimeter wave (MMW) systems constitute a very attractive solution due to the fact that there is a several GHz unlicensed frequencies range available around 60 GHz, almost worldwide. In Europe, the frequency ranges 62 - 63 GHz and 65 - 66 GHz are reserved for wideband mobile networks (MBS, Mobile Broadband System), whereas 59 - 62 GHz range is reserved for wideband wireless local area networks (WLAN). In the USA and South Korea, the frequency range 57 - 64 GHz is generally an unlicensed range. In Japan, 59 - 66 GHz is reserved for wireless communications (Nesic et al., 2001). This massive spectral space enables densely situated, non-interfering wireless networks to be used in the most bandwidth-starving applications of the future, in all kinds of short-range (< 1 km) wireless communication. Also in this band, the oxygen absorption reaches its maximum value (10-15 dB/km), which gives an additional benefits of reduced co-channel interference. Hence, it is a promising candidate for fulfilling the future needs for very high bandwidth wireless connections. It enables up to gigabit-scale connection speeds to be used in indoor WLAN networks or fixed wireless connections in metropolitan areas.

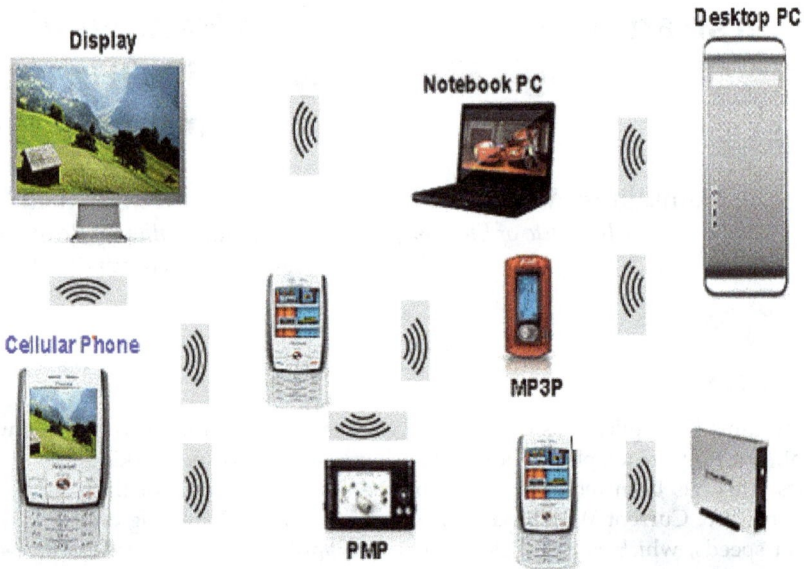

Fig. 1. Short range communication.

These new systems will need compact and high efficient millimeter wave front-ends including antennas. For antennas, printed solutions are often demanding for the researchers because of its low profile, lightweight and ease of integration with active components (Zhang et al., 2006). High gain and high efficient antennas are needed for 60 GHz communication due to high path losses at this range of frequencies. Conventional antenna arrays are used for high gain applications. But in these cases for achieving high gain, a large number of elements are needed, which not only increases the size of the antenna but also decreases its efficiency (Lafond et al., 2001), (Kärnfelt et al., 2006) & (Soon-soo oh et al., 2004). It has been reported that for high gain, a superstrate layer can be added at a particular height of 0.5 λ_0 above the ground plane (Choi et al., 2003) , (Menudier et al., 2007) & (Meriah et al., 2008).

2. Superstrate antenna technology

In this chapter the authors are explaining how to develop a wideband, high gain and high efficient antenna sufficient for 60 GHz communications using superstrate technology. Also explains the importance of different sources on antenna performance in terms of bandwidth, gain and efficiency.

2.1 Microstrip fed parasitic patch antenna with superstrate

Here the antenna configuration consists of a microstrip feed, patch and a parasitic patch, as the source, which are loaded by a superstrate. Fig. 2 shows the 3D view (a) and side view (b) of the microstrip fed stacked patch antenna with superstrate. It consists of a lower patch with an optimised dimension of 1.63 mm x 1.6 mm on a substrate RT Duroid 5880 (ε_r = 2.2, t_1 = 0.127 mm). The upper patch with an optimised dimension of 1.63 mm x 1.63 mm is printed on the lower side of a parasitic substrate RT Duroid 5880 (ε_r = 2.2, t_2 = 0.254 mm).

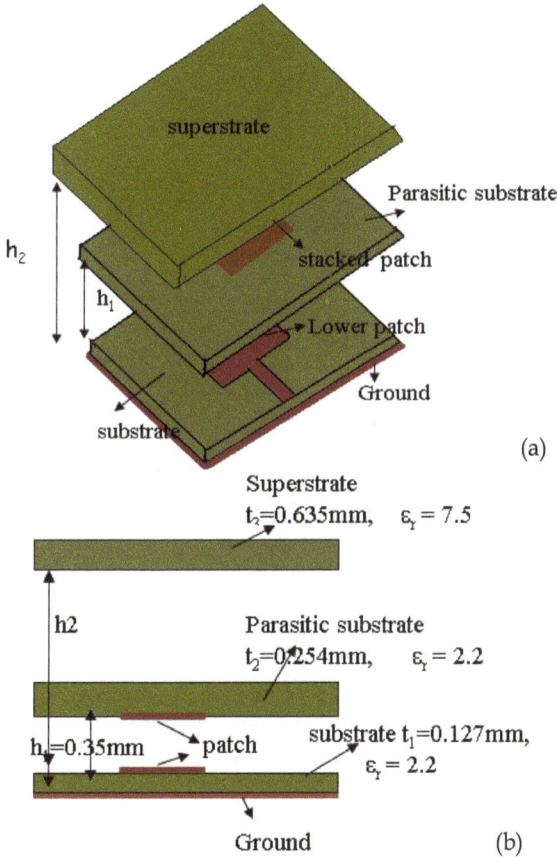

Fig. 2. Cutting plane of stacked patch antenna with superstrate.

The distance between the lower patch and the upper patch is optimized, by simulation using CST Microwave Studio®, as h_1 = 0.35 mm for a resonance at 60 GHz for a larger bandwidth and gain. This antenna is then loaded with superstrate. The material used for the superstrate is Roger substrate RT6006 (ε_r = 7.5 at 60 GHz, t = 0.635 mm). The dimension of the superstrate and the height from the ground plane are optimized as explained below. The variation of gain and VSWR bandwidth with the variation of superstrate dimension ($0.73\lambda_0$, $1.1\lambda_0$, $2\lambda_0$) for different heights ($0.5\lambda_0$, $0.6\lambda_0$, $0.7\lambda_0$) is shown in Figs. 3 (a-c). It is noted that the maximum gain with good bandwidth is achieved for a superstrate dimension of $1.1\lambda_0$ with a height = $0.6\lambda_0$, and is equal to 13.6 dB with almost flat over a frequency range of 59 GHz to 64 GHz (Vettikalladi et al., 2009). In all other cases either the gain is less than the above value or the VSWR bandwidth is poor. Also noted that when the superstrate dimension is higher than $1.1\lambda_0$, the gain goes down when the height h_2 varies from $0.5\lambda_0$ to 0.7 λ_0.

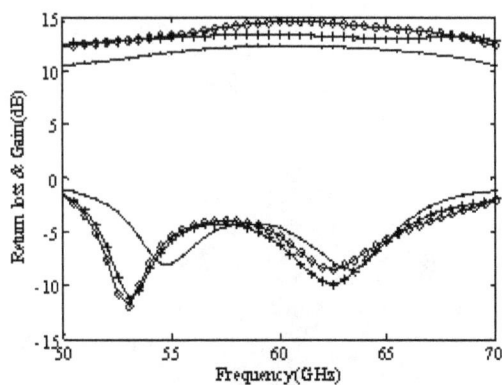

(a)

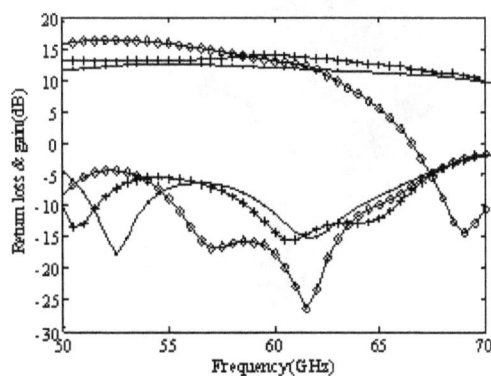

(b)

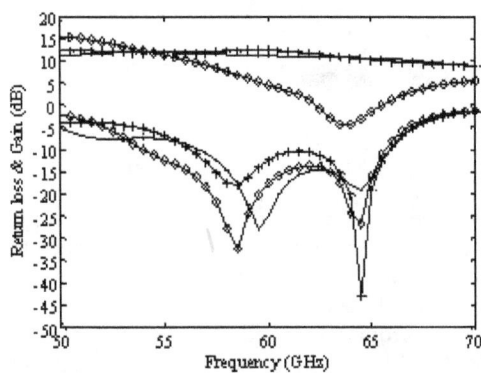

(c)

Fig. 3. variation of s11 and gain with superstrate dimension and height from ground plane.
a) h_2=0.5 λ_0 b) h_2= 0.6 λ_0 c) h_2=0.7 λ_0 for size =.73 λ_0 ——●—— size =1.1 λ_0 ——+——
size =2 λ_0 ——◇—— .

From the literature (Gupta et al., 2005), the theoretical height between the superstrate and ground plane is $0.5\lambda_0$, but in this work it is found to be $0.6\ \lambda_0$ for maximum gain, it may be due to the stacked patch. Fig. 4 shows the return loss, simulated directivity with theoretical values, and simulated and measured gain of the prototype with superstrate. It is found that there is a gain reduction in the measurement, which is due to the variation of exact heights from the theoretical values as shown in Table I.

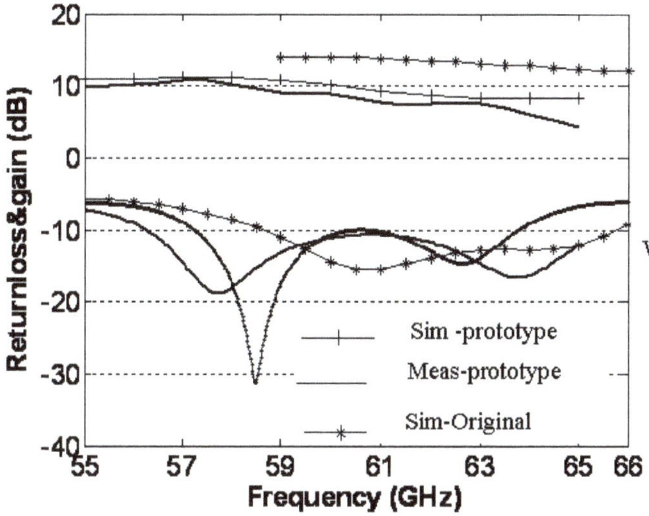

W=2mm, t=0.38mm, εr =2.2

Fig. 4 (b)

Fig. 4. (a) Comparison of return loss and gain with superstrate.

	Original (mm)	variation while implementation (mm)
Lower Patch	1.6x1.63	1.62x1.65
Upper Patch	1.63x1.63	1.65x1.65
h1	0.35	0.38
h2	3	3.48

Table I. Variation of prototype parameters from exact values.

It is very difficult to maintain the exact thickness h1, hence we inserted a substrate cut in the form of a rectangular U shape (Fig. 4(b)), with width 2mm, thickness 0.38mm and permittivity 2.2. Also the thickness h2 is varied to 3.48mm instead of 3mm ($0.6\ \lambda_0$) and hence is the reason for the reduction of gain to nearly 10 dB. The E and H planes radiation patterns at 57 GHz and 58 GHz are shown in Figs. 5(a-b). The radiation patterns are found to be broad. There is a cross polar level of less than -20 dB on both the planes. The measured half-power beam widths are found to be 37° for E and H planes at 57 GHz, and 38° and 41° for E and H planes respectively at 58 GHz.

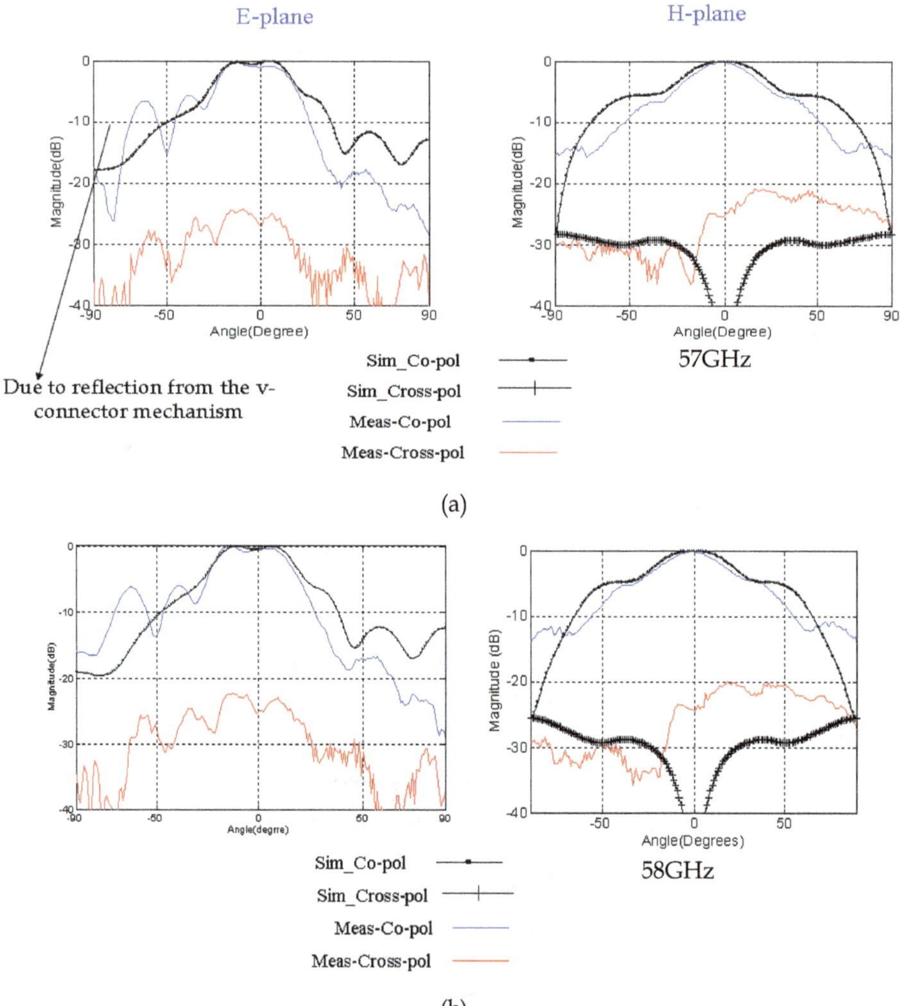

E-plane   H-plane

Due to reflection from the v-
connector mechanism

Sim_Co-pol   ━●━
Sim_Cross-pol   ━┼━
Meas-Co-pol   ————
Meas-Cross-pol   ————

57GHz

(a)

Sim_Co-pol   ━●━
Sim_Cross-pol   ━┼━
Meas-Co-pol   ————
Meas-Cross-pol   ————

58GHz

(b)

Fig. 5. Measured and simulated E-plane & H-plane radiation patterns of the parasitic patch Superstrate antenna (a) 57 GHz, (b) 58 GHz.

It is noted that for microstrip fed stacked patch antenna, the optimized superstrate size is 1.1 λ_0 for getting maximum gain and broad pattern. This value is considered as the limitation of size in this case. It is also observed from Fig. 3, that when the superstrate size is higher than 1.1 λ_0, and when the height varies from 0.5 λ_0 to 0.7λ_0 the broad nature of the gain decreases and starts coming down at 60 GHz .I.e. the pattern changes from broadside to sectorial and then to conical for different frequencies in the band as shown in Fig. 6 (for a superstrate size of 2 λ_0 & h_2=0.6 λ_0), which may suitable for some other application (Vettikalladi et al., 2009b). Here, the small superstrate size is due to the presence of the parasitic patch that disturbs the field in the cavity (thickness = 0.6 λ_0).

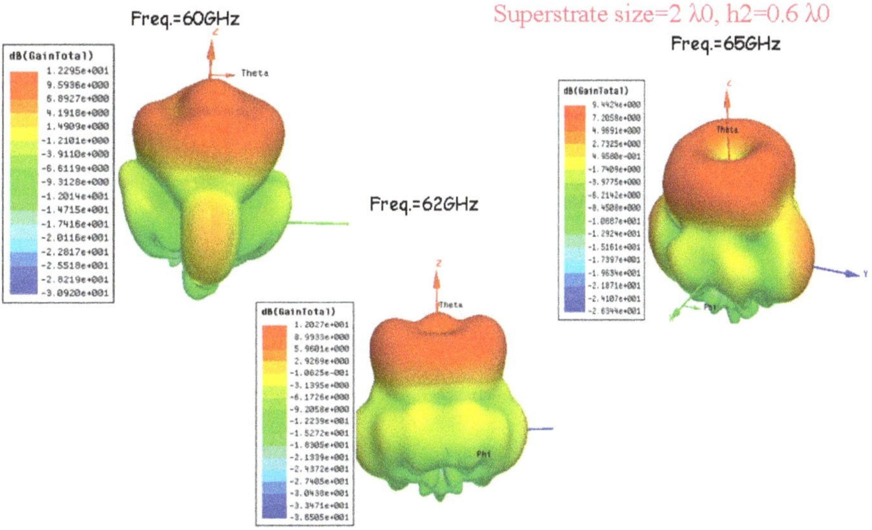

Fig. 6. Gain pattern for a microstrip fed parasitic patch superstrate with a superstrate size = $2 \lambda_0$, for different frequencies in the band.

Since this kind of prototype is very difficult to manufacture and hence we are going to discuss with other kind of technology.

2.2 Slot coupled superstrate antenna

In this section, we are explaining a superstrate antenna with aperture coupled source as the excitation. We are also showing the importance of the size of the superstrate for getting maximum gain and also for getting consistent radiation pattern all over the frequency range of interest. Fig.7 shows the side view and the 3D view of a slot coupled patch antenna with superstrate. The slot is optimised to 0.2 mm x 1 mm for maximum coupling with a stub length of 0.75 mm. In order to consider the easiness of implementation; we used a thick ground plane of thickness t=0.2 mm. The antenna consists of a patch with optimised dimension 1.3mm x 1.3mm on a substrate RT Duroid 5880 of permittivity 2.2 and a loss tangent $\tan\delta$ = 0.003 with a thickness t_1 = 0.127 mm. Low thickness and low permittivity substrate are used for reducing surface waves. A dielectric superstrate is added above the slot coupled patch antenna (Vettikalladi et al., 2009a). Here we used only one layer to avoid the technological manufacturing problems when many layers are used at 60 GHz. The material used for the superstrate is Roger substrate RT6006 with a relative permittivity of 7.5 at 60 GHz. Theoretically the thickness of superstrate must be $\lambda_g/4$ (0.456 mm), but here we took the thickness (t_2 =0.635 mm) close to the theoretical thickness available in market for good antenna performance. The distance between the superstrate and ground plane is 0.5 λ_0 as per the theory (Gupta & Kumar, 2005). A Rohacell foam layer of permittivity 1.05 is sandwiched between base antenna and superstrate to fix all the layers.

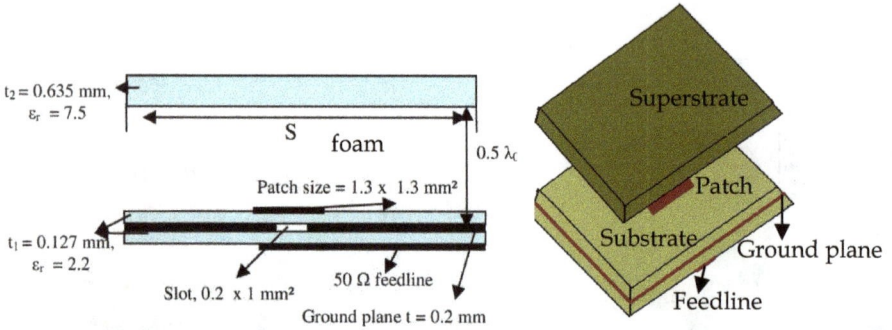

Fig. 7. Cutting plane and 3D view of aperture coupled antenna with superstrate, ground plane size = 30 x 30 mm².

Usually in all the known superstrate antennas large superstrates are used for improving the gain which not only increases the size of the antenna but also decreases the S11 bandwidth. But our objective is different, we want to use a small superstrate for obtaining high stable gain and consistant radiation pattern all over the frequency band of interest. To study the effect of superstrate size ' S ' and hence to optimize, we considered four square sizes ($1 \lambda_0, 2 \lambda_0, 4 \lambda_0$ and $6 \lambda_0$). Simulations are done using CST Microwave studio®. Fig. 8 shows the CST results of S11 and gain variations of the slot coupled antenna without superstrate and with varying superstrate size. It is observed that the S11 and gain vary with various size of the superstrate. When there is no superstrate, the antenna radiates at 60 GHz with a bandwidth of 3.7% over a frequency range of 58.9 to 61.1 GHz with a maximum gain of 5.9 dBi. It is noted that with superstrate the gain is highest for a superstrate size of $2 \lambda_0$. The 2:1 VSWR bandwidth is noted to be BW = 58.7 - 62.7 GHz i.e. 6.7% with a maximum gain of 14.9 dBi. It is also noticed that the gain decreases when the size of the superstrate is above or below $2 \lambda_0$.

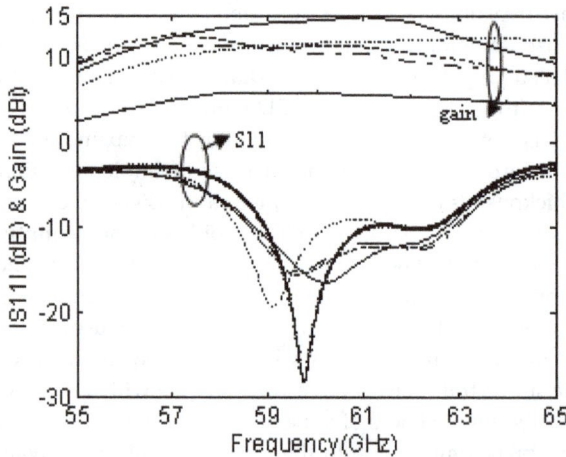

Fig. 8. Variation of S11 and gain without superstrate and with various superstrate dimensions. without superstrate ──•── $1 \lambda_0$ ·············· $2 \lambda_0$ ──── $4 \lambda_0$ ──── $6 \lambda_0$ ──·──·── .

There is a gain enhancement of 9 dB with the superstrate. Fig. 9 shows the comparison of measured and simulated S11 and gain for the optimised superstrate size of 2 λ_0. Table II gives the comparison of measured and simulated S11 and gain for the optimised superstrate size. It is noted in S11 that there is a frequency band shift of 2.8% (1.7 GHz), when a V-connector is used and a frequency band shift of 1.5% when a V-band test fixture is used. These frequency shifts are maybe due to the combined effect of connectors and the inaccuracy of the distance between patch and superstrate for the experimental prototype.

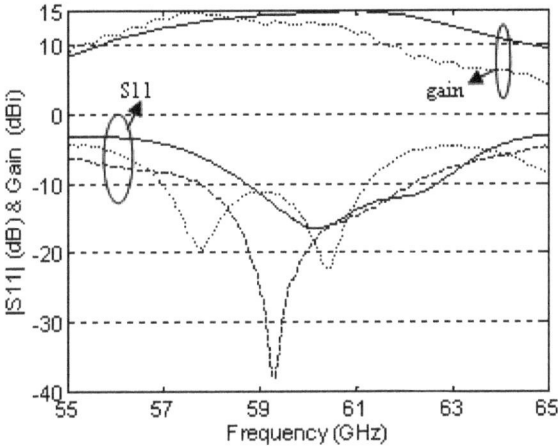

Fig. 9. Variation of S11 and gain with a superstrate dimension of 2 λ_0. Simulation ——— ; measured with V coaxial mounting connector - - - - ; measured with V test fixture — — · .

Return loss bandwidth (simulated)	Return loss bandwidth (measured)	Maximum Gain (simulated)	Maximum Gain (measured)	Efficiency Estimated η
58.7 - 62.7 GHz (6.7%)	57 - 61.1 GHz (6.8%)	14.9 dBi	14.6 dBi	76%

Table II. Comparison between simulated and measured results of aperture coupled superstrate antenna.

Also the gain measured and simulated are in good agreement but with a frequency shift as explained. The gain is measured using comparison technique with a standard horn of known gain. For calculating the efficiency, we compared the measured gain with the simulated directivity. The measured and simulated E plane radiation patterns are shown in Fig. 10a for the optimised superstrate dimension. It is clear from Fig. 9 that the measured S11 and gain are shifted; the measured gain is maximum between 57 to 59 GHz and simulated gain is from 59 to 61 GHz. Hence the radiation patterns are plotted by taking in account of this frequency shifting (e.g.; that is radiation pattern plotted is, 60 GHz simulation and 58 GHz measurement, and so on). It is noted that the radiation patterns are found to be broad and in good agreement with measurements, and there is a cross polar level of less than -28 dB at all frequencies. The radiation patterns are verified to be the same in all the frequencies in the band of interest. The measured half-power beam width is found to be 23° at 58 GHz. Also verified by simulation that the back radiation in this case is below -22 dB as compared to the antenna without superstrate (-12 dB).

Fig. 10a. Measured and simulated E-plane radiation patterns of superstrate antenna.

The measured and simulated H plane radiation patterns are shown in Fig. 10b for the optimised superstrate dimension. The radiation patterns are also plotted by taking in account of shifting as explained in E plane radiation pattern. It is noted that the radiation patterns are found to be broad and in good agreement with measurements, and there is a cross polar level of less than -28 dB at all frequencies. The measured half-power beam width is found to be 22°at 58 GHz.

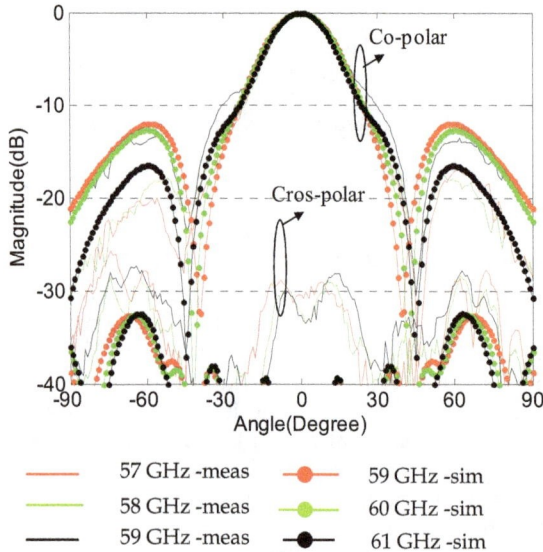

Fig. 10b. Measured and simulated H-plane radiation patterns of superstrate antenna.

When the superstrate size is higher than 2 λ_0, the broad nature of the pattern disappeared at 60 GHz. Fig. 11 shows the simulated (60 GHz) and measured (58 GHz) H plane radiation patterns of the antenna with a superstrate dimension of 6 λ_0. It is noted that the radiation patterns change from broad side to sectorial / null at 60 GHz, which is also useful for some other applications. It concludes that the dimension of the superstrate is critical for the optimum performance of the antenna. To conclude, the dimension of the superstrate is very important in order to get the consistent radiation pattern for the entire frequency band and it is found to be 2 λ_0 in this case. This is the main difference from the already developed superstrate antennas published in the literature.

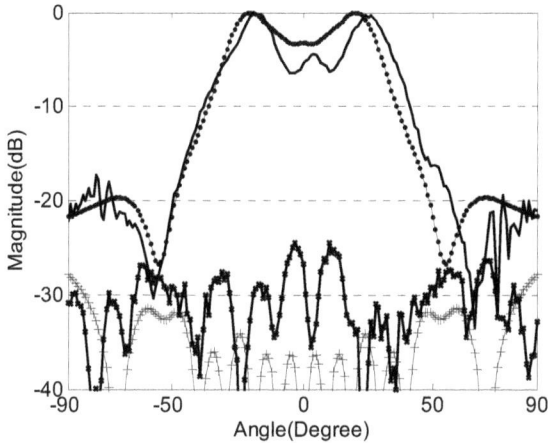

Fig. 11. Measured and simulated H-plane radiation pattern for a superstrate dimension of 6 λ_0. Co-simulated ——•—— Co-measured —— Cross-simulated ———+——— Cross-measured ——×—— .

2.2.1 Slot coupled 2x2 superstrate antenna array

Fig.12 (a) & (b) show the side view of an aperture coupled 2 x 2 patch antenna array with superstrate and the feeding network. The distance between the elements in the array are optimized to be d = 1.3 λ_0 for obtaining maximum gain and to minimize coupling. All the base antenna parameters and substrate and superstrate are same as explained in section 2.2.

As explained in section 2.2, usually, large superstrates are used for improving the gain. But our objective is different: we want to use the smallest superstrate for obtaining high stable gain and consistant radiation pattern in the frequency band. To study the effect of superstrate size ' S ' and hence to optimize it, here also we considered four square sizes (2.4 λ_0, 3.2 λ_0, 4 λ_0 and 6 λ_0). Simulations are done using CST Microwave studio. Fig. 13 shows the CST results of S11 and directivity variations of the 2 x 2 slot coupled antenna array with varying superstrate size. The S11 and directivity are affected by the size of the superstrate: the highest directivity of 18 dBi is obtained for a superstrate size of 3.2 λ_0. The resulting 2:1

VSWR bandwidth is 5% from 58.6 to 61.6 GHz. It is also noticed that the directivity decreases when the size of the superstrate is above or below 3.2 λ_0 and hence the optimised size of the 2 x 2 superstrate antenna array is 3.2 λ_0 x 3.2 λ_0. Fig. 14 shows the comparison of measured and simulated S11, and measured gain with simulated directivity for the optimised superstrate size of 3.2 λ_0.

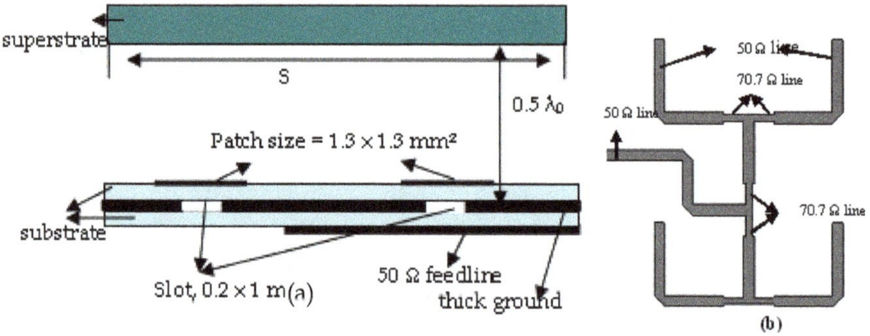

Fig. 12. a) Cutting plane of an aperture coupled 2 x 2 antenna array with superstrate, ground plane size = 6 λ_0 x 10 λ_0, for connecting V band connector and for ease of measurement purpose, b) 2x2 feeding network.

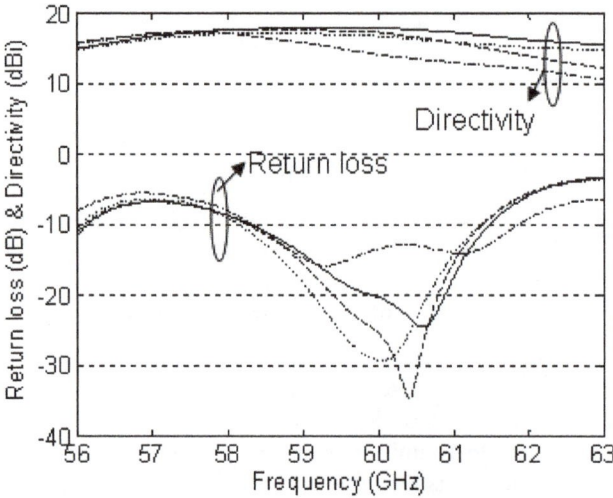

Fig. 13. Variation of return loss and directivity with various superstrate dimensions.
2.4 λ_0 ················   3.2 λ_0 ———   4 λ_0 – – – –   6 λ_0 – · – · – · .

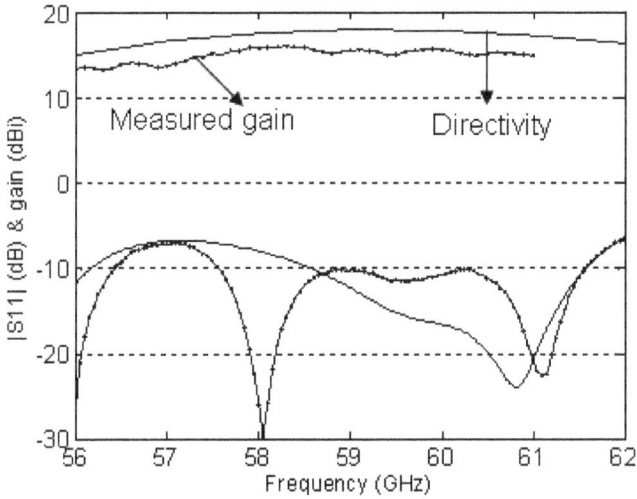

Fig. 14. Variation of S11 and gain with a superstrate dimension of 3.2 λ_0.
Simulation ——— measured ——•—— .

Table III gives the comparison of measured and simulated results for the optimised superstrate size (Vettikalladi et al., 2010a). It is found that the measured maximum gain is 16 dBi with S11 bandwidth of 6.7% and an estimated efficiency of 63%. With superstrate there is a gain enhancement of 4 dB compared to the classical 2 x 2 array (Liu et al., 2009, Book chapter 5, O. Lafond & M. Himdi). The measured gain is maximum at 58 GHz while the simulated directivity is maximum at 59 GHz, which corresponds to 1.7% frequency shift.

Return loss bandwidth (simulated)	Return loss bandwidth (measured)	Maximum Directivity (simulated)	Maximum Gain (measured)	Efficiency Estimated η
58.6 - 61.6 GHz (5%)	57.6 - 61.6 GHz (6.7%)	18 dBi	16 dBi	63%

Table III. Comparison of simulated and measured 2 x 2 superstrate antenna array.

The simulated and measured E-plane radiation patterns are shown in Fig. 15 for the optimised superstrate dimensions. It is clear from Fig. 14 that the measured gain is maximum between 58 to 59 GHz and simulated is from 59 to 60 GHz. Hence the radiation patterns are plotted by taking in account of this 1.7% shift (e.g. ; that is radiation pattern plotted is, 60 GHz simulation and 59 GHz measurement, etc). It is noted that the radiation patterns are found to be broad and in good agreement with measurements. The measured half-power beam width (HPBW) is found to be 17° at 59 GHz.

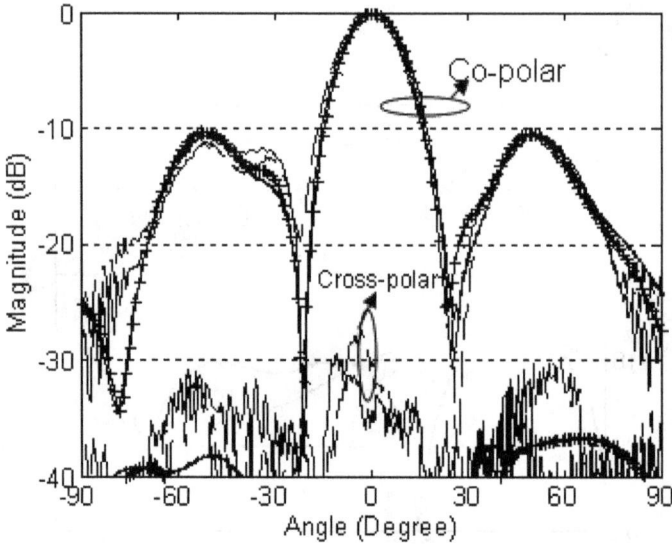

Fig. 15. Measured and simulated E-plane radiation patterns of 2 x 2 superstrate antenna array.
58 GHz - measured – – – –   59 GHz - simulated   ——•——
59 GHz - measured ——————   60 GHz - simulated   ——+—— .

The measured and simulated H-plane radiation patterns are shown in Fig. 16 for the optimised superstrate dimension. The radiation patterns are also plotted by taking into

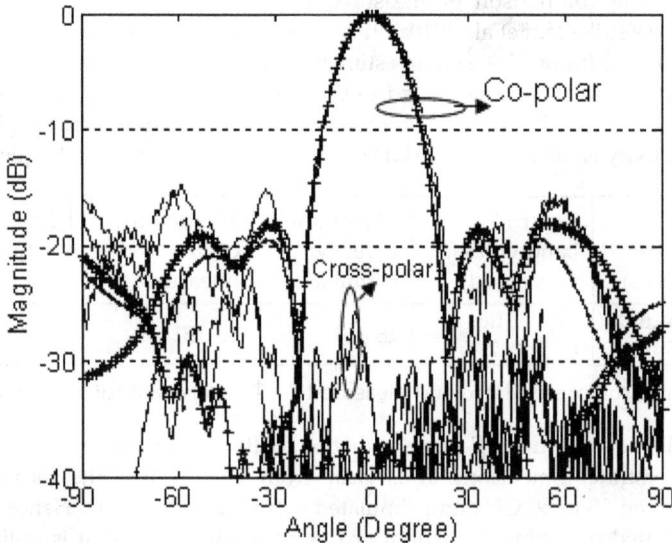

Fig. 16. Measured and simulated H-plane radiation patterns of 2 x 2 superstrate antenna array.
58 GHz - measured ——————   59 GHz - simulated   ——+——
59 GHz - measured – – – –   60 GHz - simulated   ——•—— .

account the shift as explained for E-plane radiation patterns. It is noted that the radiation patterns are found to be broad and in good agreement with measurements. The measured HPBW is found to be 16° at 59 GHz. The cross polarisation level is lower than -26 dB on both the E and H-plane, and is lower than -19 dB at 45° cut plane in 3D pattern.

2.2.2 Slot coupled 4x4 superstrate antenna array

Fig. 17 shows the photograph of a 4 x 4 array antenna array with superstrate. The antenna parameters and the distance between the elements are the same as explained for 2 x 2 superstrate antenna array in Section 2.2. The same substrate for the superstrate is used. Here also the superstrate should be optimised and it is found to be 6 λ_0 x 6 λ_0 which is the total size of the antenna. For manufacturing this prototype, because of the 4x4 array, two metal wedges of 3 mm width are used to position the superstrate at 2.3 mm (~ $\lambda_0/2$ - 0.127 mm) above the patch array. This mechanical solution is found to be better in this case than foam due to the relation between superstrate position sensitivity and gain increase.

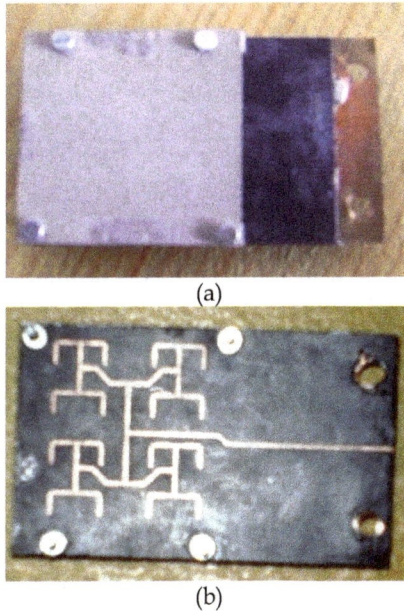

(a)

(b)

Fig. 17. Photograph of 4 x 4 superstrate antenna array prototype (a) Top view of superstrate antenna (b) view of antenna feed line network from bottom side. Ground plane taken is 6 λ_0 x 10 λ_0, for connecting V band connector and for ease of measurement purpose.

Fig. 18 shows the measured and simulated S11, and measured gain with simulated directivity. It is found that the maximum gain measured is 19.7 dBi with an efficiency of 51% (simulated directivity = 22.6 dBi) which is far better than a classical 6 x 6 array antenna of gain 17.5 dBi with an efficiency of 40% at 59 GHz (Lafond et al. 2001), and an 8 x 8 array antenna (size = 6.5 λ_0 x 6.5 λ_0) of gain 19.7 dBi with an efficiency of ~ 40% as explained in (Nesic et al., 2001). The S11 bandwidth measured for the 4 x 4 superstrate antenna array is 57.9 GHz to 61.3 GHz (5.7%).

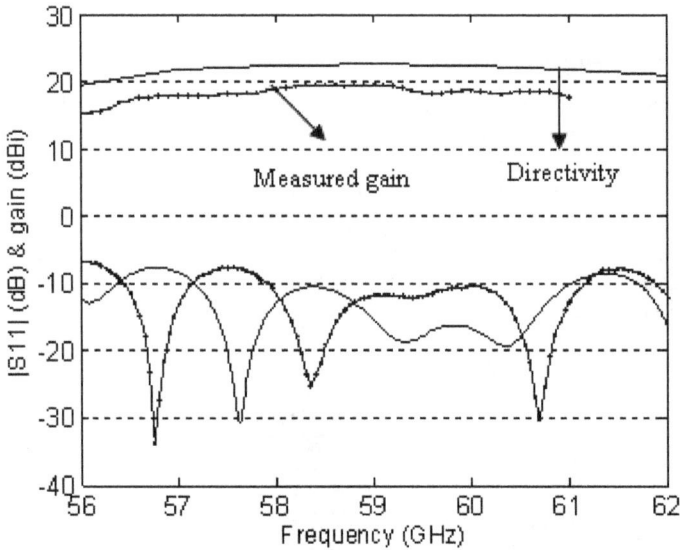

Fig. 18. Variation of S11 and gain with a square superstrate dimension of 6 λ_0. Simulation ———— measured ——•— .

It is clear from Fig. 18 that the gain measured and simulated are maximum from 58 GHz to 60 GHz. The simulated and measured H-plane radiation patterns are shown in Fig. 19 (a). It is noted that the simulated patterns are in good agreement with the measured results. The measured HPBW is 8° at 60 GHz.

The measured and simulated E-plane radiation patterns are shown in Fig. 19(b) for the optimised superstrate dimension. The radiation patterns are broad and the agreement between measurement and simulation are quite acceptable. The measured HPBW is found to be 10° at 60 GHz. In this case, the cross polarization level is lower than -25 dB on both the E and H-plane, and is lower than -16 dB at 45° cut plane in 3D pattern.

For both the presented antennas, a distance between the elements of 1.3 λ_0 is used for the source array, which induces high ambiguity side lobes for both cases when there is no superstrate: -2 dB for a 2 x 2 array and -1.9 dB for 4 x 4 array as shown in Fig. 20. Adding a superstrate will strengthen the main lobe while suppressing the ambiguity side lobes to less than -10 dB for both the arrays without affecting the back radiation as shown in Fig. 20. It also strengthens the front to back ratio as shown in the figure. It is to be underlined that the size of the superstrate is a key point of the design of such structures. In fact the nature of the pattern is conditioned by the choice of this parameter: a broad pattern is obtained for a size limited to 3.2 λ_0 x 3.2 λ_0 for 2 x 2 array and 6 λ_0 x 6 λ_0 for 4 x 4 arrays.

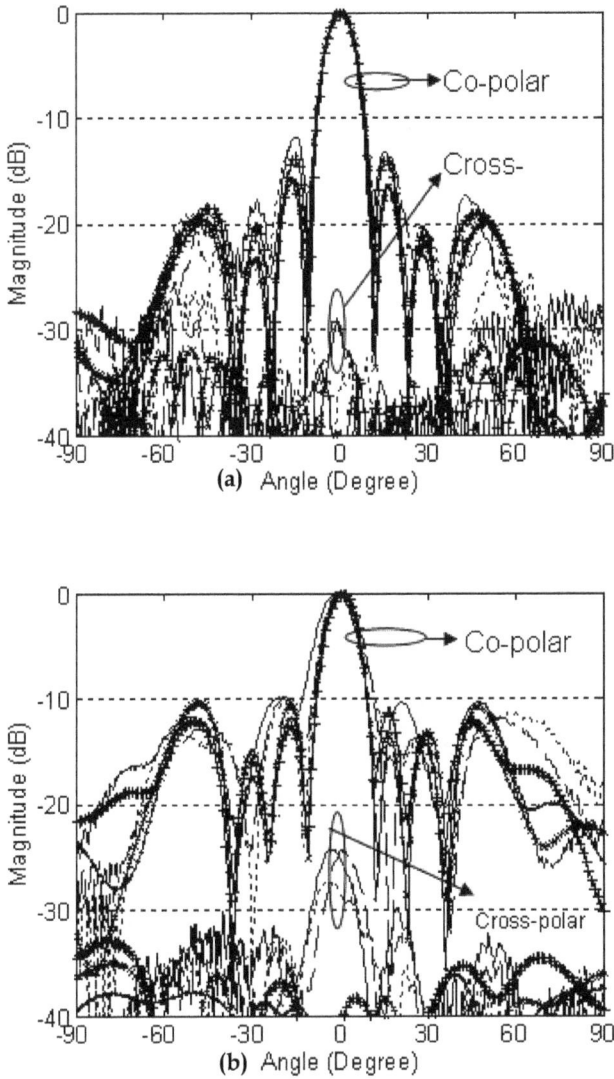

Fig. 19. Measured and simulated (a) H plane & (b) E plane radiation patterns of 4 x 4 superstrate antenna array.

58 GHz - measured ················· 58 GHz - simulated ——✕—
59 GHz - measured — — — — 59 GHz - simulated ——•—
60 GHz - measured —————— 60 GHz - simulated ——+—— .

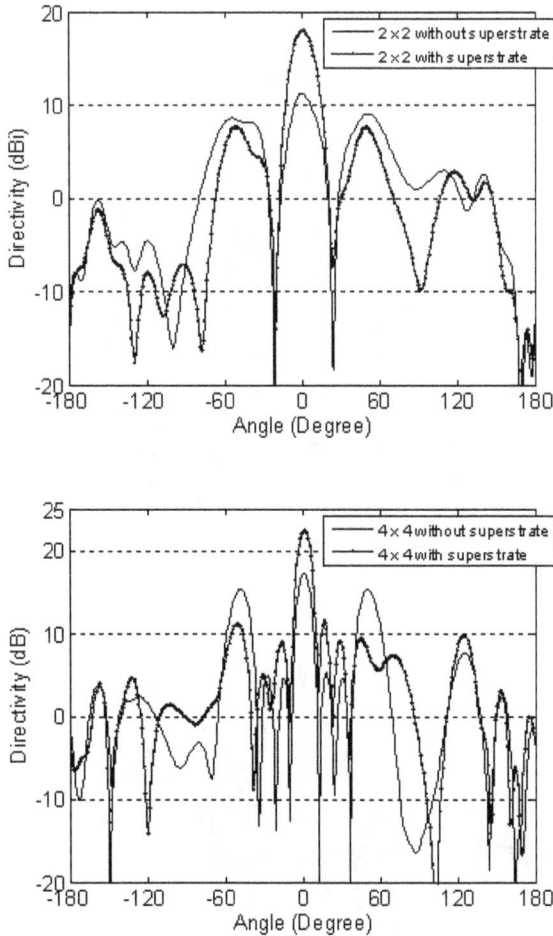

Fig. 20. Comparison of simulated results of 2 x 2 and 4 x 4 arrays without and with superstrate in terms of main beam, side lobe and back radiation.

2.3 Superstrate aperture antenna

In this section we are using another source, as aperture, for exciting the antenna. The side view of an aperture antenna with superstrate is shown in Fig. 21(a). The aperture is optimised to 4.4 mm x 1 mm for maximum coupling with a stub length of 0.4 mm (Fig. 21(b)). To improve the rigidity of antenna, a ground plane of thickness t = 0.2 mm is used. For maintaining the exact air thickness in practical prototype, the superstrate is inserted within an air pocket realized in a Rohacell foam block of permittivity 1.05, as shown in Fig. 21(c). All the substrate and superstrate material used are the same as explained in section 2.2.

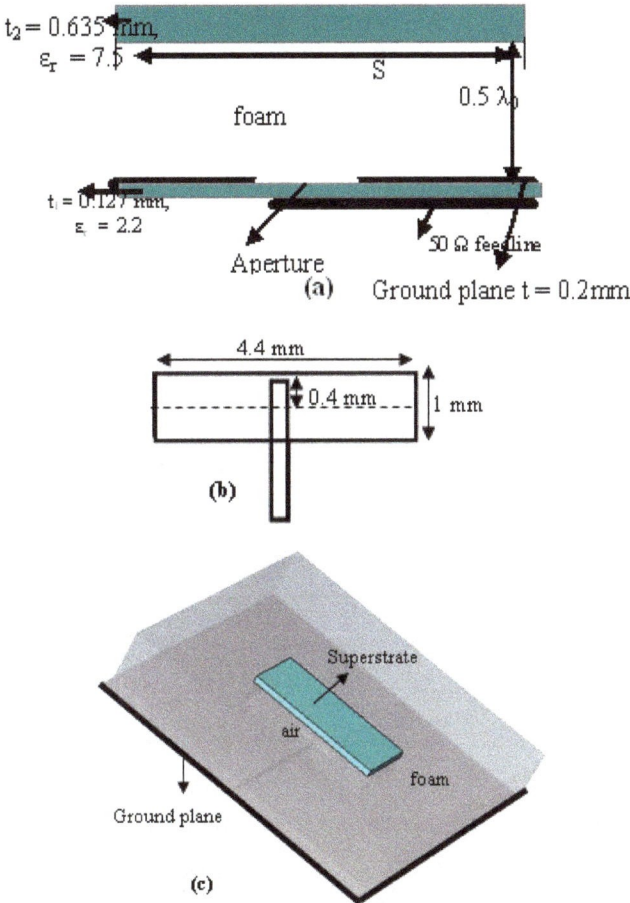

Fig. 21. (a) Cutting plane of aperture antenna with superstrate, ground plane size = 6 λ_0 x 6 λ_0. (b) Aperture and stub in details. (c) Overview of the Prototype: Details of the superstrate and air gap within foam.

In this case also we want to study the effect of superstrate size on antenna performance. To study the effect of superstrate size ' S ' and hence to optimize it, a parametric study has been performed, using commercial electromagnetic software CST Microwave studio. To highlight the effects of this parameter, results obtained for four sizes (1 λ_0 x 2 λ_0 , 2 λ_0 x 2 λ_0, 1.2 λ_0 x 2.7 λ_0 and 3 λ_0 x 3 λ_0) are reported in Fig. 22. It is observed that both the S11 and the directivity vary according to the size of the superstrate. A maximum directivity of 14.5 dBi is obtained for a superstrate size of 1.2 λ_0 x 2.7 λ_0. The corresponding 2:1 VSWR bandwidth is noted to be equal to 57.5 - 71 GHz i.e. 22.5%. It is also noticed that the directivity decreases when the size of the superstrate is above or below this optimized value. We plotted directivity only up to 65 GHz because of the decline in the values after that. When the superstrate size is higher than the optimized value, then there is a plunge in directivity as shown in Fig. 22. Hence the broad nature of the pattern moved out at 60 GHz as explained in (Vettikalladi et al., 2009a), i.e the radiation patterns change from broad side to sectorial / null at 60 GHz, which is also useful for some other applications.

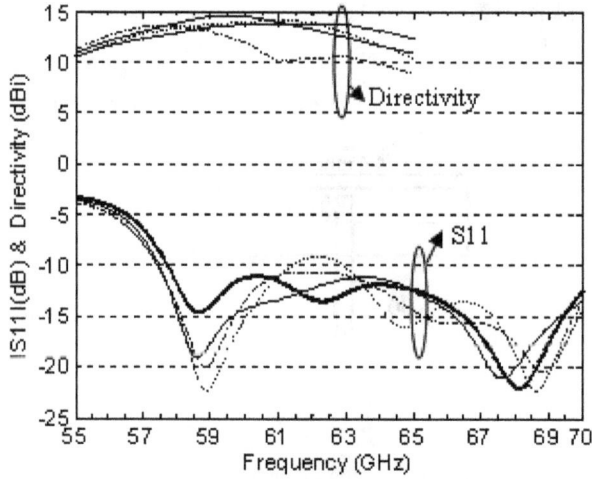

Fig. 22. Simulated results of S11 and directivity with various superstrate dimensions.
$1 \lambda_0 \times 2 \lambda_0$ ——•—— ; $2 \lambda_0 \times 2 \lambda_0$ ·············· ; $1.2 \lambda_0 \times 2.7 \lambda_0$ ——— ; $3 \lambda_0 \times 3 \lambda_0$ —·—·— .

It concludes that in this case the dimension of the superstrate is critical for the optimum performance of the antenna. Also we can control the shape of the pattern by changing the dimension of the superstrate from broadside to sectorial / null. The comparison of measured and simulated S11, and measured gain with simulated directivity for the optimized superstrate size is shown in Fig. 23 (a). Table IV gives the summary of these results (Vettikalladi et al., 2010b). It is noted that the measured 2:1 VSWR bandwidth is 15% which is larger compared to the superstrate slot coupled antenna (Vettikalladi et al., 2009a), where the bandwidth was only 6.8%. The gain is measured using comparison technique with a standard gain horn. It is found to be 13.1 dBi. Moreover, it is almost flat (ripple ~ 0.5 dB) over a bandwidth of 5 GHz. To determine the efficiency, we compared the measured gain with the simulated directivity. The estimated efficiency is 79%. In order to highlight the effect of the superstrate for this configuration, the simulated comparison of E-plane radiation pattern of aperture antenna with superstrate and without supertstrate is shown in Fig. 23 (b). The ripples in the pattern without superstrate are due the diffraction from the edges of the limited ground plane. Also it is clear that aperture antenna is a bidirectional antenna, superstrate technology make this antenna to unidirectional without adding any reflector, which is a highlight of the superstrate with this kind of source. I.e. Superstrate makes the antenna pattern directive and there is a gain enhancement of 8 dB compared to its basic aperture antenna.

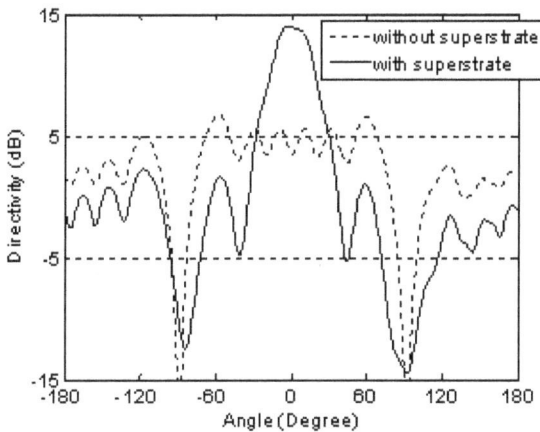

Fig. 23. (a) Results of S11 and gain with a superstrate dimension of 1.2 λ_0 x 2.7 λ_0.
Simulation ——•—— ; measurement ——— . (b) Simulated comparison of E-plane antenna radiation pattern with and without superstrate.

The measured and simulated H- and E-plane radiation patterns at 57 GHz, 60 GHz and 62 GHz are shown in Fig. 24 for the optimized superstrate dimension. It is noted that the radiation patterns are found to be broad and in agreement with simulations. The cross polar level is less than -25 dB for H-plane and -20 dB for E-plane respectively, for all the frequencies in the band. The measured half-power beam widths (HPBW) at 60 GHz are 26° for H-plane and 30° for E-plane respectively. The measured cross polarization level is lower than -17 dB at 45° cut plane in 3D patterns of both planes.

Return loss bandwidth (simulated)	Return loss bandwidth (measured)	Maximum Directivity of the prototype (simulated)	Maximum Gain (measured)	Efficiency Estimated η
57.5-71 GHz (22.5%)	55 - 64GHz (15%)	14.1 dBi	13.1 dBi	79%

Table IV. Comparison between simulated and measured results superstrate aperture antenna.

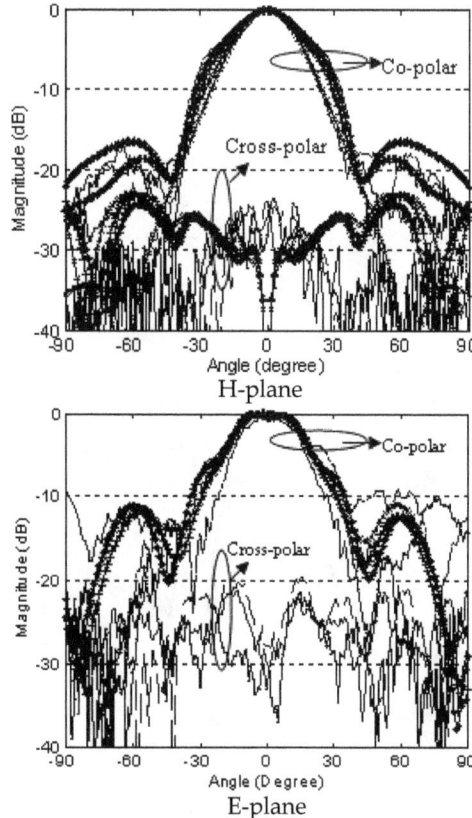

H-plane

E-plane

Fig. 24. Measured and simulated H-plane & E- plane radiation patterns of superstrate antenna (Co and Cross polarisation).

57 GHz -Simulated ——×—— ; 57 GHz - Measured ——•—— ;
60 GHz -Simulated ———+——— ; 60 GHz -Measured ——— ;
62 GHz - Simulated ——✳—— ; 62 GHz - Measured – – – – .

2.3.1 2x2 superstrate aperture antenna array

The side and 3D view of a 2 x 2 aperture antenna array with superstrate are shown in Figs. 25(a) and (c). All the parameters of the antenna are the same as explained in section 2.3. For maintaining the exact air thickness, the superstrate is inserted within an air pocket realized in Rohacell foam as shown in Fig. 25(c). The distance between the elements in the array is optimized as d = 1.3 λ_0 for obtaining maximum gain and to minimize coupling. The 2 x 2 feeding network is exposed in Fig. 25(b). In a classical array (without superstrate), when the distance between the patches is d = 1.3 λ_0, high ambiguity side lobes appear with almost the same level as the main lobe. Adding a superstrate strengthens the main lobe while reducing the ambiguity side lobes to less than -10 dB (E-plane) as shown in Fig. 26. It also strengthens the front to back ratio as shown in the figure. It has to be underlined that the size of the superstrate is a key point for the design of such structures as explained in previous cases.

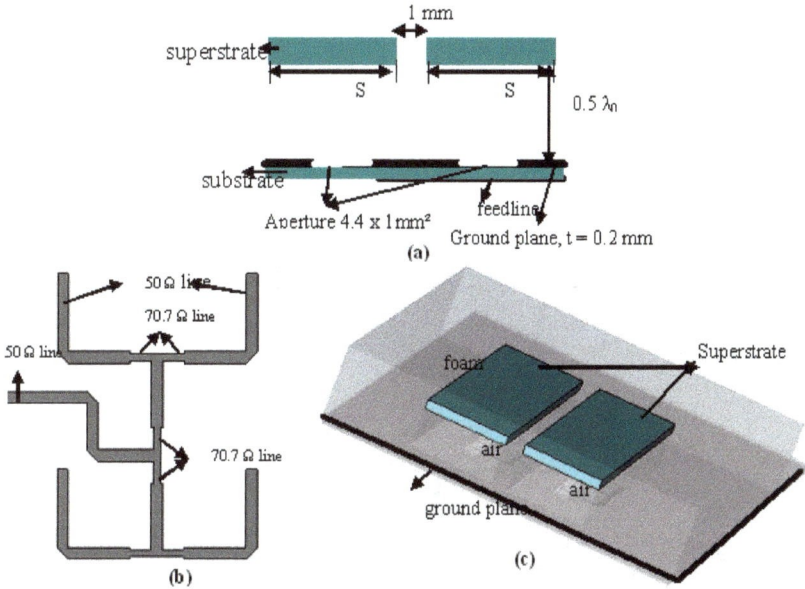

Fig. 25. (a) Cutting plane of a 2 x 2 aperture antenna array with superstrate, ground plane size = 6 λ_0 x 10 λ_0 for connecting V band connector and for ease of measurement purpose. (b) 2 x 2 feed network. (c) Overview of 2 x 2 array prototype: details of the two separate superstrate sheets and air gap within foam.

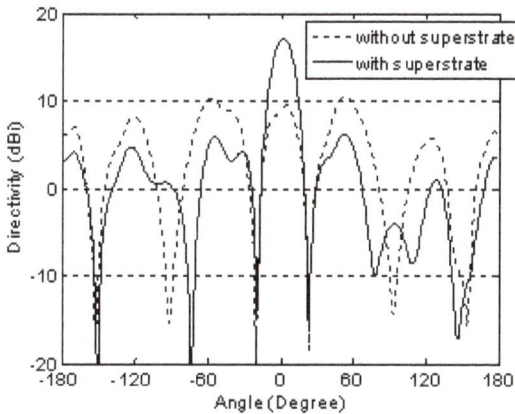

Fig. 26. Simulated comparison of 2x2 antenna array pattern (d = 1.3 λ0) with and without superstrate (E-plane).

As did in previous sections, we studied the effect of superstrate size ' S ' by simulating different sizes and the optimized solution is found to be two pieces of dimension $1.2 \lambda_0 \times 4 \lambda_0$, one sheet for two aperture antenna, with a spacing of 1mm as shown in Fig. 25(b). If we use a single piece with a size of $2.6 \lambda_0 \times 4 \lambda_0$, then the gain is little lower than in the previous case. The comparison of measured and simulated S11, and measured gain with simulated directivity for the optimized superstrate size is shown in Fig. 27. The highest directivity of 17.9 dBi is obtained for the optimized superstrate size. The resulting simulated 2:1 VSWR bandwidth is 11.3% from 57.2 to 64 GHz.

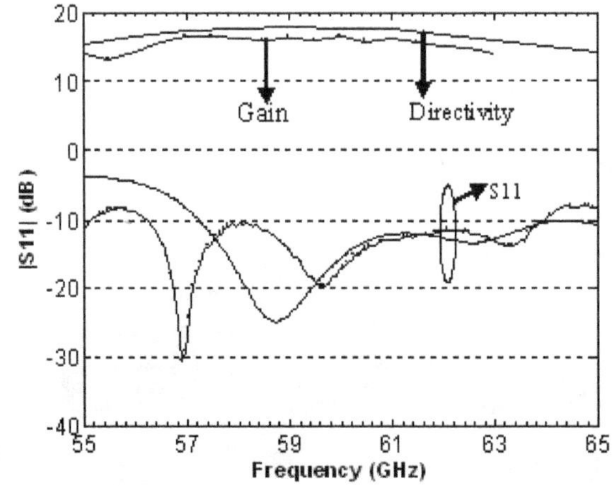

Fig. 27. Variation of S11 and gain for a 2 x 2 superstrate aperture antenna array.
Simulation ———; measurement ——•— .

Table V gives the comparison of measured and simulated results for the optimized superstrate size. It is found that the maximum measured gain is 16.6 dBi with S11 bandwidth of 13.3% (56 GHz - 64 GHz), and an estimated efficiency of 74%. This gain is comparable to a classical 4 x 4 array at 60 GHz but with better efficiency (Lafond 2000). Also the measured gain is almost stable (ripple ~ 0.8 dB) over 5 GHz (57 GHz - 62 GHz) in the band of interest (Vettikalladi et al., 2010c).

Return loss bandwidth (simulated)	Return loss bandwidth (measured)	Maximum Directivity (simulated)	Maximum Gain (measured)	Efficiency Estimated η
57.2 - 64 GHz (11.3%)	56 - 64 GHz (13.3%)	17.9 dBi	16.6 dBi	74%

Table V. Comparison between simulated and measured results of a 2 x 2 superstrate aperture antenna array.

The measured and simulated H- and E-plane radiation patterns at 57 GHz, 60 GHz and 62 GHz are shown in Figs. 28 (a) & (b) respectively for the optimized superstrate dimension. It is noted that the radiation patterns are found to be broad and in good agreement with simulations. The measured cross polar levels are -26 dB for H-plane and -20 dB for E-plane respectively. The radiation patterns are verified to be the same in all the frequencies in the band of interest. The measured HPBWs are 17° for H-plane and 16° for E-plane respectively at 60 GHz.

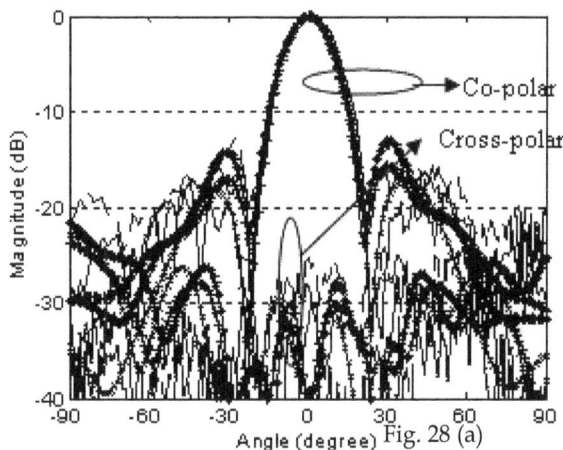

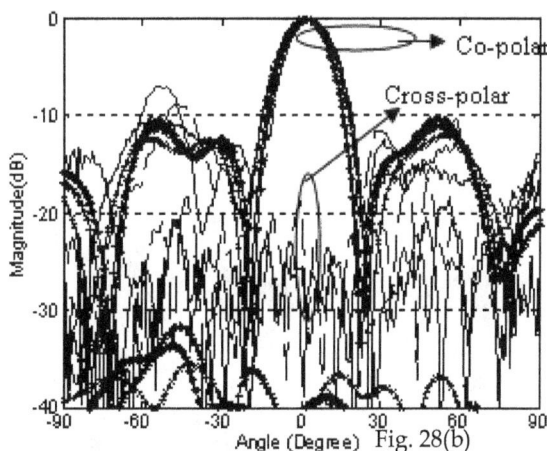

Fig. 28. Measured and simulated H-plane (a) & E-plane (b) radiation patterns of the 2 x 2 superstrate antenna array (Co and Cross polarisation).
57 GHz - Simulated ———×——— ; 57 GHz - Measured ———•—— ;
60 GHz - Simulated ———+——— ; 60 GHz - Measured ——— ;
62 GHz - Simulated ———✳——— ; 62 GHz - Measured – – – – .

2.3.2 16 x 16 superstrate aperture antenna array

Finally we developed a big array to obtain very high gain of nearly 30 dBi for 60 GHz outdoor communication, for example from one department to another department inside a university (< 1km). Fig. 29 (a) depicts the 3D side view of the 16 x 16 array prototype. The antenna parameters and the distance between the elements are all same as explained in Section 3.2. For maintaining the exact air thickness, the superstrate is inserted within an air pocket realized in Rohacell foam as shown in Fig. 29(a). The 16 x 16 feeding network is showing in Fig. 29(b).

As pointed out in previous section , we want to use the smallest superstrate for obtaining high stable gain and consistent radiation pattern in the frequency band. We studied the effect of superstrate size ' S ' by simulating different sizes and the optimized size is found to be 16 pieces of dimension 1.2 λ_0 x 21.8 λ_0, one sheet for 16 aperture antenna, with a spacing of 1mm as shown in Fig. 29(a). If we use a single piece with a size of 20.6 λ_0 x 21.8 λ_0, then the gain is little lower than in the previous case.

(a)

(b)

Fig. 29. (a) Side overview of 16 x 16 array prototype: details of the 16 separate superstrate sheets and air gap within foam, total size = 20.6 λ_0 x 21.8 λ_0. (b) 16 x 16 feed network.

The comparison of measured and simulated S11, and measured gain with simulated directivity for the optimised superstrate size is shown in Fig. 30. The highest simulated directivity of 33.3 dBi is obtained for the optimised superstrate size. The resulting simulated 2:1 VSWR bandwidth is 22 %. It is found that the maximum measured gain is 29.4 dBi (at point 'P' in Fig. 29 (b)) with S11 bandwidth of 16.7 % (54 GHz - 64 GHz), and an estimated efficiency of 41%. The measured and simulated E- and H-plane radiation patterns at 57 GHz, 60 GHz and 62 GHz are shown in Figs. 31(a) and (b) respectively for the optimised superstrate dimension. It is noted that the radiation patterns are found to be broad and in good agreement with simulations. The measured cross polar levels are -28 dB for H-plane and -26 dB for E-plane respectively. The radiation patterns are verified to be the same in all the frequencies in the band of interest. The measured HPBWs are 2.5° for H-plane and E-plane respectively at 60 GHz.

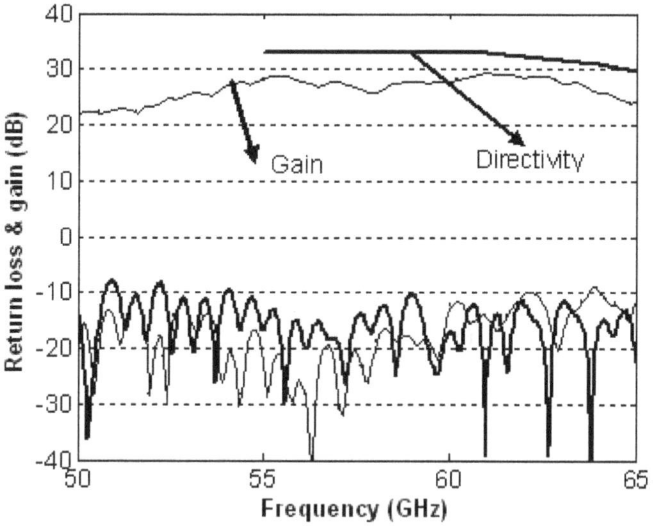

Fig. 30. Variation of S11 and gain for a 16 x 16 superstrate aperture antenna array. Simulation ———— ; measurement —•— .

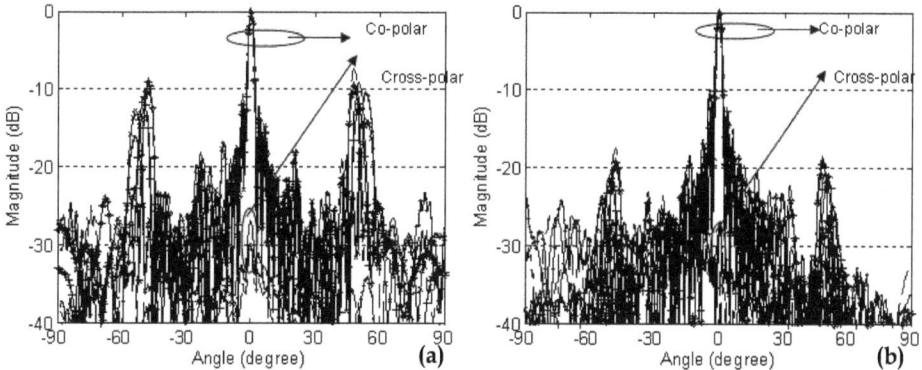

Fig. 31. Measured and simulated E-plane (a) & H-plane (b) radiation patterns of the 2 x 2 superstrate antenna array (Co and Cross polarisation).
57 GHz - Simulated ——×—— ; 57 GHz - Measured —•— ;
60 GHz - Simulated ——+—— ; 60 GHz - Measured ———— ;
62 GHz - Simulated ——✳—— ; 62 GHz - Measured – – – – .

It is noted from the study that single/small array superstrate antenna technology is very good for high gain, wide bandwidth and high efficiency, but is not suggestive for big arrays because of the gain go up is not upto the point but is good in terms of efficiency.

3. Comparison of superstrate slot coupled antenna with superstrate aperture antenna

Table VI gives the comparison of the superstrate aperture antenna element or antenna array presented with slot coupled superstrate antenna element or antenna array. It is clear that superstrate aperture antenna element / antenna array give broad S11 bandwidth of 15% / 13.3% with better efficiency as compared to superstrate slot coupled antenna element / antenna array. Also the antenna size is smaller compared to the other in both the cases. I.e. superstrate aperture antenna element or antenna array gives sufficient bandwidth, gain and efficiency for 60 GHz applications.

Antenna	Return loss bandwidth (measured)	Maximum gain (measured)	Efficiency Estimated η	Size
Single superstrate Aperture	15%	13.1	79%	6 mm x 13.5 mm x 3.48 mm
Single superstrate slot coupled	6.8%	14.6	76%	10 mm x 10mm x 3.48 mm
2x2 superstrate Aperture	13.3%	16.6	74%	13 mm x 20 mm, x 3.48 mm
2x2 superstrate slot coupled	6.7%	16	63%	16 mm x 16 mm x 3.48 mm

Table VI. Comparison between superstrate aperture s antenna and slot coupled superstrate antenna explained in section 2.

4. Conclusion

In this chapter we explained about the significance of superstrate on antenna performance at millimeter wave frequencies. It is found that the size of the superstrate is critical, which is not the case in lower frequencies. Also we studied the antenna performance with different source of excitation. It is noted that superstrate technology is very good for a single patch and also for small array but not that much good for big arrays. As a conclusion, it is found that superstrate aperture antenna element / antenna array is a good candidate for wideband, high gain and high efficiency antenna design in millimetre wave range. Moreover it is easy to integrate with electronics by placing the feed on backside of the substrate where the electronic components are integrated, and the radiating aperture and superstrate (i.e. the radiation part) are on the other side.

5. References

Cho, W; Yong Hei Ccho; Cheol-Sik Pyo & Jae_Ick Choi. (2003). A high gain microstrip patch array antenna using a superstrate layer, *ETRI journal*, 2003, vol. 25, pp.407-411.

Gupta, R. K. & Kumar, G. (2005). High gain multilayered antenna for wireless applications, *Microw. Opt. Technol. Lett.*, vol. 50, no. 7, pp. 152-154, Jul. 2005.

Julio-Navarro. (2002). Wide-band, low-profile millimeter wave antenna array, *Microw. Opt. Technol. Lett.*, 2002, vol. 34 , pp. 253-255.

Kärnfelt, C; Hallbjörner, P; Zirath, H. & Alping, A. (2006). High gain active microstrip antenna for 60-GHz WLAN/WPAN applications, *IEEE Trans. on Microw. Theor. and Techniq.*, Jun. 2006, 54 (6), pp. 2593-2602.

Liu, D; Gaucher, B; Ullrich, P. & Janusz, G. (2009). *Advanced Millimeter-wave technologies*, (Wiley, 2009), pp. 170-172.

Lafond, O. (2000). Conception et Technologies D'antennes imprimees Multicouches a 60 GHz, *PhD thesis, University of Rennes1*, France, Dec. 2000, pp. 52-54

Lafond, O; Himdi, M. & Daniel J. P. (2001). Thick slot-coupled printed antenna arrays for a 60 GHz indoor communication system, *Microw. Opt. Technol. Lett.*, 2001, vol. 28, pp. 105-108.

Meriah, S. M; Cambiaggio, E; Staraj, R. & Bendimerad, F. T. (2008). Gain enhancement for microstrip reflect array using superstrate layer, *Microw. Opt. Technol. Lett.* , 2008, vol. 46, pp. 1923-1929.

Menudier, C; Thevenot, M, Monediere, T & Jecko, B . (2007). Ebg resonator antennas state of the art and prospects, *International Conference on Antenna Theory and Techniques*, 17-21 September, 2007, Sevastopol, Ukraine.

Nesic, A; Nesic, D; Brankovic, V; Sasaki, K. & Kawasaki, K. (2001). Antenna solution for future communication Devices in mm-wave range, *Microwave Review*, pp. 9-17, Dec. 2001.

Soon-soo oh; John Heo; Dong-Hyeon Kim; Jae-Wook Lee; Myung-sun song & Yung-sik kim. (2004). Broadband millimeter-wave planar antenna array with a waveguide and microstrip feed network, *Microw. Opt. Technol. Lett.*, 2004, vol. 42, pp. 283-287.

Vettikalladi, H; Lafond, O. & Himdi, M. (2009a). High-Efficient and High-Gain Superstrate Antenna for 60 GHz Indoor Communication, *IEEE Antennas and Wireless Propagation Letters* vol. 8, pp. 1422-1425, 2009.

Vettikalladi, H; Lafond, O. & Himdi, M. (2009b). High-Gain Broad-band Superstrate Millimeter wave Antenna for 60 GHz Indoor Communications, *5th ESA Workshop on Millimetre Wave Technology and Applications and 31st ESA Antenna Workshop*, 18 - 20 May 2009, ESTEC, Noordwijk, The Netherlands.

Vettikalladi, H; Le Coq, L; Lafond, O. & Himdi, M. (2010a). Efficient and High-Gain Aperture Coupled Superstrate Antenna Arrays for 60 GHz Indoor Communication Systems, *Microwave and Optical Technology Letters*, 2010.

Vettikalladi, H; Le Coq, L; Lafond, O. & Himdi, M. (2010b). Wideband and High Efficient Aperture Antenna with Superstrate for 60 GHz Indoor Communication Systems, *2010 IEEE AP-S International Symposium on Antennas and Propagation and 2010 USNC/CNC/URSI Meeting* in Toronto, ON, Canada, on July 11- 17, 2010.

Vettikalladi, H; Le Coq, L; Lafond, O. & Himdi, M. (2010c). Broadband Superstrate Aperture Antenna for 60 GHz Applications, *European Microwave Week*, 26th sept. to-1st October 2010, Paris, France.

Zhang, Y.P & Wang, J.J. (2006). Theory and analysis of differentially-driven microstrip antennas, *IEEE Trans. Antennas Propag.*, 2006, vol. 54, pp.1092-1099.

Part 2

Wireless Communication Hardware

Hardware Implementation of Wireless Communications Algorithms: A Practical Approach

Antonio F. Mondragon-Torres
Rochester Institute of Technology
USA

1. Introduction

Wireless communication algorithms are implemented using a wide spectrum of building blocks such as: source coding; channel coding; modulation; multiplexing in time, frequency and code domains; channel estimation; time and frequency domain synchronization and equalization; pre-distortion; transmit and receive diversity; combat and take advantage of fading and multi-path channels; intermediate frequency (IF) processing in software defined radio, etc.

Due to this breadth of different algorithms, the traditional approach has been to create a system model in a high level language such as Matlab (Mathworks, 2011), C/C++ and recently in SystemC (SystemC, 2011). Usually these models use floating point representations, are architecture agnostic, and are time independent, among others characteristics. After the system model is available, then based on the specifications it is manually converted into a fixed point model that will take care of the finite precision required to implement the algorithm and compare its performance against the "Golden" floating point model. The reason to perform this conversion is due to cost and performance. While it is possible to program the algorithm on a floating point Digital Signal Processor (DSP) or using floating point hardware on application specific integrated circuit (ASIC) technology, the resulting: complexity; signal throughput; silicon area and cost; and power consumption among others, usually prohibits its implementation in floating point arithmetic. This is one of the reasons most of the wireless communications algorithms are implemented using a finite precision fixed point number representation.

In the last decade several technologies have made the conversion from floating point to fixed point seamless to a certain point. These technologies rely either on either a high level language such as C or C++ or a set of hardware model libraries for a particular field programmable gate array (FPGA) or ASIC technologies. In addition to these, there are some other electronic system level (ESL) design tools that can take a floating point algorithm and even preserve the same floating point testbench and transform the algorithm into a fixed point representation, where different architectural trade-offs can be made based on the area/power/latency/throughput requirements are in the system specifications.

In this chapter we do not propose a one solution fits all applications methodology, rather we will navigate through the author's encounters with different technologies at different stages in his career and how different applications have been and are currently approached. This is a summary of the last ten years of working with different tools, methodologies and design flows. What has prevailed due the level of integration of current Systems on a Chip (SoC) has been for example: component and systems reusability; fast algorithm and architecture exploration; algorithm hardware emulation; and design levels of abstraction.

2. System level design

By system level design (SLD), we refer to the modeling of the wireless communications systems based solely on the specifications or target standard. At this stage, individual and collective block level performance can be evaluated and also interconnects with other components in the system can be specified. There are two major known approaches for system design, top-down and bottoms-up methodologies.

System level design calls for a top-down methodology. In sophisticated systems such as SoCs, their complexity can be very large and it is a common practice in system level design to create a set of high level specifications with a complete vision of the system including their complete set of interconnects. The next phase is to divide the system into functional blocks, specify all internal interconnects and design each block in the subsystem. This allows the complete system to be simulated using for example a system level language such as SystemC and then be able to replace each block with its Register Transfer Level (RTL) functional equivalent. These techniques are also being heavily used to speed up system verification in which it is not possible to perform in a reasonable amount of time a complete RTL or gate level simulation due to time to market (TTM) constraints or because it is not computationally feasible. SLD methodologies allow performing a complete system level simulation at a higher level of abstraction by just including the key blocks required at the gate level to test interconnectivity and performance.

A system level simulation is in the order of tenths to thousands times faster than gate level simulations, thus assuring that all individual blocks or combinations of blocks will work after being interconnected. In Figure 1 it is shown an ideal case where a system level model or commonly referred as the "running specification" is first generated and creates a "golden" model against all performance implementations will be compared against. Ideally we would like to keep the original testbench for all modeling, design, implementation, simulation and verifications tasks, but this is not always possible. The problem arises when manual or automatic translations could change the behavior of the original testbench. One of the most critical problems in SLD development is that once you descend in the level of abstraction, the system level testbench and models are no longer updated and maintained, then deviating from the original running specification reference.

2.1 System modelling

SLD has been traditionally been done using C language, therefore it is common to refer in industry to the "C-model" as the running specification or "golden" model. The advantage is that C language is particularly fast, runs on all platforms and can represent fixed point precision easily after taking care of the fixed point operations such as rounding, truncation, saturation, etc. One disadvantage of this methodology is that it is not very straight forward to couple C simulations with RTL simulations and then obtain the complete benefits of system level modeling.

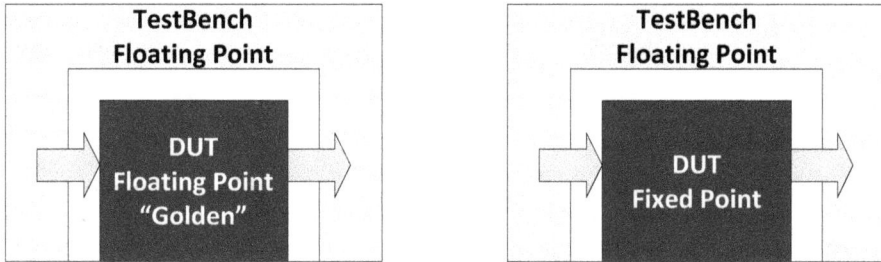

Fig. 1. System Level Modeling approach. Testbench should ideally be reused while verifying the Device Under Test (DUT) at different level of hierarchy. E.g. Behavioral, RTL, gate level netlist and parasitic extracted netlist.

More recently, SLD has been done using C++, since the level of abstraction can be taken one level further and the interfaces and testbenches can be encapsulated and reused. SystemC is a set of libraries that extend C++ to brings capabilities such as fixed point types, transaction level modeling (TLM), parallel event driven simulation compatibility, and testbench reutilization among several other features. Recently SystemC have been used to create complex reusable testbenches that interface directly with RTL code and can be executed using most of the high performance RTL event driven simulators.

A relatively new player in the SLD is System Verilog which in addition to have unique properties to perform verification and design tasks, it can also be used for system level design due to its enhancements comparable with SystemC features. The current belief is that System Verilog can be the "one size fits all" language due to its system and blocks level modeling, system and block level verification, synthesis constructs, and simulation capabilities. One company working in this space is Bluespec that provides high level system modeling, architecture exploration, verification and synthesis using a System Verilog (Bluespec, 2011).

So far, we have talked about languages that are capable of performing SLD, but the drawback of these languages is that they rely on the user knowing the architectural constraints of the design. There is also another very popular complete set of SLD languages that also allow to perform system level modeling at the same time that its users are closer to the algorithm development rather than the language options we just mentioned above. The primary SLD system language for modeling is Mathwork's Matlab and it's time-driven block-based tool Simulink. There are also other tools that also used for system level modeling such as Agilent's SystemVue (Agilent, 2011) and Synopsys' SPW (Synopsys, 2011a) to cite a few used previously by the author.

The author has been exposed to more SLD projects done in Matlab, and in some cases the complete running specification has been kept in Matlab m-code, even the fixed point implementation and test vector generation. Other projects, had Matlab as the main algorithm verification driver, followed by a C model implementation and then by an RTL implementation. Each tool/language translation can potentially introduce errors in the system level design and verification stages. In an ideal world, we should only deal with one system level language, one system level testbench and multiple implementations at different levels of abstraction. By having different models at different levels of abstraction we can have a different model to resolve efficiently different problems such as interconnection, timing, programming, functional verification, synthesizability, and feasibility of implementation.

2.2 Algorithmic focused system level design

The focus on this chapter will center around Matlab and especially on Simulink. The two major FPGA providers Xilinx (Xilinx, 2011b) and Altera (Altera, 2011a), make available libraries that allow efficient block level modeling of wireless communications algorithms and its automatic conversion to RTL. The code can be either downloaded to the FPGA for standalone algorithm implementation or used with hardware in the loop (HIL) functionality that allows a particular block of the system to be emulated using an FPGA device, this is with the purpose of performing hardware acceleration.

Nowadays the common first step taken by researchers is to test their ideas in Matlab's m-code. Matlab as a system level platform allows a very fast and efficient algorithm implementation of complete systems. Matlab does not include the conception of time; it is more comparable to high level programming languages; has a vast set of libraries or toolboxes in many disciplines; and it is not limited to math or engineering. Matlab has become an indispensable tool in modern electronic design and engineering in general.

If the designer would like to model the system including time as another design dimension, Simulink could be used to design complete dynamic systems that are time aware and also include a large number of libraries or toolboxes for a large number of disciplines.

2.3 System architecture

When evaluating an algorithm, the designer is mostly concerned on modeling a system. One of the problems is that the final implementation cannot be readily extracted from this system level modeling easily. There are different levels of system models, some models can be bit accurate and/or cycle accurate.

In a bit accurate model, the system traditionally has been modeled using floating point precision, and then the algorithm has been converted into fixed point precision for efficient implementation. At this stage the main concern is that the signal to quantization noise ratio (SQNR) will dictate the losses due by the effects of for example: quantization, rounding and saturation. This transformation stage can be performed in Matlab/Simulink, SystemC and C/C++. A bit accurate model will have a very close representation of the final implementation in terms of hardware cost and performance. One problem here is that the internal precision of the operation is difficult to model until the final architecture has been decided.

In a cycle accurate model, the systems are architected such that the generated hardware corresponds one to one to the behavioral model in terms of time execution. The advantage is that a true bit accurate and cycle accurate simulation can be obtained, but at much higher simulation speed to their RTL or gate level simulations. In the author's experience, this model has not been used much in the past, since it is tied up with a fixed architecture so the conversion to RTL is straightforward with no ambiguities.

After the fixed point precision has been proposed, it is traditionally coded either in a high level language or in a hardware description language. Of course at this stage the model can continue to be modeled in Simulink. Typically and architectural description is being pursued at this level and the model should closely represent the hardware to be implemented.

What is interesting is that at this stage, there are at least from two or more "system models." One very common error is to not update the higher level with architectural changes once high level modeling stage has "finished", this could lead to inaccuracies on the implementation since it is no longer compared with the "golden" model anymore. As we mentioned, the models can get out of synchronization due to lack of communication between the system's team and the implementation's team. It is of extreme importance throughout the life of the project to have all models updated to reflect the latest changes in both SLD and RTL since each one represents a running specification of the system at different levels of hierarchy.

2.4 System testbench

A testbench is created at the behavioral level, what this means is that the testbench is not to be synthesized, that is why the testbench can include language constructs that represent stimulus and analysis rather than processing and are not directly synthesizable. The testbench is designed to test a "black box" or commonly known as the Device Under Test (DUT), generate inputs, measure responses and compare with known "golden" vectors. One very useful feature in Verilog HDL is to be able from the testbench to descend into the design hierarchy and probe on internal signals that are not available at the interface level. VHDL 2008 includes hierarchical names for verification as well.

A rule of thumb says that when a design is "finished", it is just 30% complete and the validation and verification (V&V) stages will start to cover the remaining 70% effort to have a verified finished design. There are different methodologies to accomplish this and unfortunately Verilog HDL and VHDL have not been robust enough to allow complete and efficient design verification. Due to the later, several proprietary verification languages evolved and recently several methodologies such as Open Verification Methodology (OVM)(Cadence, 2011), Verification Methodology Manual (VMM)(Accellera, 2011) and Universal Verification Methodology (UVM)(Synopsys, 2011c) have been developed to fill the gap between HDL and proprietary verification languages including a common framework for verification. The common denominator in all these methodologies is the use of System Verilog as the driver of all three. System Verilog is evolving as the verification and design solution language since it contains the best of design, synthesis, simulation and verification features, the versatility of the HDLs, and it is designed for system level verification.

Talking about levels of design abstraction, another very common approach is to use the popular C and C++ languages to describe algorithms to be implemented in hardware. We have found that several Electronic System Level (ESL) design tools generate SystemC testbenches that could be used as standalone applications as well as integrated into event driven simulators that are the core when designing hardware implementations. Some examples are Pico Extreme from Synfora (acquired by Synopsys and now is SynphonyC compiler)(Synopsys, 2011b) and CatapultC from Mentor Graphics (CatapultC is more like C++rather than SystemC) (MentorGraphics, 2011).

We have talked about Matlab/Simulink being used at the system level design phase. In order to take full advantage of a common testbench, a hardware design could rely entirely on this platform for rapid prototyping by accomplishing transformations at the level of modeling hierarchy.

Once a design is transformed for example from Matlab m-code to Simulink, or perhaps the design was started in Simulink directly, there are a series of custom libraries that allow the designer to transform their design directly into hardware and keep the original Simulink testbench to feed the hardware design. The design could be verified by generating HDL RTL and by running event driven simulations side by side the Simulink engine and compare with the original Simulink model to verify that the RTL code generated matches the desired abstracted model. Not only a standalone simulation is conceivable, it is possible to download the application directly into an FPGA and generate excitation signals and receive the data back in Simulink. This allows to verify hardware performance at full speed or to accelerate algorithm execution that will take a long time on an event driven simulator. There are several products with similar capabilities such as National Instrument's LabView (NI, 2011) that also allows the option to have "Hardware In the Loop" (HIL) as a way to accelerate computing performance by implementing the algorithm directly in hardware.

The philosophy at this level is to try to reuse the testbench as much as possible to verify correctness of the design at a very high level of abstraction and to code a single testbench that could be used at the system level, while still being able to run the components at single levels of abstraction, namely behavioral, RTL and gate level.

3. Fixed point number representation

This section will cover the different formats used to represent a number using fixed point precision. In addition, the effects of truncation, rounding, and saturation will be covered. SystemC provides a standard set of fixed point types that have been also adopted and adapted by electronic system level (ESL) tools. We will talk about SystemC's fixed number representation. We will talk also about traditional RTL fixed point implementations and the required hardware, complexity and performance.

3.1 SystemC fixed point data types

SystemC includes the *sc_fixed* and *sc_ufixed* data types to represent fixed point signed and unsigned numbers the syntax to include these in a SystemC program is the following:

sc_fixed<wl, iwl, q_mode, o_mode, n_bits>
sc_ufixed<wl, iwl, q_mode, o_mode, n_bits>

where

wl: total word_length
iwl: integer word length
q_mode: quantization mode
o_mode: overflow mode
n_bits: number of saturated bits

Quantization modes: SC_RND, SC_RND_ZERO, SC_RND_INF, SC_RND_MIN_INF, SC_RND_CONV, SC_TRN, SC_TRN_ZERO

Overflow modes: SC_SAT, SC_SAT_ZERO, SC_SAT_SYM, SC_WRAP, SC_WRAP_SM

For example if we would like to declare a signed integer variable with 16 total bits of which 8 bits are integer, we declare:

sc_fixed<16,8>   number;

As can be observed in Figure 2, the 16 bit number will contain 8 integer bits and 8 fractional bits. The maximum number that can be represented is $2^7 - 2^{-8} \approx 128$ and the minimum number will be $-2^7 = -128$ with a $2^{-8} \approx 3.9 \times 10^{-3}$ resolution. By default, the number will have a quantization mode of *q_mode* = *SC_TRN* which means that the number precision will be truncated after each mathematical operation or assignment, and the number will have an overflow mode o_mode=SC_WRAP which means that the number will wrap from approximately 128 to -128. The different modes allow for flexibility in the rounding and saturation operations that are useful to limit the number of bits enhance the SQNR and also to allow infrequent numbers to be saturated and save on the total number of bits. Of course, the price is additional hardware and probably timing to perform these operations.

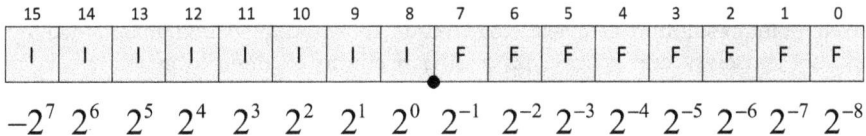

15	14	13	12	11	10	9	8	7	6	5	4	3	2	1	0
I	I	I	I	I	I	I	I	F	F	F	F	F	F	F	F

$$-2^7 \quad 2^6 \quad 2^5 \quad 2^4 \quad 2^3 \quad 2^2 \quad 2^1 \quad 2^0 \quad 2^{-1} \quad 2^{-2} \quad 2^{-3} \quad 2^{-4} \quad 2^{-5} \quad 2^{-6} \quad 2^{-7} \quad 2^{-8}$$

Fig. 2. *sc_fixed<16,8>* representation of a fixed point number.

There are too many ways to describe fixed point notations and representations, but we think that this represents a commonly used format in most of ESL tools that we have explored.

4. Floating to fixed point design considerations

A practical implementation of a wireless communication algorithm involves the conversion of a floating point representation into a fixed point representation. This process is related to the optimum number of bits to be used to represent the different quantities through the algorithm. This process is performed to save complexity, area, power, and timing closure. A fixed point implementation is the most efficient solution since it is customized to avoid waste of resources. The tradeoffs against a floating point implementation are noise, non-linearities and other effects introduced by the processes of: quantization, truncation, rounding, saturation and wrapping among the most important.

Both the floating point and the fixed point solutions have to be compared against each other and one of the most common measure of fixed point performance is the signal to quantization noise ratio (SQNR) (Rappaport, 2001).

Several tools are available to allow the evaluation of a fixed point implementation against a floating point implementation. One of the most important factors is the dynamic range of the signal in question. Floating point adapts to the signal dynamic range, but when the conversion is to be done, a good set of statistics has to be obtained in order to get the most out of the fixed point implementation. The probability density function of the signal will give insight on the range of values that occur as well as their frequencies of occurrence. It may be acceptable to saturate a signal if overshoots are infrequent. We need to carefully evaluate the penalty imposed by this saturation operation and the ripple effects that it could have. This process allows to use just the necessary number of bits to handle the signal most of the time, thus saving in terms of area, power and timing. In section 5.3, we talk about some of the little steps that have to be taken throughout the design in order to save in power consumption. As mentioned, power consumption savings start at the system level architecture throughout the ASIC and FPGA methodologies.

Sometimes the processed signal could be normalized in order to have a unique universal hardware to handle the algorithm. It is very important to take into consideration the places where the arithmetic operations involve a growth in the number of bits assigned at each operation. For example, for every addition of two operands, a growth of one bit has to be appended to account for the overflow of adding both signals. If four signals are added, only a growth of two bits is expected. On the other hand a multiplication creates larger precisions since the number of bits in the multiplication result is the addition of the number of bits of the operands and also it has to be taken into account if the numbers are signed or unsigned.

The fixed point resolution at every stage needs to be adapted and maintained by the operations themselves and specific processing needs to be done to generate a common format. These operations are the truncation, rounding, saturation and wrapping covered briefly for SystemC data type in section 3.

A nice framework of the use of fixed point data types that could be incorporated into C/C++ algorithm simulations are the SystemC fixed point types available in the IEEE 1666™-2005: Open SystemC Language Reference Manual (SystemC, 2011). There are some other alternatives to fixed point data types such as the Algorithmic C Datatypes (Mentor-Graphics, 2011) that claim to simulate much faster than the original SystemC types and used in the ESL tool CatapultC. The ESL tool Pico Extreme uses the SystemC fixed point data types as the input to the high level synthesis process.

Matlab/Simulink also has a very nice framework to explore floating to fixed point conversion. When hardware will be generated directly from Simulink, it is very natural to alternate between floating point and fixed point for system level design. Designs that are targeted for Xilinx or Altera FPGAs could naturally use this flow and reuse the floating point testbench to generate the excitation signals that could be used within the Matlab/Simulink environment in for example Hardware in the Loop (HIL) configuration or fed externally to the FPGA using an arbitrary waveform pattern generator.

Another very useful tool for creating executable specifications in C++ is to use IT++ (IT++, 2011) libraries available for simulation of communication systems.

Each EDA vendor has a different set of tools that allow designers to make the implementation of floating point to fixed point as seamless as possible. This conversion process is a required step that cannot be avoided and traditionally it has been done manually and by matching the results of the Golden model against HDL RTL simulation. Sometimes this comparison is bit accurate, but in some cases the comparison is just done at the SQNR level due to the difficulty to model all the internal operations and stages of a particular hardware implementation.

5. Register transfer level design

Once a system has been verified for performance and has been converted from a floating to a fixed point representation, the specifications are passed to the register transfer level (RTL) design engineer to come up with an architecture that will achieve the desired performance, while consuming minimum power at the right frequency of operation, using minimum area, sharing resources efficiently, reusing as much components as possible, and coming with an optimum tradeoff between hardware and software implementations. We can see that this is not usually an easy task to perform, even for experienced designers.

5.1 Architecture

In this section we will give an overview of the importance of the architecture in RTL design. Examples of different architectures for complex multipliers, finite impulse response (FIR) filters, fast Fourier transforms (FFT) and Turbo Codes will be given comparing their complexity, throughput, maximum frequency of operation and power consumption.

When an efficient architecture is sought, each gate, each register, each adder and each multiplier counts. Sometime it is a good approximation at the system level to count the number of arithmetic operations to get an initial estimate of the silicon area that will be used for the algorithm. While this is a crude approximation it is a very good start point to allocate resources on the System on a Chip (SoC). Many companies have spreadsheets that contain average values for different operations in a particular technology; based on hundredths of designs. The architecture task is to find the optimum implementation of a particular algorithm while accomplishing all the above referred design parameters.

When an algorithm is implemented, what will be the final underlying technology for implementation? ASIC or FPGA; or will it be driven by software and just primitive building blocks will be used as coprocessors or hardware accelerators. Whenever a product needs to be designed on an application area that continues to grow and generate new algorithms and implementation such as video processing, sometime an analytics engine must be architected that will provide co-processing or hardware acceleration by implementing the most common image processing algorithms. This idea could be applied to any communications system or signal processing system where a solution could include a common set of hardware accelerators or coprocessors that realize functions that are basic and will not easily change. One very good example is the TMS320TCI6482 Fixed-Point Digital Signal Processor (Texas-Instruments, 2011) that is used for third generation mobile wireless infrastructure applications and contains three important coprocessors: Rake Search Accelerator, Enhanced Viterbi Decoder Coprocessor and Enhanced Turbo Decoder Coprocessor.

So the question is: When implementing a particular algorithm, how can we architect it such that it is efficient in all senses (are, power, timing) as well as versatile? The answer depends on the application. That is why hardware/software partitioning is a very important stage that has to be developed very carefully by thinking ahead of possible application scenarios. In some cases there is no option, and the algorithm has to be implemented in hardware, otherwise the throughput and performance requirements may not be met. Let's explore briefly some practical examples of blocks used in wireless communication systems and just brainstorm on which architectures may be suitable.

Finite Impulse Response Filters

An FIR filter implementation can be thought as a trivial task, since it involves the addition of the weighted version of a series of delayed versions of an input signal. While it seems very simple, we have several tradeoffs when selecting the optimum architecture for implementation. For an FIR filter implementation we have for example the following textbook structures: Transversal, linear phase, fast convolution, frequency sample, and cascade (Ifeachor, 1993). When implementing on for example on FPGAs, then we found for example the following forms: Standard, transpose, systolic, systolic with pipelined multipliers(Ascent, 2010).

Most of the FPGA architectures are enhanced to make more efficient the implementation of particular DSP algorithms and the architecture selection may fit into the most efficient configuration for a particular FPGA vendor or family. If we are targeting ASIC, then the architecture will be different depending on the library provided by the technology vendor. When implementing an FIR or any other type of filter or signal processing algorithm, we need to evaluate the underlying implementation technology for tuning the structure for efficient and optimum operation.

Turbo Codes

One interesting example is on Turbo Codes, while the pseudo-random interleaver is supposed to be "random", there has been a pattern defined on how the data could be efficiently accessed. Some interleavers are contention free, while some others have contentions depending on the standard. For example, one of the major differences on the third generation wireless standards namely 3GPP(W-CDMA) and 3GPP-2 (CDMA2000) is on the type of interleaver generator used, this means that to a certain degree it would be possible to design a Turbo Coder/Decoder that could easily implement both standards.

The purpose of an efficient implementation of an interleaver hardware is to have different processing units accessing different memory banks in parallel, some examples on the search for common hardware that could potentially be used for different standards are shown in (Yang, Yuming, Goel, & Cavallaro, 2008), (Borrayo-Sandoval, Parra-Michel, Gonzalez-Perez, Printzen, & Feregrino-Uribe, 2009) and (Abdel-Hamid, Fahmy, Khairy, & Shalash, 2011). The architecture is a function of the standard and sometimes it is very difficult to find a "one architecture fits all" type of solution and in some case to make the interleaver compatible with multiple standards, on-the-fly generation is the best approach, but there can be irregularities or bubbles inserted into the overall computation. This is one of the challenges in mobile wireless that sometimes is easier to implement complete different subsystems performing efficiently one particular standard, rather than having an architecture that could perform all. This is the case in mobile cellular second generation GSM (Global System for Mobile Communications, originally Groupe Spécial Mobile) and third generation cellular W-CDMA (wideband code division multiple access) that minimum reusability could be achieved and to a certain extent there are two complete wireless modems implemented for each standard.

Fast Fourier Transform

Many of the modern wireless communications algorithms migrated from the CDMA to the Orthogonal Frequency Division Multiple Access (OFDMA) technologies. One of the main reasons to transfer to a completely new technology might have been that the current state of the art on integrated circuit design allowed the efficient implementation of algorithm architectures that were not previously convenient to implement in hardware. This is the case of the Fast Fourier Transform (FFT) which is the core of Orthogonal Frequency Division multiplexing (OFDM) and its derivatives such as OFDMA (Yin & Alamouti, 2006).

OFDM and FFT techniques are not new, as a matter of fact they have been around longer that many of the current wireless technologies. What it is new, is the feasibility of the algorithms to be implemented on silicon. An efficient architecture implementation for a pipelined FFT (Shousheng & Torkelson, 1998) has been used as a benchmark for hardware implementation of the FFT algorithms, this technique allows all hardware units to be used at all times once the pipeline is full and is very convenient for FPGA or ASIC implementation.

We will just briefly talk about this on section 10, since it is one example that comes with the FPGA libraries and the purpose of this chapter is not to develop a new FFT form, but rather to see how it can be implemented.

5.2 Maximum operating frequency

While it could be easy to convert an algorithm from floating point to fixed point and to identify architectures for its implementation, the final underlying technology should be taken into account to determine the maximum operating frequency and in some cases the required level of parallelism and/or pipelining. It can be true that an algorithm designed for FPGA will run without major modifications on ASIC, but the reverse is not always true. FPGAs are used widely to perform ASIC emulation, but it does not make much sense to have two different versions of the algorithm running on either technology, since this could invalidate the overall algorithm validation. Sometimes the same code could be run, but in slow motion on FPGAs if real time constraints are not required. If real time is a factor, only some of the low throughput modes could be run on the FPGA platform and simulated for ASIC.

5.3 Power consumption

Power consumption in mobile devices is a crucial part of the algorithm selection and it is tightly coupled to architecture's implementation, frequency of operation, underlying technology, voltage supply, and gate level node toggle rates to give some examples. In this section we will cover some of the important features to be considered when designing power optimized algorithms implementations.

When designing digital systems we all know that a magic button exists that reduces power consumption to the minimum. Unfortunately this is not the case, the magic button does not exist and power savings start at the system level design, the architecture selection, the RTL implementation, the operating frequency, the integrated circuit technology chosen, the gate clocking methodology, use of multi-V_{dd} and multi-V_{th} technologies, and leakage among some of the most important factors. In reality power savings are being done in small steps starting from efficiency at the system and RTL level design. One power saving criteria is: if you do not have to toggle a signal, don't do it! Power consumption is a function of the frequency of operation, the load capacitance and the power supply voltage. On average, the gate level nodes switch at around 10% to 12%, while an RTL level simulation could have toggles close to 50% meaning that all units are being used all the time and there is no waste in terms of hardware resources.

When deciding the fixed point representation, every bit in the precision counts towards the total power consumption, the number of gate levels between registers the load capacitance of each node. If we decide to include saturation and/or rounding, there are additional gates required to perform these operations. The cost of additional hardware can be worth the gates if the bit precision is reduced from a system with a wide dynamic range that takes into account no overflow for signals that can have very large excursions but are very infrequent. So what could be the best tradeoff between complexity, fixed point precision, internal normalizations, and processing? There is not a single solution to the problem, the best will be to statistically characterize the signals being handled to find out their probability distributions and then based on these determine the dynamic range to be used and if saturation/wrapping and truncation/rounding could be used and within these which methods to apply as mentioned in section 3.

Power consumption depends on the circuit layout as well, while old technologies used to be characterized in terms of gate delays, input capacitance and output load driving capacitance, the end game has changed and modern technologies have to take into account the effects of interconnection delays due to distributed resistance, inductance and capacitance. The solution to the power consumption estimate is not final until the circuit has been placed and routed and transistors are sized. If an FPGA implementation is sought, a similar approach is taken but control is coarser due to the huge number of paths that the signals have to flow in order to be routed among all resources.

Another important factor are the power supply V_{dd} and the threshold voltage V_{th} of the transistors. These two factors control the voltage excursion of the signals and most important the operation region of the transistor. Most of the digital logic design rules assume that the transistors are operating in saturation, power is consumed while transitioning through the active region and this is the region where you want to get out as fast as possible. A transistor operating under saturation regime has a quadratic transconductance relation of the current I and the input gate voltage V_g. When a transistor is not in saturation, it could be in linear region or even in sub-threshold. A transistor in the latter does not have a quadratic, but an exponential transconductance relation. While this is the most power efficient operating regime, it is also the slowest. Many circuits that need very low power consumption can work in sub-threshold, but there is a huge variability and precision constraints. Most of these designs involve linear analog mode operations.

So what is the secret formula to design power efficient devices? The answer is discipline! Try to save as much as possible at each level in the design hierarchy. If it is in software, set the processor to sleep if there is nothing important to do. If it is hardware, do not toggle nodes that do not require to be toggled, gate the clocks so you can lower power consumption in blocks not used, reduce powers supply V_{dd} to the minimum allowed for efficient operation of the algorithm and design using just the right number of bits. More techniques for low-power CMOS design have been published and good overviews are given in (Chandrakasan & Brodersen, 1998) and (Sanchez-Sinencio & Andreou, 1999).

6. Electronic System Level Design

Electronic System Level Design (ESL) design has come from a promising technology to a reality. Companies such as Cadence, Mentor Graphics and Synopsys have their own ESL tools and have integrated these into their System on a Chip (SoC) design flows. In this section we will address some of the most important features of ESL which are architecture exploration, power consumption estimation, throughput, clock cycle budgets allocated, and the overall integrated verification framework from untimed C/C++ golden model, all the way to gate level synthesis.

One of the advantages of ESL tools is that the same testbench used to design a block could be reused at all levels of abstraction thus minimizing the probability of introducing errors at different levels of the implementation. While RTL design requires thinking very carefully on a target architecture, ESL allows exploring different architectures and taking tradeoffs using a high level description of the algorithm, and avoids the designer to go to the RTL level to verify block's performance. We will go through examples of an OFDM FFT implementation as well as MIMO signal processing. ESL niche applications are hardware accelerators that traditionally are hooked to a microcontroller platform such as an ARM processor and handle data processing intensive operations. This is a common practice in SoC design, several intellectual property (IP) vendors concentrate their products in offering very high performance blocks that interface with a common bus architecture such as AMBA.

7. FPGA implementation

For FPGA implementations we could always resort to the traditional RTL implementation of the algorithm. For this section we will resort to Mathwork's Matlab/Simulink implementations of particular algorithms by the automatic generation of RTL code to be either downloaded to the FPGA and to be tested standalone or to the Matlab/Simulink testbench that could be used to drive the simulation and the actual RTL code will be executed in the FPGA. The latter is referred as hardware in the loop (HIL).

We will give examples of: converting a chaotic modulator/demodulator from Matlab code to a Simulink model; to a Simulink model using Altera DSP builder blocks; and demonstrating the algorithm working on a development board after digital to analog and analog to digital conversions.

In FPGAs the pool of resources is fixed. Depending on the particular algorithm, it could be better placed in one of the different families of FPGAs available by different vendors. Datapath architectures can be very efficiently instantiated on FPGAs since most of building blocks included in these devices are designed for very high performance digital signal processing algorithms. We will talk about the tradeoffs when FPGA utilization is low and high and the effort to place and route (P&R) as well as timing closure.

8. ASIC implementation

Most of the wireless communication algorithms would have two versions: one for wireless infrastructure that needs high performance and power is important but not critical since it is always connected to an external power source, and another for mobile wireless devices in which performance is a requirement but power has to be optimized in order to make the device usable, power efficient and competitive. In this section we will explore these two types of implementation in applications specific integrated circuits (ASIC). We will give an example of a turbo code interleaver/de-interleaver that had been implemented and verified using simulation and an FPGA platform and the changes required to take it to an ASIC implementation.

9. Hardware acceleration

Sometimes it is not possible to evaluate an algorithm using regular simulation techniques due to the computing power that is required to perform these tasks. SoC designs are a good examples of these constraints, not all block could be implemented and verified at the gate level in simulation due to the fact that it will take from hours to weeks to perform these simulations. For these cases it is common to use FPGAs as hardware accelerators or ASIC emulators. ESL tools are very efficient in generating these type of blocks that can be either instantiated for FPGA or ASIC and the only real difference is on the characterized libraries used as well as the system clock frequency.

The basic requirements while designing custom datapath components is to create hardware accelerators that could work as standalone blocks. Normally these components will become part of a large SoC. Many of the current embedded products recently designed are composed of a microcontroller such as an ARM core, a standard bus such as AMBA, and a series of Intellectual Property (IP) blocks that realize specific functions that require high performance and low-power. This is mostly true on cellular mobile devices, while for base stations a dedicated Digital Signal Processor (DSP) could be used since throughput is a more important constraint than power consumption. It is worth mention that these designs could be done in the same technology geometry, but with different characteristics: base station would most likely use a high performance, higher threshold voltage and large leakage process while the mobile device will be constrained to medium performance, very low leakage process and low and probably variable threshold voltages.

Some examples of systems that are designed as hardware accelerators in cellular technologies are:

- Equalizers
- Viterbi, Turbo and LDPC decoders

- OFDM Modems
- Rake receivers
- Correlators
- Synchronizers
- Channel estimators
- Matched filters
- Rate matching filters
- Encryption/decryption
- Modulator/demodulator
- Antenna diversity and MIMO processing

The question is which functions will run on software and which functions will run on hardware. This lies in the gray area of hardware/software partitioning. There are different specifications that need to be considered before taking an educated decision. In theory, anything that could be done in hardware could be done in software and vice versa (of course having an infinitely fast processor with a humongous bus bandwidth and a large number of I/Os). We must carefully evaluate the hardware components to be implemented since no field upgradeability will be possible once an ASIC has been manufactured; we need to find the equilibrium where a firmware patch could potentially get rid of any anomaly not detected at verification and validation time.

In particular, the author worked for many years in teams concentrated on hardware accelerators, but all these components were part of a SoC where traditionally an ARM processor was used with a standard interconnect such as AMBA(ARM, 2011) or OCP (OCP, 2011) and the hardware accelerators were mapped as peripherals in the processor memory space. The ASIC design was first simulated, then emulated on a large FPGA platform at a constrained speed and then the ASIC could finally be developed.

In academia we are more involved with FPGA designs and in particular the platforms being used for teaching include the possibility of a soft core processor. For the author's particular case the platform is Altera and the soft core processor is the Nios II. It is interesting to find that a C to RTL application program exists that allows functions implemented in software could be converted into hardware accelerators. The application is C2H (Altera, 2011b) and even that the author has not been able to test it, it looks promising since it allows the exploration of different hardware/software partitions that could impact the total silicon area, performance, power and cost of a particular application (Frazer, 2088). In the case of FPGA design it could lead to be able to reduce costs or performance by moving back and forth different FPGA migration devices that are pin compatible, but vary in the number of logic elements available, the number of I/O pins available and cost. An equivalent tool exist from Xilinx called Auto-ESL (Xilinx, 2011a) that generates code from C/C++/SystemC.

10. Hardware implementation examples

10.1 MOC digital communications system implementation

In this design example, we will walk through the steps required to implement a mutually orthogonal chaotic (MOC) digital communications system (Glenn, 2009) algorithm architected in Simulink to run on FPGA hardware and the constraints imposed by these steps that were not considered in the original design, that affect the systems performance.

The MOC algorithm was coded first in m-code and later converted into a Simulink model. This is shown in Figure 3. The model allows following what the algorithm does without going deep into the details and the model is time dependent. The data rates at the input and output of each block are not shown and this is one of the most important features to consider in a datapath Simulink model.

After looking at the architecture presented for implementation, each of the blocks was substituted by the equivalent Altera DSP Builder available blocks. Some of the blocks have a direct equivalent while some others have to be converted into an equivalent hardware component. This is shown in Figure 4.

Since this block is originally excited by a binary signal, some digital components were used to group the bitstream into a fixed number of bits that will be used to select the modulation waveform. The original Simulink model does not have time restrictions and could potentially generate a waveform with a very large precision, but for practical reasons the implementation is restricted to a particular clock frequency and thus the number of samples to choose for the modulation waveform has an impact on the algorithm performance. A study of the optimum number of samples and the optimum number of bits to represent each modulation waveform had to be done. Each modulated waveforms also could change in sign and or magnitude, for Simulink the operation is just a simple multiplication, but for a hardware implementation it is more efficient to allocate ROM tables and access the correct magnitude and phase. This is similar to storing one quarter of the phase of a sine wave and generates sine and cosine waveforms out of this reduced table. The difference is that the basis functions for this algorithm are chaotic waveforms, then it is difficult to exploit any symmetry property.

In Simulink it is very convenient to add very high level functions such as the modulators and demodulators observed in Figure 3b and Figure 3c, while this may not be required for a baseband algorithm like the one that are implemented on FPGAs. For implementation and testing we decided to work just at the baseband level.

After the model was converted, we compared the values generated by the Simulink blocks simulation against the one generated by using the Altera DSP Builder blocks. The signals were matched and SQNR was computed to validate the approach as well as rate matching was performed to match the samples. The bit sequence and the modulated waveforms are shown in Figure 4d.

The next step is to generate HDL RTL out of the Altera DSP Builder blocks. This is shown in Figure 5a where RTL code is generated, a Simulink simulation is run, followed by a Modelsim RTL simulation and both simulations are compared and the differences are noted. The generated HDL RTL now can be synthesized and programmed into the FPGA for further development. Since for this particular system the excitation is being generated in the test bench by using a Bernoulli random number generator, we decided to use a pseudo random noise (PRN) sequence generator to embed into the FPGA for standalone testing.

The results for the transmitter are shown in Figure 6, where a) is the Altera Cyclone II FPGA testing board with two 14-bit resolution and data rate up to 65 MSPS analog to digital converters and two 14-bit resolution and data rate up to 125 MSPS digital to analog converters. This configuration is suited for testing communication transceiver applications, digital signal processing algorithms and as a platform for various modulation techniques such as the presented in this implementation example.

Figure 6b shows the modulation operation when an all zero pattern is generated. Figure 6c shows the PRN sequence excitation modulation waveforms and Figure 6d shows a screen capture of the MOC modulated waveforms.

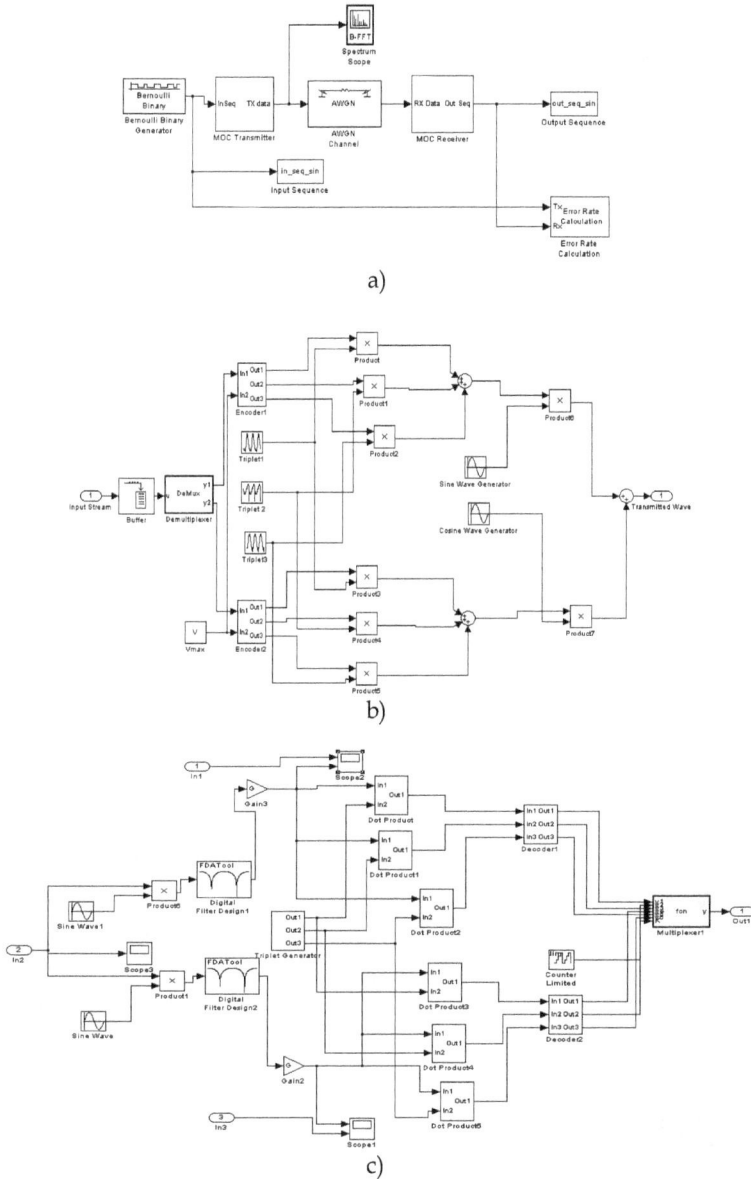

a)

b)

c)

Fig. 3. MOC algorithm architecture implemented as Simulink models.
a) Complete MOC communications system block diagram including channel modeling.
b) MOC transmitter block diagram. c) MOC receiver block diagram.

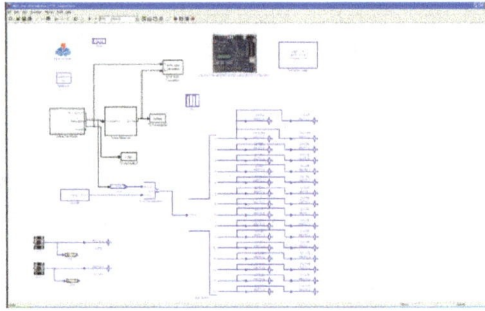

a)

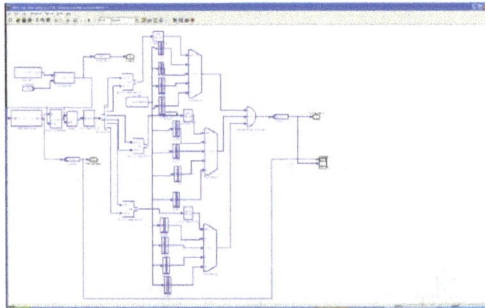

b)

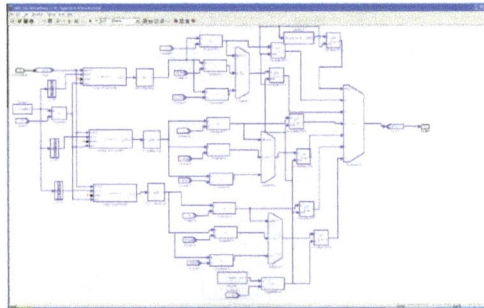

c)

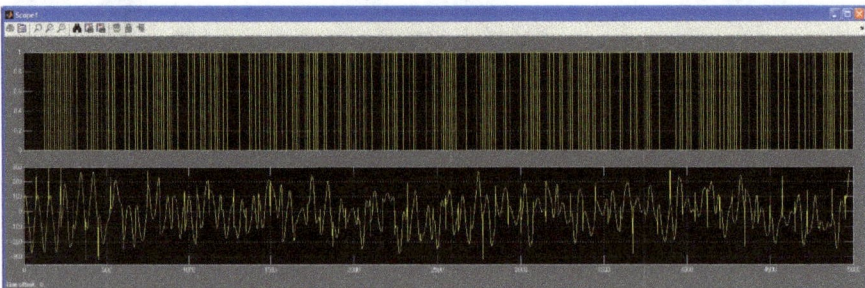

d)

Fig. 4. MOC algorithm transformed to use Altera DSP Builder blocks to automatically generate HDL for FPGA implementation. a) Testbench and interface signals to FPGA. b) Transmitter sub-system. c) Receiver subsystem. d) Simulink simulation waveform.

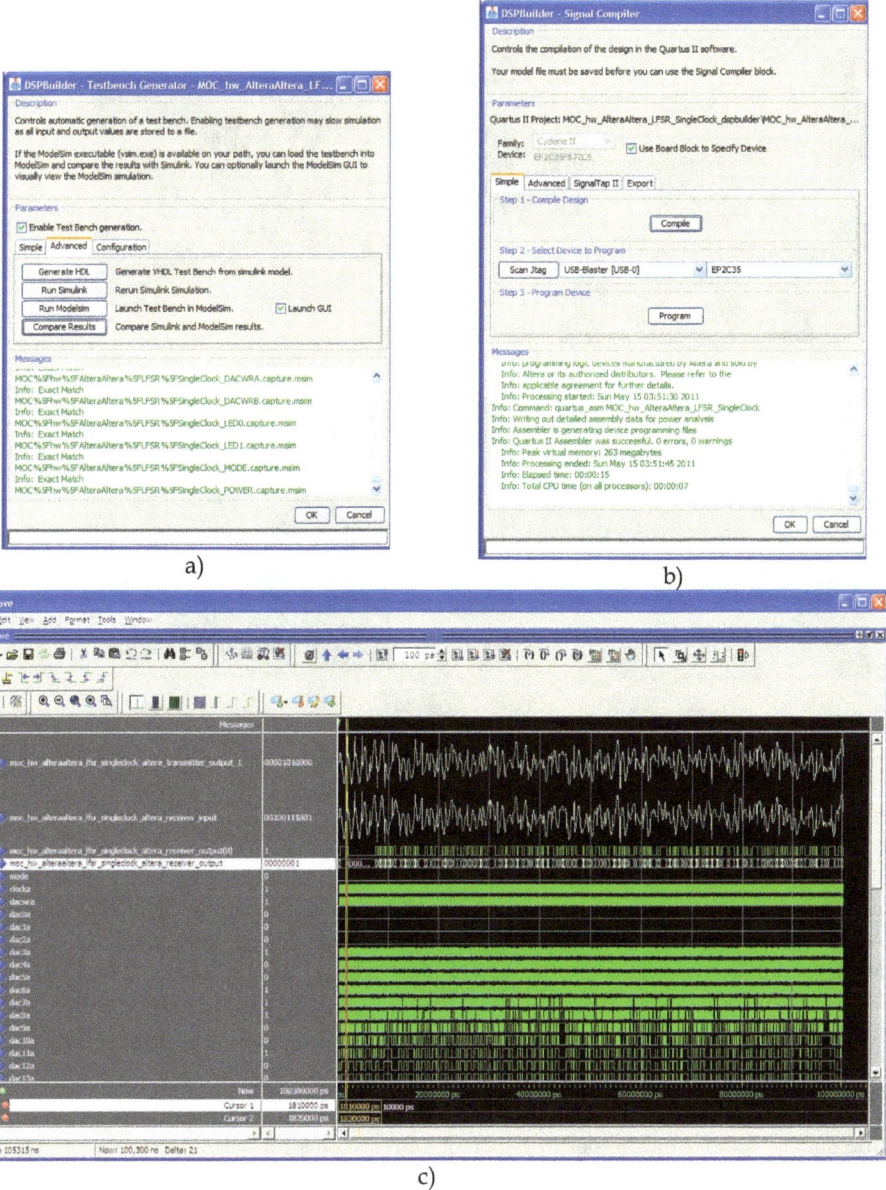

a)

b)

c)

Fig. 5. In addition to a system level simulation within Simulink, it can also control an RTL simulation of the generated HDL code and compare against the system level simulation. a) Test bench generator for RTL simulation. b) RTL HDL simulation of the code generated by DSP Builder. c) Signal compiler for synthesis, place and route, and FPGA programming.

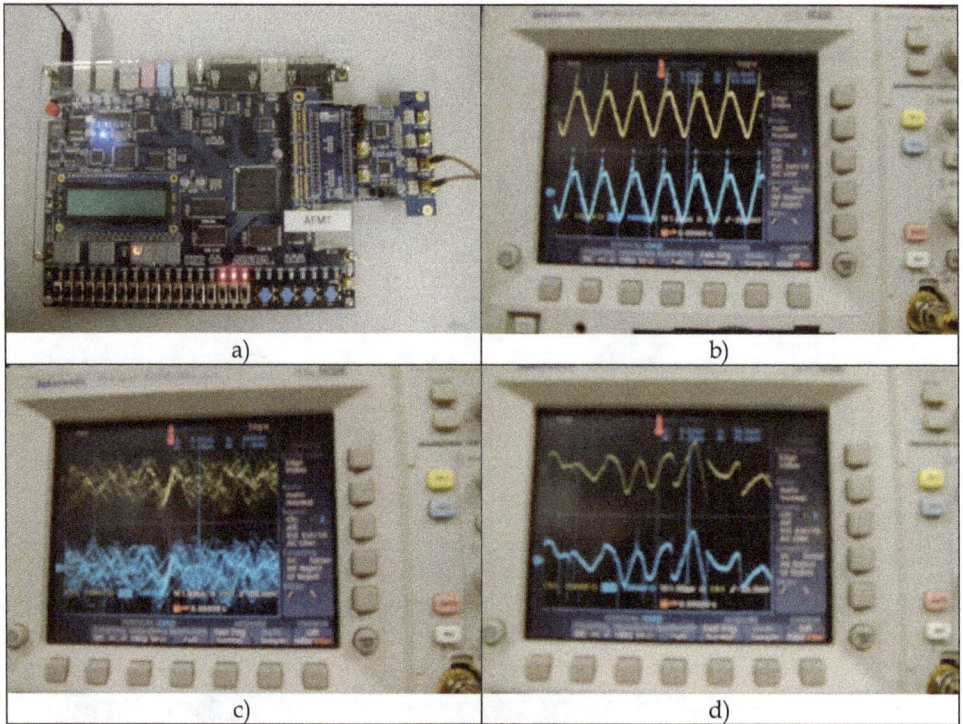

Fig. 6. MOC hardware implementation on an Altera Cyclone II FPGA.
a) Altera DE-2 with daughtercard dual AD channels with 14-bit resolution and data rate up to 65 MSPS and dual DA channels with 14-bit resolution and data rate up to 125 MSPS.
b) MOC modulation output when the input is a stream of constant zeros.
c) MOC modulation output when the input is driven by a PRBN sequence generator.
d) MOC modulation output snapshot when the input is driven by a PRBN sequence generator.

10.2 Improving the performance of DSP systems for MIMO processing

In the paper "Improving the performance of DSP systems for MIMO processing" (Horner, Kwasinski, & Mondragon, 2011), we explored the efficient implementation of select Multiple Input Multiple Output (MIMO) communications algorithms. Two implementation approaches were considered: adding new instructions to the DSP instruction set and adding a hardware accelerator to the DSP system. Of the two approaches, the second was concluded to be best, as it resulted in notable processing speedups and a more efficient use of the computational resources.

While the research into MIMO algorithms have reached levels of development that show important wireless systems performance improvements, the development of DSP systems to implement them has limited the realization of these algorithms to the simplest and least performing ones. This example addresses this technological gap by studying how to design DSP systems to better handle the increased complexity arising from the particular operations typical of MIMO processing algorithms.

Two hardware co-processors were designed, as shown in Figure 8 one for a Householder decomposition algorithm and one for a Greville pseudo inverse algorithm. These hardware co-processors resulted in a simulated speedup of 2.7 for the Greville algorithm and between 4 and 4.7 for the Householder algorithm.

For the design of the hardware accelerator, Synfora's Pico Extreme (acquired recently by Synopsys) ESL tool was used. The author had previous experience with the tool and the task performed for this work was limited to architecture exploration and to find which ASIC implementation would result in the best compromise between throughput, area, power, and easy of interfacing. The algorithms were written in floating point C code and then converted to fixed point C code by evaluating the impact in performance due to the hardware implementation.

Pico Extreme is a very versatile tool since it is structured as a series of logical steps from running an untimed sequential ANSI C program, to single-to-multi-threaded transformations; to hierarchical block-level resource sharing & scheduling; to automatic retiming and pipelining; to performance and throughput analysis; to rapid exploration of performance impacts of loop unrolling, scheduling, and other optimizations; and to RTL verification among others. The flow methodology is shown in Figure 7.

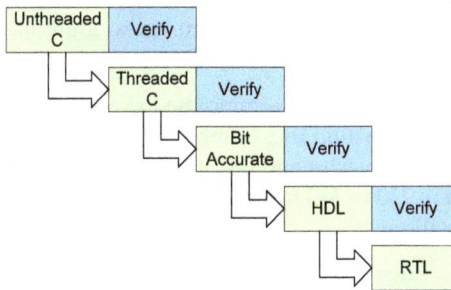

Fig. 7. PICO Extreme design flow.

While this seems to be a dream in which the system designer can implement his design by exploring architectures and trade-offs, then pushing a button and get verified RTL as an output, the reality is that the learning curve of these tools is quite steep and it is not as straight forward as it looks. Even that a very thorough architecture exploration can be performed, the designer still needs to think in terms of hardware when writing the C code to have the same effect as writing in HDL RTL. The C code has to be written in terms of functional units, pipeline stages, memory implementations, operator sharing and general hardware efficiency.

There are two basic methods to specify the design (Synfora, 2009). The number of clock cycles between iteration starts is called II (Initiation Interval) and the number of clock cycles to start all iterations is called MITI (Maximum Inter Task Interval). For this example, MITI can be as small as N*II (where N is the number of loop iterations).

The user is able to provide a target maximum number of clock cycles taken per stage MITI and the tool will select from the library of high-speed components the optimum to achieve higher levels of parallelism at the same time of sharing resources and achieving performance.

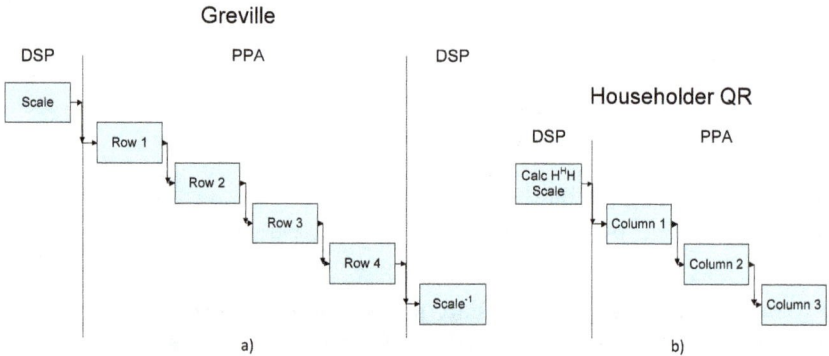

Fig. 8. Processing pipeline for Greville and Householder decomposition methods.

To provide a tradeoff between complexity and speedup, different implementations with different target MITIs were generated. It was noted that as timing constraints tightened, hardware multipliers were switched from two-cycle to one-cycle and the number of multipliers increased to be able to complete complex multiplications (requiring three multiplies) in a single cycle.

MITI timing constraints were used to determine the lowest complexity implementation for each algorithm. The constraints within these ranges of target clock cycles were then used to produce a tradeoff between complexity and resulting speedup. Resulting ranges of targeted number of clock cycles were 230 to 330 for the Householder implementation and 130 to 210 for the Greville implementation.

The resulting speedup was calculated as the ratio of cycles on the DSP-only implementation to the cycles of the DSP-PPA implementation. The resulting silicon area was calculated based on the estimated number of gates given by Pico Extreme and using a characterized CMOS 65nm technology library with an estimate of 854,000 gates per mm². This technology was selected, given that is the one in which the DSP was manufactured and can provide an estimate of the growth of the silicon area for the DSP to enable MIMO processing. A plot of speedup vs. complexity for both clocks and both simulators is shown in Figure 9.

The resulting maximum speedups were close to 2.75 for the Greville algorithm and between 4 and 4.7 for the Householder QR decomposition algorithm. This speedup would result in a large reduction (129 μs for the Greville implementation and 521 μs for the Householder implementation) in the amount of time required to compute the channel equalization matrices for an entire OFDM channel in MIMO communication. There is an upper limit to the speedup, however. Because the DSP is still required for some pre-processing operations, there is an asymptotic limit on the actual speedup achieved. Once the PPA unit is able to compute one stage of the processing pipeline in the same amount of time as the software pre-process, there is little added benefit to faster clock or higher complexity. There is also not a major advantage in the 1 GHz clock over the 500 MHz. While the slower clock would require the more complex implementations to compute faster than the DSP software, the savings on power consumption could outweigh the cost of higher complexity.

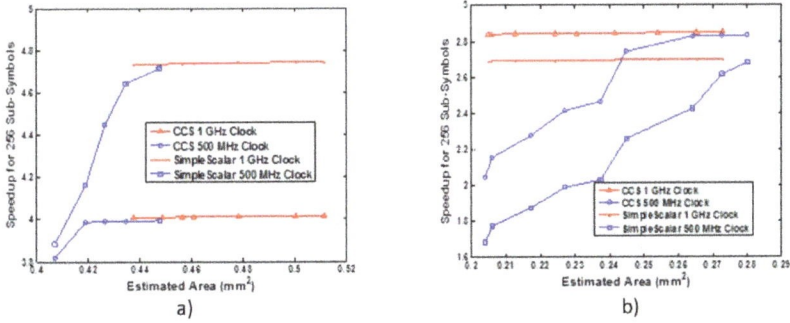

a)   b)

Fig. 9. a) Speedup vs. Complexity for Householder implementation b) Speedup vs. Complexity for Greville implementation.

10.3 OFDM – FFT example

In (Mondragon-Torres, Kommi, & Bhattacharya, 2011), the author proposes the development of an OFDM educational platform that will make use of all the methodologies and tools presented in this chapter with the objective of creating a single system that will allow students to explore different levels of abstraction on hardware design as well as to quantify the effects of the decisions taken on the fixed point precisions as well as all the intermediate signal processing and conditioning through the datapath.

The heart of the OFDM modulation technique lies in the use of the Fast Fourier Transform (FFT), which is a very structured algorithm to convert a time domain signal into the frequency domain and by taking the inverse FFT (IFFT) can be transformed back into the time domain. In Figure 10, a complete digital communication system that employs OFDM modulation is shown (Cho, 2010).

The approach in OFDM systems is to have digital information encoded by traditional phase modulation techniques such as Quadrature Amplitude Modulation (QAM). This modulation technique maps a series of bits into QAM modulated symbols. The number of symbols used for each OFDM frame is traditionally a power of two. Then the IFFT of a block is performed on the frame to convert it back into a time domain representation that can be further processed and sent through the transmitter chain and through the antenna. On the receiver side the process is reversed after frame synchronization by taking the FFT of the received block and obtaining an estimate of the QAM symbols which are mapped back into a series of bits. This sounds pretty straightforward but there are many subtle details that could be investigated in terms of the effects of: quantization, distortion, channel noise, multipath propagation, fading, Doppler shift, synchronization, etc.

A very simple implementation of a 256 point FFT is presented in this section as shown in Figure 12. No architectural decisions were performed and a regular textbook implementation is used just to demonstrate some of the capabilities of CatapultC. In Figure 11, technology parameters and some common definitions are shown as reference for the reader. Based on the above definitions, we started to change the system parameters to get a feel of their implications.

In Figure 13 it is shown how by unrolling and pipelining the input and output operations we can drastically reduce the latency. What is the price for this? Answer: Memory bandwidth. We can observe that the area has been maintained constant and this is due to the fact that no memories have been considered in these solutions.

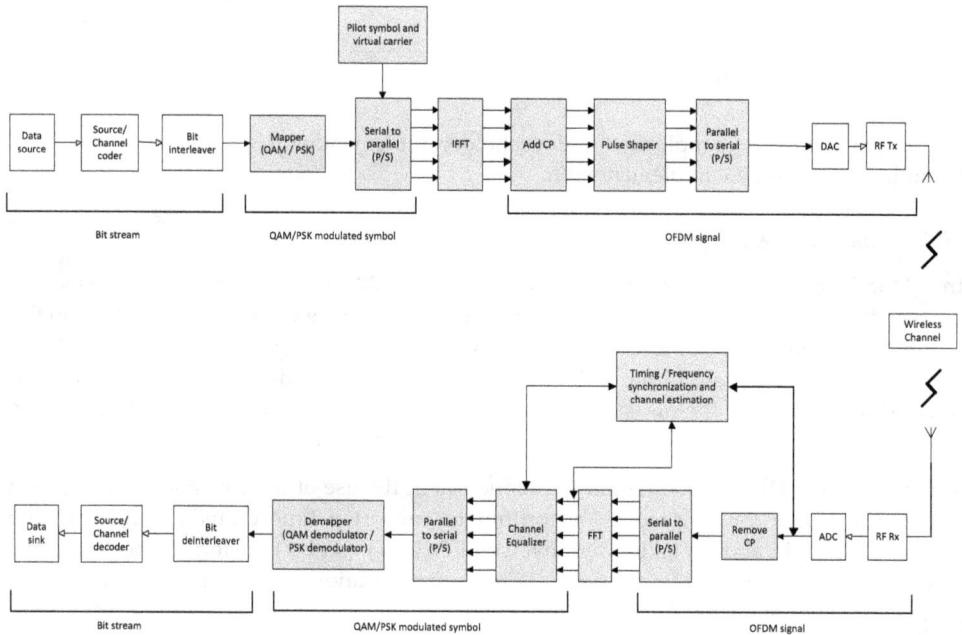

Fig. 10. Digital communications system using OFDM modulation.

Figure 14 and 15 shows the complexity of the solution and we can observe that most of the area is being used in multiplexers to route the signals. On the other hand, more memory will be required for unrolling *printing* and pipelining *reading*. So far we have not touched a single line of code and just by modifying the outer input and output loops we have been able to reduce the latency by 2x at the cost of 2x memory. This is a simple illustration of using the same code to tradeoff performance vs. complexity.

Technology used: Generic CMOS ASIC 90 nm, 200MHz

Definitions

Loop unrolling: Loop unrolling can be used to compute multiple loop iterations in parallel.

Partial unrolling: Computes 'n' copies in parallel

Pipelining: Starts the next loop iteration before the current iteration of the data path contained in the loop has completed

Initial Interval: indicated how often to start a new loop iteration

Latency: Latency refers to the time, in clock cycles, from the first input to the first output

Throughput: Throughput, not to be confused with IO throughput, refers to how often, in clock cycles, a function call can complete.

Fig. 11. Technology used and some common definitions.

```
1    #include <iostream>
2    # include "FixedButterfly.h"
3    # include "Twiddle.h"
4    using namespace std;
5
6        int      Bsep, p, Bwidth;
7        int      topval, Botval;
8        float1   pi=3.141593;
9        float1   Theta, wnr, wni;
10       float1   Tempr, Tempi;
11       float1   xr[N], xi[N];
12
13   #pragma hls_design top
14   void FixedButterfly ( ac_channel<float1> &data_inR, ac_channel<float1> &data_inI,
15                          ac_channel<float1> &data_outR, ac_channel<float1> &data_outI)
16   {
17   //Reading data from the channels bit by bit
18       reading: for(int i=0;i<N;i++)
19       {
20           data_inR.read(xr[i]);
21           data_inI.read(xi[i]);
22       }
23       Stage: for(int s=1;s<=m;s++)
24       {
25           Bsep=Bsep1[s];
26           p=p1[s];
27           Bwidth=Bwidth1[s];
28           coefficients: for(int j=0;j<=Bwidth-1;j++)
29           {
30               wnr=twiddle_real[s][j];
31               wni=twiddle_img[s][j];
32               finalvalues: for(int topval=j;topval<N;topval=topval+Bsep)
33               {
34                   Botval=topval+Bwidth;
35                   Tempr=xr[Botval] * wnr - xi[Botval] * wni;
36                   Tempi=xi[Botval] * wnr + xr[Botval] * wni;
37                   xr[Botval]=xr[topval]-Tempr;
38                   xi[Botval]=xi[topval]-Tempi;
39                   xr[topval]=xr[topval]+Tempr;
40                   xi[topval]=xi[topval]+Tempi;
41               }
42           }
43       }
44       printing: for(int i=0;i<N;i++)
45       {
46           data_outR.write(xr[i]);
47           data_outI.write(xi[i]);
48       }
49   }
50
```

Fig. 12. Program to compute 256 point FFT.

The FFT algorithm itself has not been optimized due to the data dependency among inner and outer loops. Additional pipe stages will need to be implemented in order to break the loop dependency implicit in the direct implementation of the FFT. This probes the point that there the designer has to guide the tool by writing the C code in such a way that the hardware can be inferred.

Another simple tradeoff was executed by increasing the frequency of operation from 100 MHz to 500 MHz as shown in Figure 16. We can observe that the area remained almost constant, while the latency cycles increased by 3% with respect to the 200 MHz implementation baseline, the latency cycles increased by 19%. We can interpret these numbers as the logic required to implement the FFT had a larger critical path, but since the clock was increased 2.5x, the latency time was reduced by 2.0x demonstrating that there is not a linear relationship between the parameters and depends on the implementation given by the particular constraints.

Talking about power, increasing the frequency by 2.5x will have an impact on the power, but at the same time if it is 2.0x faster, we can think for example on reusing the FFT for some other part of the OFDM processor such as computing the IFFT and FFT using the same hardware and sharing it on the time domain rather than have two cores to perform both operations independently.

Solution /	Latency ...	Latency ...	Through...	Through...	Total Area
NoConstraints.v1 (allocate)	1415	7075.00	1417	7085.00	291555.47
UnrollingRead.v1 (allocate)	1176	5880.00	1177	5885.00	292156.30
UnrollingRead & Printing.v1 (allocate)	666	3330.00	667	3335.00	291849.43
Unrolling Print pipeling Read.v1 (allocate)	**650**	**3250.00**	**652**	**3260.00**	**291555....**

Fig. 13. Different solutions by selecting different architectural constraints.

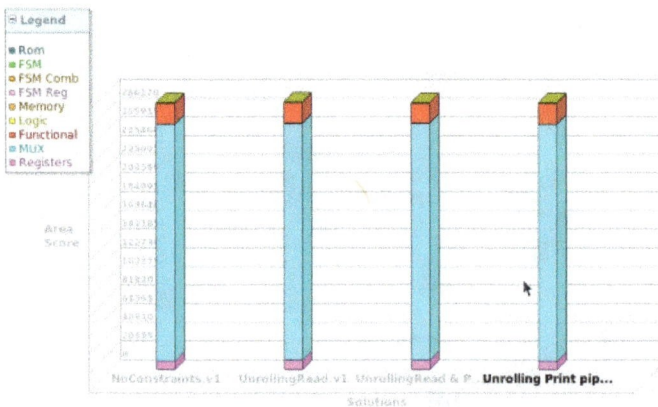

Fig. 14. Graphical view plotting Area.

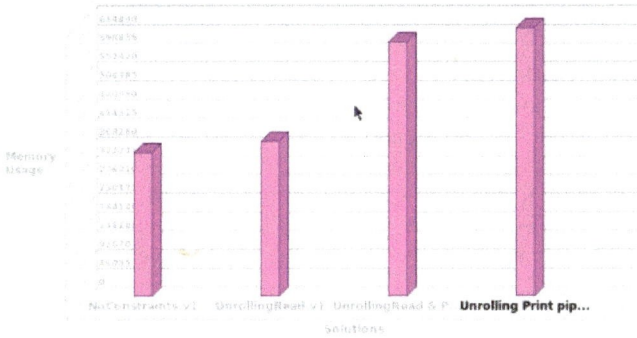

Fig. 15. Graphical View plotting memory usage.

Solution	Latency Cycles	Latency Time	Throughput Cycles	Throughput Time	Total Area
100MHz.v1 (allocate)	1391	13910.00	1393	13930.00	289966.67
200MHz.v1 (allocate)	1415	7075.00	1417	7085.00	291555.47
300MHz.v1 (allocate)	1415	4711.95	1417	4718.61	304308.54
400MHz.v1 (allocate)	1423	3557.50	1425	3562.50	303989.20
500MHz.v1 (allocate)	1695	3390.00	1697	3394.00	300547.22

Fig. 16. Change in performance with change in frequency.

10.4 Hardware In the Loop (HIL)

Hardware in the loop has become a buzz word when designers want to run their algorithm at full speed or at least hundredths or thousands times faster than an RTL or gate level simulation. In SoCs, simulation can take days, weeks and sometimes months, and that depends on the level of detail that is included in the top level simulation. That is why it is important to be able to replace each block by its behavioural, RTL and gate level models in order to refine the level of simulation control and granularity.

Rather than talking about ASIC emulators that are not traditionally available for small companies or universities, we will take a poor's man approach and show how we can integrate hardware in our computations to able to speed up the testing and processing of algorithms.

Let's take a closer look at the first level of implementation which is generating automatic HDL code from a Simulink model. Each block or a set of few blocks of the entire communication system can be implemented on hardware this was demonstrated in Section 10.1. So far, we have used an Altera Stratix III FPGA to do system level hardware testing of the Fast Fourier Transform block in the OFDM communication model. For this purpose we have used Hardware in Loop (HIL) block provided by the DSP builder Altera library. This block acts as a link between Simulink and the actual hardware we want to configure.

In modern digital communication systems, the current trend is to implement a pipelined FFT to generate orthogonal sub-carriers. A pipelined FFT generate an output every clock cycle which helps in real-time applications like digital communication systems where data is being continuously fed. We have designed Simulink models to implement FFT using butterfly diagrams which use simple Simulink blocks as well as pipelined FFT which use the advanced block set from DSP Builder. In this section we are going to talk more about the pipelined FFT for the above mentioned reasons. For more information on the architecture of the pipelined FFT implemented refer to (Shousheng & Torkelson, 1998).

The hardware implementation was done using the Altera's Quartus II version 10.1 and DSP Builder version 10.1. Care must be taken to properly design a Simulink model which would involve block sets from both advanced and standard block sets of DSP Builder. We created this model in layers. The lower level consists of the device block which has the information about the FPGA available in the hardware platform (Stratix III) and the functional blocks that essentially form the FFT. However, on the top level we could only use the signal and control blocks from the advanced block set and other blocks have to be at the lowest level in the design hierarchy.

We make use of the signal compiler and testbench from the standard block set on the top level. The signal compiler is used for creating a Quartus II project, start synthesis, to launch place and route after generating the HDL code. The testbench is used to compare the block level simulations in Simulink and the HDL simulations using Modelsim. Input and output blocks are inserted before and after the subsystem that contains the advanced block set. These blocks have external type parameters to convert from floating or other format handled by Simulink to fixed point as FPGA implementations can only be configured for fixed point. These blocks act as boundaries to the advanced and basic block sets. The procedure to convert the FFT model to HDL, configure the FPGA with the HDL code, and running it from Simulink is detailed below.

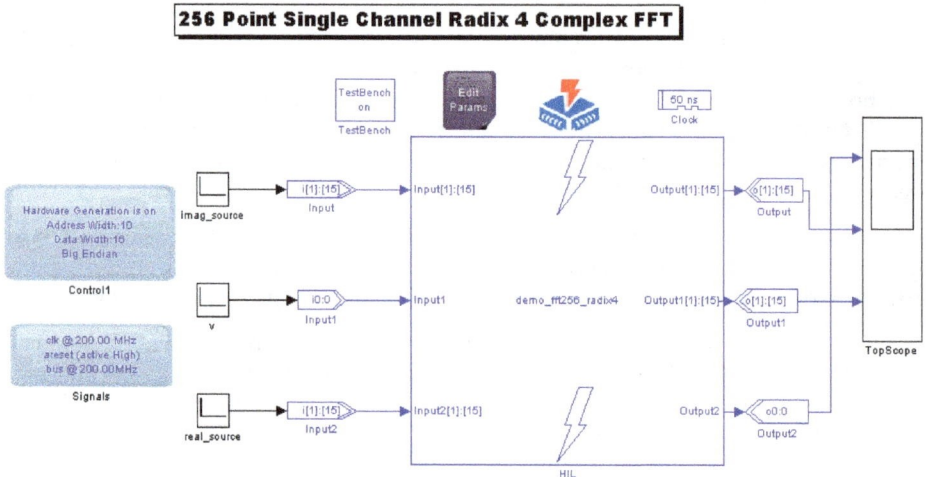

Fig. 17. Hardware In the Loop (HIL) Simulink simulation, actual code runs on the FPGA.

We first run the signal compiler block on the top level to generate HDL code and create a Quartus II project. Then compile the design with Quartus II using the compile option in the signal compiler block. We have now created a Quartus II project for the model and synthesized the HDL code for the same. Now save a copy of this model and instantiate a HIL block on the top layer of the new model from the Altera DSP Builder library found in the standard block set. Open the HIL block and copy the Quartus II project that was earlier created into the file path. This would generate proper ports for the HIL block. Connect these ports to the appropriate signals. Configure the simulation in burst mode to observe high speed of simulation. In the next menu entry of the HIL block, compile the Quartus II project again, scan JTAG in order to recognize the FPGA device and program it. If we simulate this model it runs at a remarkable speed when compared with the native Simulink simulation. Figure 17 above shows the model which has the advanced block set replaced with a HIL block. This example was modified from the one supplied by Altera to run the FFT on the FPGA platform and to be controlled by the Simulink simulation. We are in the process of converting some other algorithms into hardware following the same methodology to be able to create custom hardware acceleration blocks (Altera, 2007).

11. Conclusions

In this chapter we summarized a few of the methodologies, technologies, tools and approaches that can be taken to convert a wireless communications algorithm into a feasible hardware implementation.

While this chapter is far from being a single methodology to be followed when designing for hardware implementation of wireless communication circuits, we explored many of the practical aspects on how to achieve quick results and also to have a baseline where the final design may compare with.

Push button methodologies are still far from being a reality and even that ESL tools can achieve impressive results and can verify all the way from system level down to gate level against a golden model, there is still some reluctance from the backend teams to rely on automatic tools to do the job. While this approach has been done in automatic place and route in digital systems, ESL has been pushed the level of abstraction one level above RTL design.

What are the advantages of ESL system level design? The most valuable for the author is the ability to explore different architectures and the possibility of generating very complex datapath designs easily with simple constraints and with high hardware reusability.

Can a good RTL designer do it better? The answer is yes if he has all the time to select the best architecture for implementation. SoC design methodologoes rely on IP reutilization and to spend the valuable design time just on those blocks that will make the product differentiation.

Due to time to market constraints, design teams cannot spend much time trying to find the best and optimal architecture to implement, sometimes the task are reduced to get the job done on time. One important aspect to remember that most of the products, when the designer announces that the module is ready, it is still no more than 30% of the complete SoC design. Integration, verification & validation, design for testability, design for manufacturability, synthesis, automatic place and route will consume more than 70% of the SoC development time.

Another very important aspect is to be able to run an algorithm on hardware to take advantages of computational speed that for example could be obtained on an FPGA. This is a step required to prove if an algorithm is robust enough. ASIC technologies cannot be verified using FPGAs, but at least system level emulation can be performed to verify interconnectivity and overall signal flow.

12. Acknowledgements

There are many people that contributed directly and indirectly to the contents of this chapter with their algorithms, ideas for implementation, hard work and enthusiasm. I would like to recognize the following individuals and organizations that contributed in the following areas:

Name	Project
Dr. Chance Glenn Sr.	MOC Digital communications System Implementation
Padma Ragam	MOC Digital communications System Implementation
Nathaniel Horner	Improving the performance of DSP systems for MIMO processing
Dr. Andres Kwasinski	Improving the performance of DSP systems for MIMO processing
Mahesh Nandan Kommi	OFDM – FFT Hardware in the Loop (HIL)
Department of Electrical, Computer and Telecommunications Engineering Technology	Publishing funds.

13. References

Abdel-Hamid, E. M., Fahmy, H. A. H., Khairy, M. M., & Shalash, A. F. (2011, 15-18 May 2011). *Memory conflict analysis for a multi-standard, reconfigurable turbo decoder.* Paper presented at the Circuits and Systems (ISCAS), 2011 IEEE International Symposium on.

Accellera. (2011). UVM World: Universal Verification Methodology, 2011, from http://uvmworld.org/

Agilent. (2011). SystemVue ESL Software | Agilent, 2011, from http://www.home.agilent.com/agilent/product.jspx?cc=US&lc=eng&ckey=1297131&nid=-34264.0.00&id=1297131

Altera. (2007). An OFDM FFT Kernel for Wireless Applications (Vol. AN-452).

Altera. (2011a). Digital Signal Processing, 2011, from http://www.altera.com/products/software/products/dsp/dsp-builder.html

Altera. (2011b). Nios II C-to-Hardware Acceleration Compiler, 2011, from http://www.altera.com/devices/processor/nios2/tools/c2h/ni2-c2h.html

ARM. (2011). CoreLink System IP & Design Tools for AMBA - ARM, 2011, from http://www.arm.com/products/system-ip/amba/index.php

Ascent, S. (Ed.). (2010). *FPGAs for DSP and Communications Course Notes, UCLA Extension,* January 24-27, 2011 *Course Notes.*

Bluespec. (2011). Bluespec, Inc., 2011, from http://www.bluespec.com/

Borrayo-Sandoval, H., Parra-Michel, R., Gonzalez-Perez, L. F., Printzen, F. L., & Feregrino-Uribe, C. (2009, 9-11 Dec. 2009). *Design and Implementation of a Configurable Interleaver/Deinterleaver for Turbo Codes in 3GPP Standard.* Paper presented at the

Reconfigurable Computing and FPGAs, 2009. ReConFig '09. International Conference on.

Cadence. (2011). OVM-based verification flow 2011, from http://www.cadence.com/products/fv/pages/ovm_flow.aspx

Chandrakasan, A., & Brodersen, R. (1998). *Low power CMOS design / edited by Anantha Chandrakasan, Robert Brodersen*: Piscataway, NJ IEEE Press, 1998.

Cho, Y. S. (2010). *MIMO-OFDM wireless communications with MATLAB*: Singapore ; Hoboken, NJ : IEEE Press : J. Wiley & Sons (Asia), c2010.

Frazer, R. (2088). Reducing Power in Embedded Systems by Adding Hardware Accelerators, 2011, from http://www.eetimes.com/design/embedded/4007550/Reducing-Power-in-Embedded-Systems-by-Adding-Hardware-Accelerators

Glenn, C. M. (2009). MOC Technical brief (ECTET, Trans.). Rochester, NY: Rochester Institute of Technology.

Horner, N., Kwasinski, A., & Mondragon, A. (2011, 22-27 May 2011). *Improving the performance of DSP systems for MIMO processing*. Paper presented at the Acoustics, Speech and Signal Processing (ICASSP), 2011 IEEE International Conference on.

Ifeachor, E. C. (1993). *Digital signal processing : a practical approach*. Wokingham, England ; Reading, Mass. :: Addison-Wesley.

IT++. (2011). Welcome to IT++! , 2011, from http://itpp.sourceforge.net/devel/index.html

Mathworks. (2011). MathWorks - MATLAB and Simulink for Technical Computing, 2011, from http://www.mathworks.com/

Mentor-Graphics. (2011). Algorithmic C Datatypes - Mentor Graphics, 2011, from http://www.mentor.com/esl/catapult/algorithmic

MentorGraphics. (2011). Catapult C Synthesis Overview - Mentor Graphics, 2011, from http://www.mentor.com/esl/catapult/overview

Mondragon-Torres, A. F., Kommi, M. N., & Bhattacharya, T. (2011). *Orthogonal Frequency Division Multiplexing (OFDM) Development and Teaching Platform*. Paper presented at the 2011 Annual Conference & Exposition, Vancouver, BC, CANADA. http://www.asee.org/search/proceedings?fields%5B%5D=title&fields%5B%5D=a uthor&fields%5B%5D=session_title&fields%5B%5D=conference&fields%5B%5D=y ear&search=mondragon-torres&commit=Search

NI. (2011). NI LabVIEW FPGA - National Instruments, 2011, from http://www.ni.com/fpga/

OCP. (2011). OCP-IP : Home Page, from http://www.ocpip.org/

Rappaport, T. S. (2001). *Wireless communications : principles and practice* (2nd ed ed.). Upper Saddle River, N.J. : London :: Prentice Hall PTR.

Sanchez-Sinencio, E., & Andreou, A. (1999). *Low-Voltage/Low-Power Integrated Circuits and Systems: Low-Voltage Mixed-Signal Circuits, Wiley-IEEE Press*, January 1999

Shousheng, H., & Torkelson, M. (1998, 11-14 May 1998). *Design and implementation of a 1024-point pipeline FFT processor*. Paper presented at the Custom Integrated Circuits Conference, 1998. Proceedings of the IEEE 1998.

Synfora. (2009). *PICO USER MANUAL - Writing C Applications: Developer's Guide.* (PE-ASIC-UM-WCADG-VER 09.03-6). Mountain View, CA.

Synopsys. (2011a). Signal-Processing, 2011, from http://www.synopsys.com/systems/blockdesign/digitalsignalprocessing/pages/ signal-processing.aspx

Synopsys. (2011b). Synphony C Compiler, 2011, from
    http://www.synopsys.com/Systems/BlockDesign/HLS/Pages/SynphonyC-
    Compiler.aspx
Synopsys. (2011c). Verification Methodology Manual for SystemVerilog, 2011, from
    http://vmm-sv.org/
SystemC. (2011). Home - Open SystemC Initiative (OSCI), 2011, from
    http://www.systemc.org/home/
Texas-Instruments. (2011). TMS320TCI6482 Fixed Point Digital Signal Processor, 2011, from
    http://www.ti.com/product/tms320tci6482
Xilinx. (2011a). AutoESL High-Level Synthesis Tool, 2011, from
    http://www.xilinx.com/tools/autoesl.htm
Xilinx. (2011b). System Generator for DSP, 2011, from
    http://www.xilinx.com/tools/sysgen.htm
Yang, S., Yuming, Z., Goel, M., & Cavallaro, J. R. (2008, 2-4 July 2008). *Configurable and
    scalable high throughput turbo decoder architecture for multiple 4G wireless standards.*
    Paper presented at the Application-Specific Systems, Architectures and Processors,
    2008. ASAP 2008. International Conference on.
Yin, H., & Alamouti, S. (2006, 27-28 March 2006). *OFDMA: A Broadband Wireless Access
    Technology.* Paper presented at the Sarnoff Symposium, 2006 IEEE.

6

Analysis of Platform Noise Effect on Performance of Wireless Communication Devices

Han-Nien Lin

Feng-Chia University
Taiwan, R.O.C.

1. Introduction

Cloud computation and always-connected Internet attracts the most industrial attention for the past few years. Meanwhile, with the development of IC technologies advancing toward higher operating frequencies and the trend of miniaturization on wireless communication products, the circuits and components are placed much closer inside the wireless communications devices than ever before. The system with highly integrated high-speed digital circuits and multi-radio modules are now facing the challenge from performance degradation by even more complicated platform EMI noisy environment. The EMI noises emitted by unintentionally radiated interference sources may severely impact the receiving performance of antenna, and thus result in the severe performance degradation of wireless communications. Due to the miniaturization of a variety of wireless communications products, the layout and trace routing of circuits and components become much denser than ever before. Therefore, we have investigated and analyzed the EMI noise characteristics of commonly embedded digital devices for further high performance wireless communications design. Since the camera and display module is most adopted to the popular mobile devices like cellular phone or Netbook, we hence focus on EMI analysis of the built-in modules by application of IEC 61967[1][2] series measurement method.

Since the causes of reduction of throughput or coverage due to receiving sensitivity degradation of wireless system could result from decreased S/N via conducted or radiated EMI noises from nearby digital components shown in Figure 1. This chapter discusses RF de-sensitivity analysis for components and devices on mobile products. To improve the TIS performance of wireless communication on notebook computer, we investigated the EMI noise from the built-in camera and display modules as examples and analysed the impact of various operation modes on performance with throughput measurement. We also utilized the near-field EM surface scanner to detect the EMI sources on notebook and locate the major noisy sources around antenna area. From the emission levels and locations of the noisy components, we can then figure out their impact on throughput and receiving sensitivity of wireless communications and develop the solutions to improve system performance. Finally, we designed and implemented periodic structures for isolation on the notebook computer to effectively suppress noise source-antenna coupling and improve the receiving sensitivity of wireless communication system.

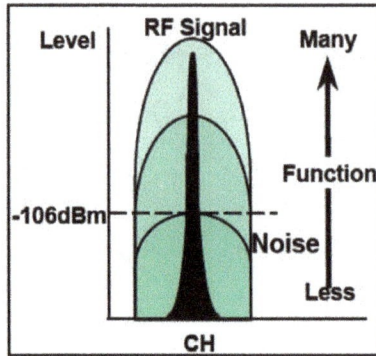

Fig. 1. S/N ratio decreases due to digital components for multi-functions.

2. The noise impact of camera and Touch Panel (TP) modules on product performance

2.1 Performance testing for Wireless Wide Area Network (WWAN) devices

There are two different purposes for the OTA (Over-The-Air) testing[3] on mobile stations. The first testing is for the carrier's cell site coverage which is relative with loss plan and link budget of the cell site. For example, the sensitivity measurement of the W-CDMA receiver is performed by the base-station simulator to determine the receiving sensitivity of EUT (Equipment Under Test) by reporting the minimum forward-link power which resulting in a bit-error-rate (BER) of 1.2% or less at the data rate of 12.2 kbps with a minimum of 20,000 bits. The second tresting is the throughput for supporting all kind of the applications for cloud computing. The minimum throughput required will depend on application. For example, the minimum throughput we need to link YOU TUBE for HD video is about 1Mbps at least. Therefore, the mobile station (Smart Phone, Tablet PC, Note book PC, etc.) is required the OTA performance testing on TRP, TIS and De-Sense.

2.1.1 Total Radiated Power (TRP)

TRP measurement is to evaluate the transmitting RF power performance of mobile device by summing the effective isotropic radiated power (EIRP) of complete Theta- and Phi-cut as shown in Figure 2. The procedure is first to measure the radiated power at each Phi degree interval for 360 degree rotation (if interval is 30 degree then it need 12 measurement), and then for the Theta axial. Finishing the 180 degree rotation along Theta axial, the TRP is obtained with following formula.

$$TRP \cong \frac{\pi}{2NM} \sum_{i=1}^{N-1} \sum_{j=0}^{M-1} \left[EiRP_{\theta}(\theta_i, \phi_j) + EiRP_{\phi}(\theta_i, \phi_j) \right] \sin(\theta_i) \qquad (1)$$

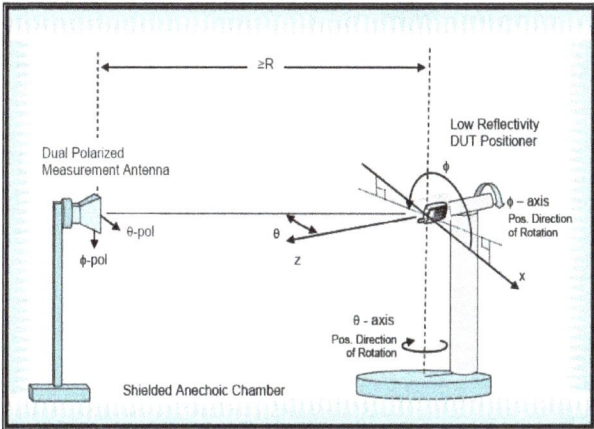

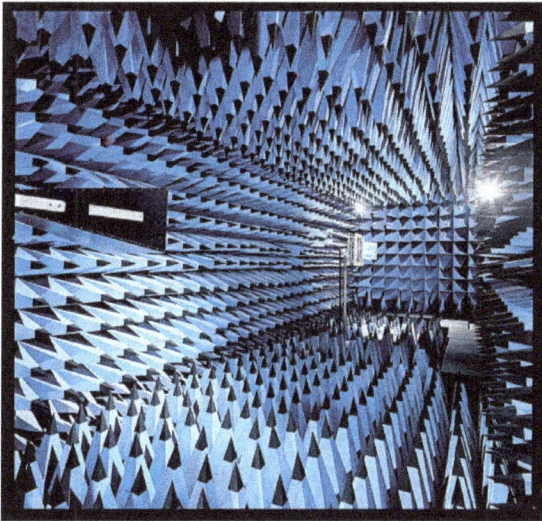

Fig. 2. Total Radiated Power measurement.

For the ideal case, the TRP should be equal to the conducted power (Watts) times mismatching Loss (%) and antenna efficiency as shown in following relationship and illustration. But the antenna efficiency measurement can actually with error resulting from coaxial cable connection as illustrated in Figure 3. When the coaxial cable is connected to the SMA connector, the surface on it could cause measurement error of the antenna efficiency.

$$TRP = \frac{1}{4\pi} \int_{\theta=0}^{\pi} \int_{\phi=0}^{2\pi} (EiRP_\theta(\theta,\phi) + EiRP_\phi(\theta,\phi))\sin(\theta)d\theta d\phi$$

$$TRP = P_A \cdot L_m \cdot eff \tag{2}$$

Transmit Power = Pc (Conducted Power) + Antenna Gain (in dB)

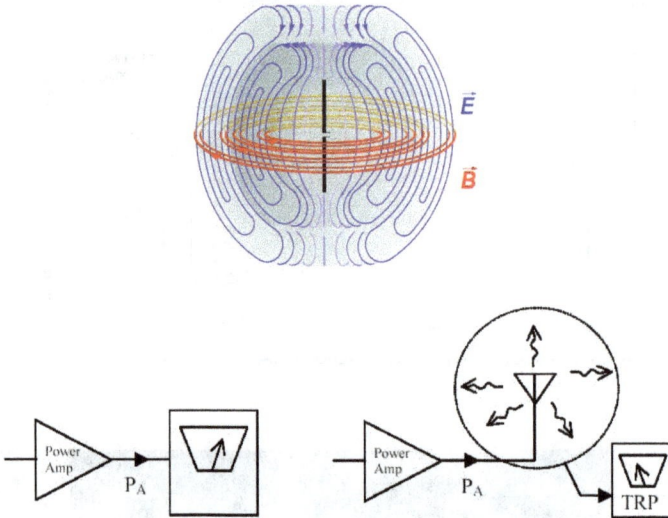

Fig. 3. Illustration of Antenna TRP.

2.1.2 Total Isotropic Sensitivity (TIS)

The measurement setup for TIS testing is the same as shown in Figure 2, except with the different calculation. It is analogous to calculate the total resistance form the parallel resistor network. The Effective Isotropic Sensitivity (EIS) is illustrated in Figure 4 and calculated with following formula.

$$TIS \cong \frac{2NM}{\pi \sum_{i=1}^{N-1}\sum_{j=0}^{M-1}\left[\dfrac{1}{EIS_\theta(\theta_i,\phi_j)} + \dfrac{1}{EIS_\phi(\theta_i,\phi_j)}\right]\sin(\theta_i)} \tag{3}$$

$$G_{x,EUT}(\theta,\phi) = \frac{P_S}{EIS_x(\theta,\phi)} \tag{4}$$

$$TIS = \frac{4\pi}{\oint\left[\dfrac{1}{EIS_\theta(\theta,\phi)} + \dfrac{1}{EIS_\phi(\theta,\phi)}\right]\sin(\theta)d\theta d\phi} \tag{5}$$

For the ideal case, TIS should be equal to conductive sensitivity divided by mismatching loss and antenna efficiency as shown in following relationship and illustration. Not only the surface current on coaxial cable would cause the antenna efficiency measurement error, but also the platform noise interference investigated here would de-sense the receiver.

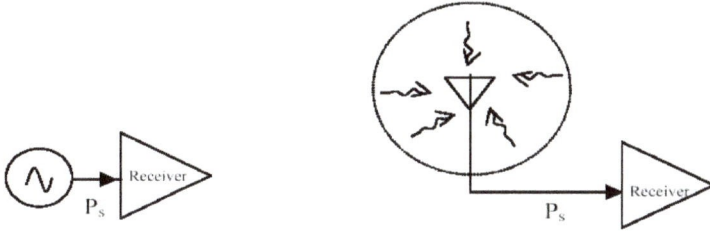

EIS: Effective Isotropic Sensitivity = Received EIRP – Antenna Gain (in dB)
TIS: Total Isotropic Sensitivity (3D Measurement)

Fig. 4. Illustration of Antenna TIS.

The relationship between receiver performance and platform noise is described by receiver bit error rate and energy per bit (Eb) /Noise(N0) as shown in Figure 5. For example, the WCDMA receiver sensitivity with QPSK modulation can be determined as following:

Bit Error Rate (BER): 1.2% BER for QPSK demodulation receiver require that $Eb/N0 = 7.5dB$

where Eb: measured at base-band output (I/Q output) for each bit.
   No: total noise power form RF front end to base-band, include LNA NF, ADC, quantize noise, PLL phase Noise,…with Gaussian system noise representation.

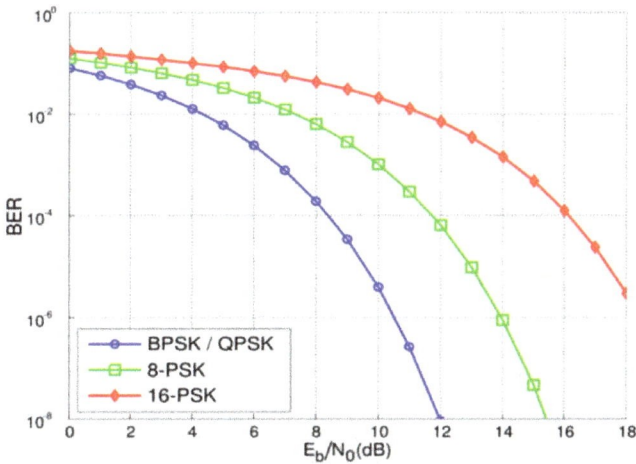

Fig. 5. Receiver Bit Error Rate vs. Energy per bit (Eb) /Noise(N0).

The BER is measured in time domain after demodulation of receiver. For WCDMA system, it needs 20k bits / 12.2Kbps = 1.64 second at each receiving channel. From communication demodulation theory, N0 is described as Gaussian noise, and it is the sum of the receiver noise (related to implementation loss) and system noise as illustrated in the following figures.

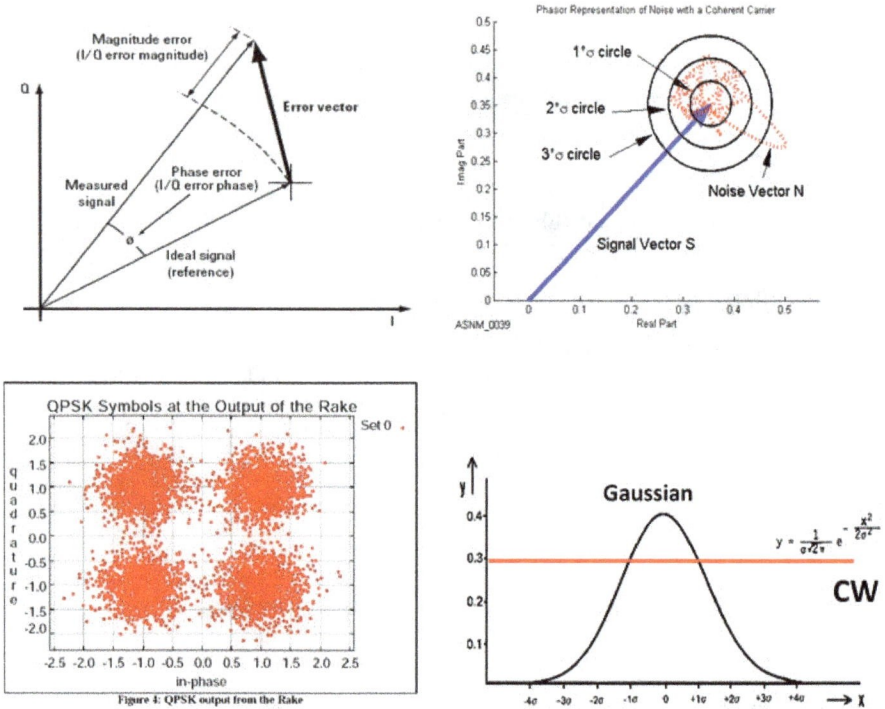

Fig. 6. Illustration of system noise effect.

Based on TIS requirement for WWAN or WLAN throughput, the noise limit of the wireless system should be set to meet the regulatory specification. Figure 7 shows the sensitivity degradation due to self-interference for GSM 1800 and WCDMA systems. The example for WCDMA system is following:

Noise Limit: TIS (dBm) + Antenna Gain (dB) – Eb/N0(dB) + Processing Gain (demodulation dependence) – System Losses (6dB) (depending on chip set, LNA NF, PLL phase nosie, ADC.....), Processing Gain (dB)=10 Log(Chip rate/Data Rate) = 10log (3840K/12.2K)= 25dB

WCDMA Noise Limit : Gaussian Probability Density Function Noise Power-103dBm + (-5dB) -7.5dB +25dB – 6dB = 96.5dBm

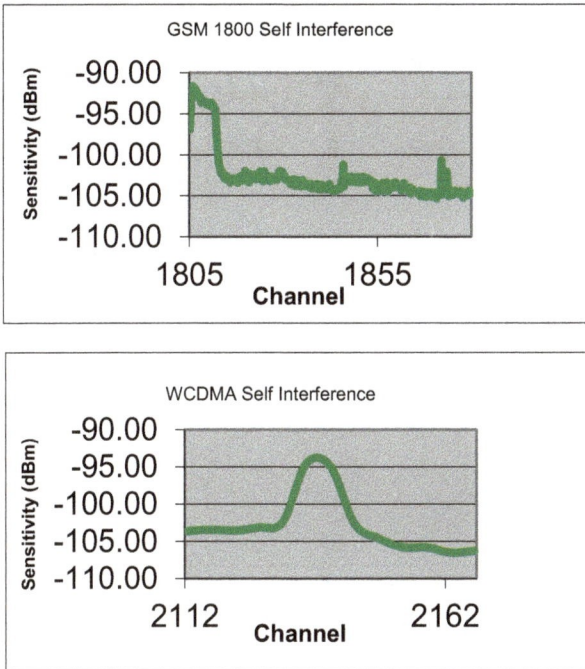

Fig. 7. Sensitivity degradation examples for GSM 1800 and WCDMA systems.

However, since there are more than one thousand Channels for GSM and WCDMA systems, we can't test all receiving channels for those WWAN devices. The alternative way for testing those intermediate channels is to measure the relative sensitivity as following steps and illustration as shown in Figure 8:

1. Move the EUT and position to the location and polarization which results in best radiated sensitivity, then measure for the closest channel (in frequency) and set as Reference Channel. After the 3D testing of the high, middle and lower channels (set as reference channel), we acn then review the 3D graph and find the best sensitivity at theta- and phi-plane of EIS along vertical or norizontal polarization (it means the best radiated sensitivity).

2. Now, there are three bset radiated sensitivity at Theta- and Phi-polarization for Low, Middle and High reference channel in one band. The rest of channels in one band should be then tested as following: The all channels in frequency range from lowest frequency channel (reference channel) to the frequency at (low + Middle)/2 should be de-sensed less than 5 dB to the reference channel. The all channels in middle band of frequency from (Low + Middle)/2 to (Middle + High)/2 should be de-sensed less than 5dB to the reference channel (middle channel). Finally, the all channels in high band of frequency from (Middle + High) / 2 to (High should be de-sensed less than 5dB to the reference channel (highest frequency channel).

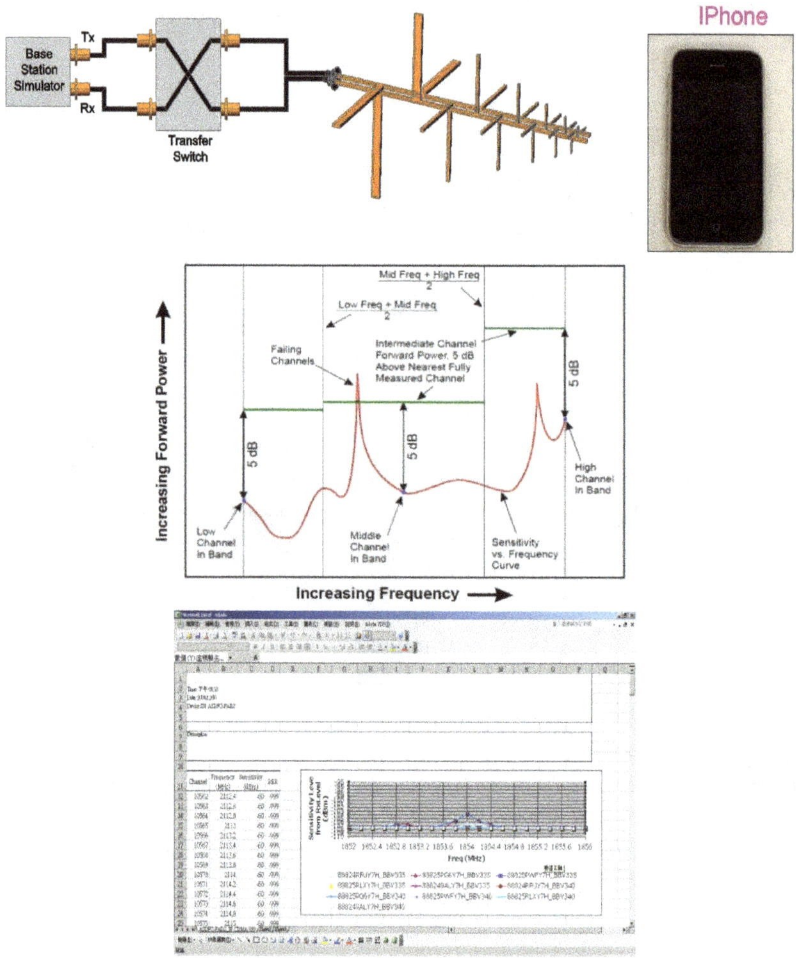

Fig. 8. Alternative way for testing relative sensitivity of intermediate channels.

2.2 De-Sense effect

DeSense is the term representing the noise impact degree on receiver sensitivity. This section will address on the popular TP and camera module about their roles on interference with embedded antenna and discuss the De-sense effect from platform noise coupling.

Nowadays, Touch Panel (TP) and camera module both occupy large part on the smartphone, Tablet PC or NB. Hence no matter where the embedded antennas are placed, the noise emitted from TP and Camera will couple to antenna and thus result in DeSense (Degradation of Sensitivity) problem. On the other hand, all the RF power transmitted by embedded antennas of the wireless products will also couple to nearby TP and camera modules. Those proximate electric (E) and magnetic (H) field coupling will affect normal operation of TP and camera modules.

Figure 9 shows the antenna locations under investigation. In addition, Figure 10 shows the Path-Loss measurement setup and test procedures as following[4]:

1. Put EUT into shielding box.
2. Connect VNA port1 to Tx antenna and laptop antenna to port 2
3. Measure for specific frequencies antenna efficiency of Tx antenna and Rx antenna

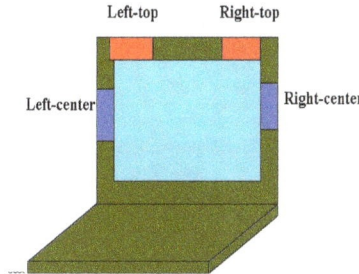

Fig. 9. Antenna locations on laptop display.

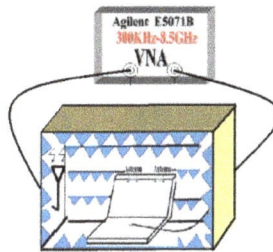

Fig. 10. Test setup to measure Path-Loss.

2.2.1 Impact of LCD EMI analysis on 802.11g throughput[5]

In a laptop, there are many interference sources which can be in the form of radiation or conduction. LCD noise is the major interference to the wireless performance. Figure 11 shows the frequency domain measurement setup and measured results for platform noise from LCD. The measurement setup is shown in Figure 11a and the test procedures are described as following:

1. Put EUT into shielding box.
2. Connect antenna cable to AMP/Spectrum analyzer via coaxial cable.
3. Power on EUT.
4. Measure noise level for the chosen target frequency.

Figure 11b shows the different antenna placements along the horizontal edge on top of a LCD panel. The measured results show that the noises at 2.400GHz, 2.450GHz, 2.490GHz (harmonics of the pixel clock) are major interference sources that fall into WLAN band. We can obtain 2-5dB noise suppression by simply moving the antenna several millimeters away from its initial location. The comparison for different antenna measurements at different locations shows that the LCD noise might have an significant impact on desensitization to

802.11g. The measurement of antenna positioned towards the left 20mm serves as a reference to quantify the impact of antenna placement on the platform noise measurement. The noise picked up by antennas would desensitizes the receiver and reduced the throughput. Meanwhile, the throughput test procedure and test setup are as follows: the setup consists of an AP (access point), EUT (laptop) and Chariot console throughput software. The AP (access point) and EUT (laptop) are connected through path-loss attenuators to control RFI strength, and the communication traffic is controlled and monitored by a desktop using Ixia Chariot® software as shown in Figure 12a. The system

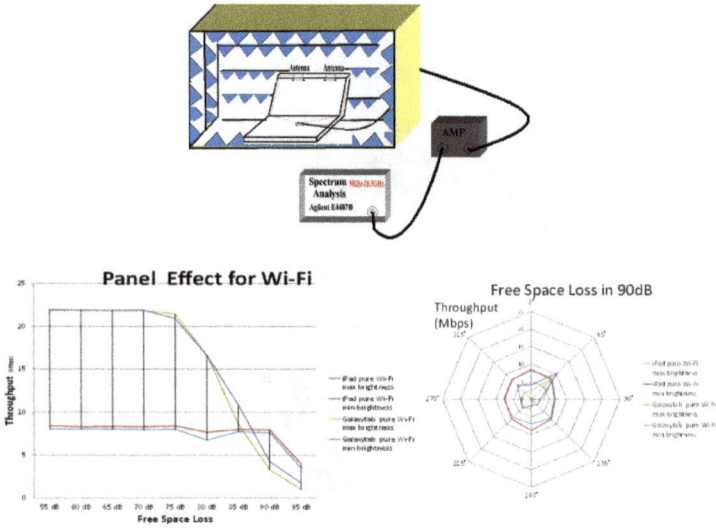

(a) Platform noise measurement setup for antenna port

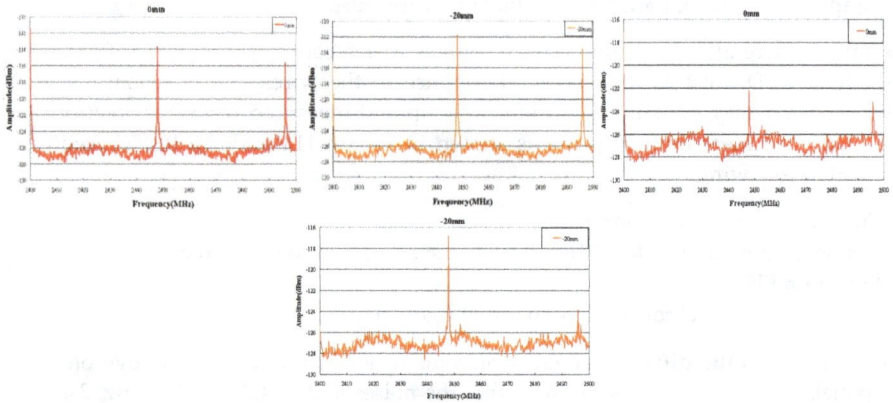

(b) Comparison of the different antennas measurement at different locations on the LCD panle.

Fig. 11. LCD noise measured at antenna port in WLAN band.

path-loss includes cable loss, space loss and attenuators. The results in Figure 12b the real line and dotted line, clearly show that the sensitivity and the throughput decrease as the LCD interference is injected to the communication link between the AP and NIC card. It is found that there is about 10dB desensitization between the two throughput measurements for different locations. It is show that when the LCD noise increases the sensitivity is reduced and performance is also degraded. Moreover, it is again shown that the location of antenna placement is significant to wireless communication performance. Figure 12c shows the photograph of the antenna integrated into a laptop for investigation.

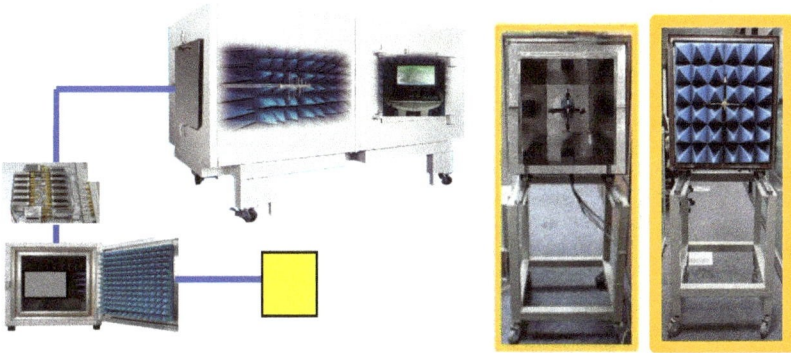

(a) The throughput test procedure and test setup

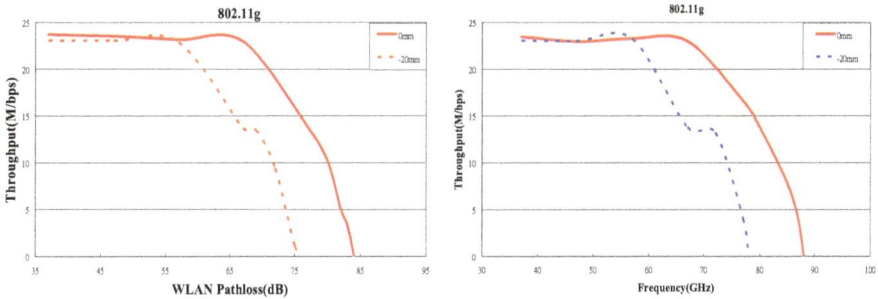

(b) Throughput comparison for in 802.11g antenna 2 and antenna 3 at two different positions.

(c) Photograph of the antenna integrate into a laptop.

Fig. 12. Throughput comparison in 802.11g for different locations on the LCD panel.

2.2.2 Platform noise analysis of Netbook system[6]

Platform noise analysis can be conducted through the noise floor measurement at antenna port of wireless device. The complete setup for noise floor measurement should at least consist of following hardware instruments: Shielded box, pre-amplifier, spectrum analyser or EMI receiver, high quality coaxial cable, and Netbook EUT.

The frequency-domain noise floor measurement setup for Netbook platform is shown as Figure 11(a). The measuring procedure is as following: Put the EUT Netbook inside the shielded box and connect its antenna port to the pre-amplifier and spectrum analyser. We first measured the ambient noise with Netbook power-off, then powered on the Netbook and measured the noise level within the selected communications bands as interfering platform noise.

Figure 13 shows the noise level captured by the integral WWAN antenna of the Netbook on GSM 850, GSM 900, and DCS 1800 bands. The platform noises on GSM 850, GSM 900, and DCS 1800 bands are shown in Figure 2(a), (b), and (c) respectively. Figure 13(a) shows that the major noise spectrum falls on 864 MHz and 888 MHz, corresponding to the 36th and 37th harmonics of the Azalia Sound Card with 24 MHz fundamental driver frequency.

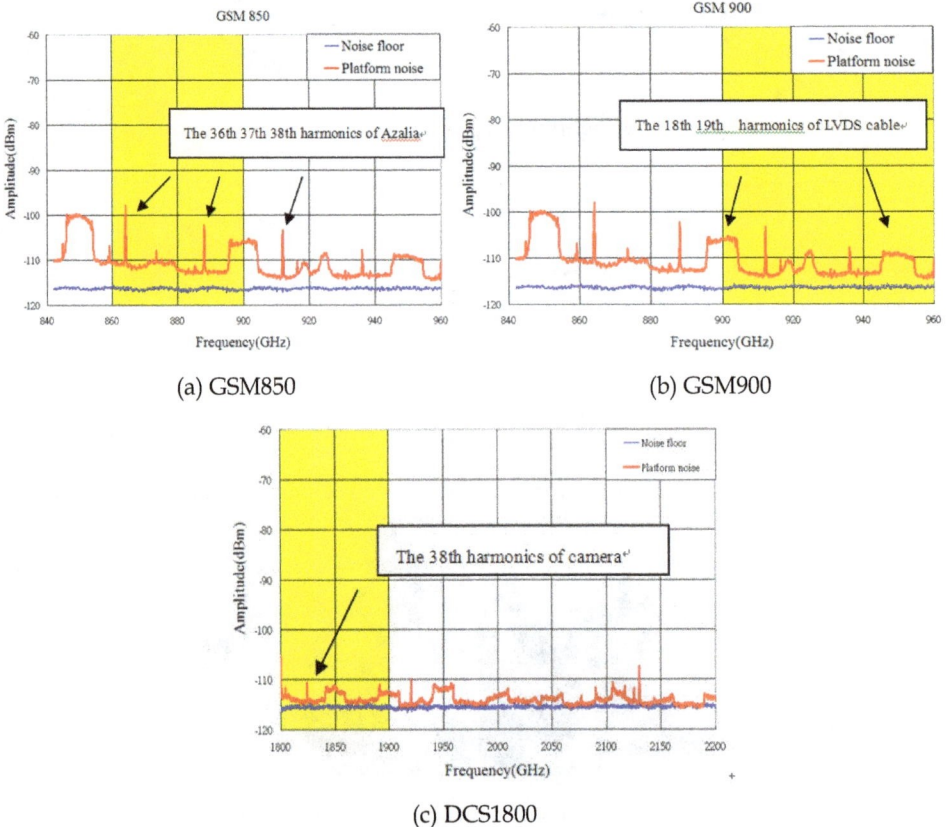

(a) GSM850   (b) GSM900

(c) DCS1800

Fig. 13. Measured noise level on WWAN bands of Netbook.

Figure 13(b) shows broadband and regularity characteristics of noise on 900 MHz band. Since the fundamental frequency of the LVDS interconnection cable is 50 MHz and noise spectrum falls between 900 MHz and 950 MHz, we can calculate from system clock map that the noise spectrum received by antenna was originated from 18th and 19th harmonics of LVDS cable. Figure 13(c) shows that the noise measured on DCS 1800 band falls on 1824 MHz, which is 38th harmonic of CCD camera. Figure 13(c) also shows noise occurring around 1850 MHz and 1900 MHz, which are the 38th and 39th harmonics of LVDS cable respectively. From the noise spectrum analysis, we found out that the noise sources mentioned above cause in-band interference on operation bands of GSM 850/950 and DCS 1800 systems frequently. Figure 14(a) shows the noise measurement result on Band-1 of WCDMA, and it appears as lower level and ambiguous. However, because Band 5of WCDMA almost operate on the same frequencies with GSM 850, it thus suffers interference from the 36th and 37th harmonics of the Azalia Sound Card.

From the platform noise measurement method and clock map analysis, we are able to establish the system design rule for related position and orientation of noise source(s) and antenna(s) placement to suppress in-band interference. We can also utilize various isolation or shielding techniques to effectively prevent the antenna port noise level from platform noise sources, and further reduce the delay time caused by lengthy product debug and speed up testing time.

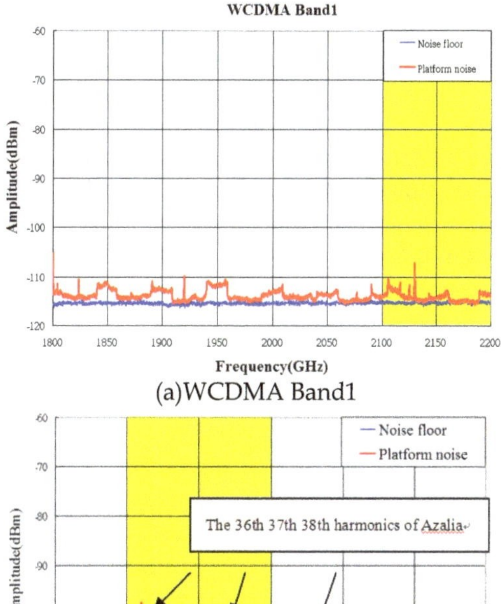

(a)WCDMA Band1

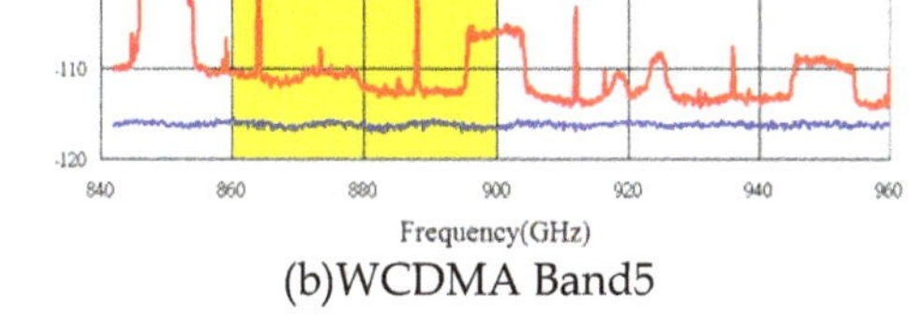

(b)WCDMA Band5

Fig. 14. Measured noise level on WCDMA bands of Netbook.

2.2.3 Impact analysis of platform noise on TIS of GSM/WCDMA systems[6]

TIS is a figure of merit for receiving performance of a mobile or wireless terminal device. Receiver performance is considered as important to over all system performance as is Transmitter performance. The downlink, or subscriber unit receive path is integral to the quality of the device's operation. The receiver performance of the Equipment Under Test (EUT) is measured utilizing Bit Error Rate (BER) or Frame Errasure Rate (FER). This test specification uses the appropriate digital error rate (as measured by the subscriber unit) to evaluate effective radiated receiver sensitivity at each spatial measurement location. All of the measured sensitivity values for each EUT test condition will be integrated to give a single figure of merit referred to as Total Isotropic Sensitivity (TIS). The BER specification of CTIA on GSM and WCDMA systems for optimal transmitted data rate are 2.44% and 1.22% respectively. TIS measurement not only measures the performance of stand-alone antenna, but also takes wireless device itself into account to realize the practical implementation. We evaluated the EIS (Effective Isotropic Sensitivity) by measuring the minimum received power that met the BER requirement on the test position. The TIS result is able to clearly show the 3-dimensional receiving performance of wireless communications device under specific mobile communications environment.

The complete setup for TIS measurement should at least consist of following hardware instruments: fully anechoic chamber, measuring antenna(s), base-station communications emulator, RF relay switch, high quality coaxial cable, control PC, and position controller. The practical setup for TIS measurement is shown as Figure 4. The operation principle of Figure 15 is as following: Connect control PC to the base-station communications emulator and then make the base-station emulator send test signal to transmit antenna. The power level of transmit antenna is set to -60 dBm for TIS measurement. The power level decreasing step specified by CTIA is 0.5 dB for transmit antenna to measure the minimum power level obtained by receiving antenna. When the transmitted power has been attenuated to some lower level and signal received from receiving antenna to base-station emulator with BER worse than 2.44% (GSM)/1.22% (WCDMA), then we have the minimum receiving power for Netbook.

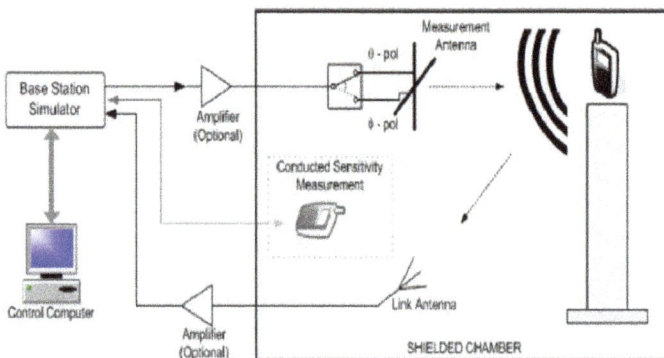

Fig. 15. Setup for TIS measurement.

The results of TIS measurement are shown in Table 1 and 2. The High/Mid/Low channels of GSM850/900 and DCS1800 systems are selected to analyse the platform noise level impact on TIS measurement as shown in Table 1. We found that TIS performance is getting worse as in-band noise increases. Table 1(c) shows the relationship between platform noise level and TIS on CH 512 and CH 698 of DCS 1800 system, it indicates that TIS has 5dB degradation as platform noise increases 2 dB. Table 2 shows the TIS measurement result of WCDMA of Netbook. From the platform noise level and TIS comparison between CH 4357 and CH 4408 of WCDMA Band-5, we found that TIS has 2dB degradation as platform noise increases 1.5 dB. From the observation above, we briefly conclude that TIS performance of the wireless product degrades 2 dB whenever intra-system platform noise level increases 1 dB. It therefore shows that the platform noise is the major factor affecting the receiving sensitivity of wireless devices.

GSM 850	TIS (dBm)	NFS (dBm)
CH128(869.2MHz)	-100.73	-111.39
CH190(882.6MHz)	-102.73	-112.17
CH251(893.8MHz)	-101.19	-112.63

(a) GSM850

GSM 900	TIS (dBm)	NFS (dBm)
CH975(925.2MHz)	-100.74	-108.97
CH037(942.2MHz)	-101.43	-113.18
CH124(959.8MHz)	-100.08	-113.72

(b) GSM900

DCS 1800	TIS (dBm)	NFS (dBm)
CH512 (1805.2MHz)	-103.76	-113.48
CH698 (1842.4MHz)	-98.91	-111.49
CH885 (1879.8MHz)	-102.49	-114.51

(c) DCS1800

Table 1. Measured TIS on WWAN bands of Netbook (a) GSM850 (b)GSM900(c)DCS1800.

WCDMA 1	TIS (dBm)	NFS (dBm)
CH10562 (2112.4MHz)	-105.44	-114.74
CH10700 (2140MHz)	-103.74	-113.12
CH10838 (2167.6MHz)	-106.32	-115.03

(a) WCDMA Band1

WCDMA 5	TIS (dBm)	NFS (dBm)
CH4357 (871.4MHz)	-102.37	-110.63
CH4408 (881.6MHz)	-104.26	-112.17
CH4458 (891.6MHz)	-104.24	-112.56

(b) WCDMA Band5

Table 2. Measured TIS on WCDMA bands of Netbook.

3. RF coexistence problems on product performance [7]

Due to the increasing add-on functions demand for consumer electronics, currently multi-radios, such as WLAN, WWAN, GPS, Bluetooth, and even DVB-H modules, have all been crowdedly embedded and highly integrated in a tiny space of wireless communications platform. Therefore the wireless devices usually have been equipped with more than one antennas, the purpose is to fit for different communication system such as cellular mobile communications, wireless local area networking, and personal area networking. Under this situation, the performance of various kinds of wireless communications is usually degraded by the mutual coupling and interference of closely arranged antennas inside the mobile device. Since the RF modules co-existence has become a critical design problem for wireless communications, we will discuss the RF coexistence problems in this section.

3.1 Isolation required for RF coexistence

Platform noise usually raises the RF receiver noise floor and dramatically degrades system performance by push the Eb/N0 to the margin when there in-band or out-of-band interference exists. A frequent cause of poor sensitivity on a single channel, or a small number of channels, is due to receiver's in-band noise from broadband digital noise or spurious signals from other coexistent transmitters. We describe in this section the potential coexistence problem for multimode and multiband RF modules, and also illustrate below in Figure 16 the example of isolation required for various RF systems to achieve better service.

Transmitter	Minimum Isolation Recommendations			
	Receiver			
	Bluetooth	802.11b/g	802.11a	GSM
Bluetooth	n/a	40dB	20dB	20dB
802.11b/g	40dB	n/a	n/a	20dB
802.11a	20dB	n/a	n/a	20dB
GSM	20dB	20dB	20dB	n/a

Fig. 16. Tx leakage in FDD system and isolation budget.

3.2 External modulation problem

Even the low-speed digital I/O traces or cables in TP may induce GFSK modulation current, when they are nearby Bluetooth module operating at 2.4GHz. The non-linear ON/OFF switching of digital signal would also play a role as pulse modulation and generate magnetic field through those traces or cables to interfere the DCS and PCS systems. Some extreme case would also happen to WCDMA. The external modulation phenomenon can't be measured by network analyzer until the TP is activated as shown in Figure 17. We also illustrate in this section how the external modulation effect could be found in the final design stage of product with WWAN and BT RF modules ready, however the re-design is needed when platform and TP operate in their normal mode.

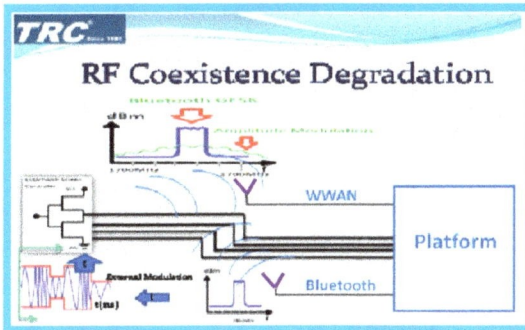

Fig. 17. External Modulation Effect.

4. Platform noise coupling mechanism

The interfering noise sources mentioned above may introduce adverse noise to the nearby wireless modules via conduction or radiation coupling or even both. The digital noise coupled from broadband digital or narrow-band RF devices may result in in-band interference and further degrade the performance of wireless communications, and vice versa. Since the digital noise would cover a wide range of frequency, we will illustrate in Figure 18 and 19 the three potential coupling mechanisms between the noise sources and victim circuit or module: conducted coupling, crosstalk coupling, and radiated coupling.

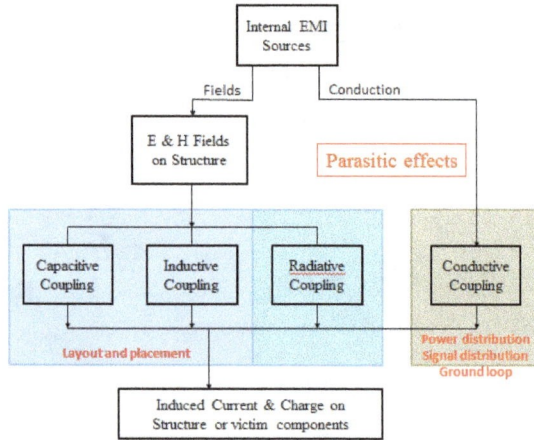

Fig. 18. Different coupling mechanisms.

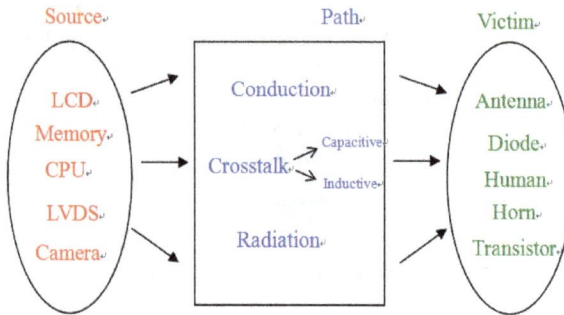

Fig. 19. Illustration of possible noise coupling for wireless device.

The Figure 20 illustrates the EM interaction between TP and embedded antenna, and the photograph of Figure 21 shows the DCS1800 and PCS 1900 transmitted power coupled to LCD panel.

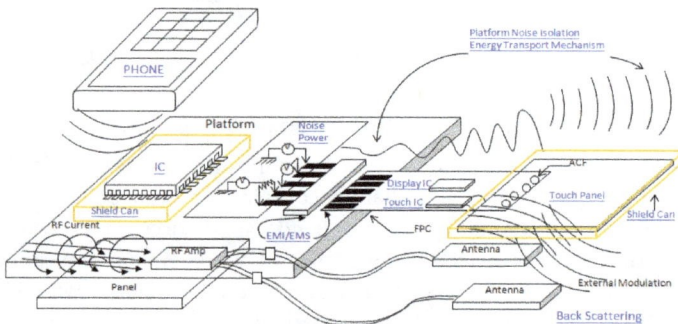

Fig. 20. EM interaction between TP and embedded antenna.

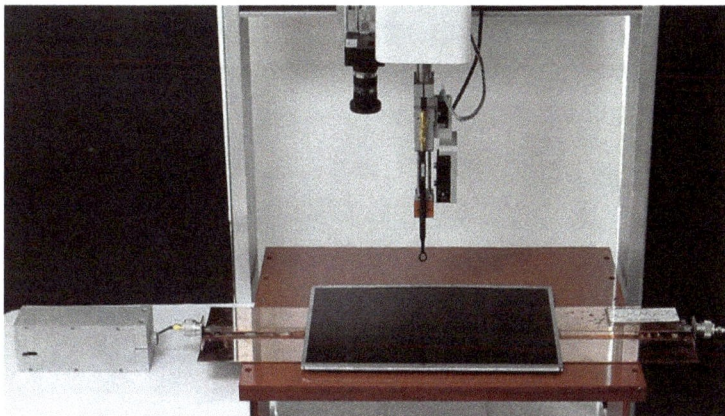

Fig. 21. Illustration of DCS1800 and PCS 1900 coupling to LCD panel.

4.1 Analysis of conductive coupling

The first step of the degradation of sensitivity (De-sense) measurement is conducted testing, because understanding how the interference platform noise conducted to the RF receiver is the most important issue for further analysis. Even the same probability distribution function of noise, there is possible to cause different De-Sense impact depending on the receiver implementation.

The best way to obtain the conducted De-Sense effect is to use the internal WWAN module or chip set of mobile device for RSSI measurement as shown later in Figure 25. The left figure shows a 50 ohm terminated at antenna connector to read the WWAN RSSI data. The figure in the center shows the dummy WWAN module with circuit ground and chassis ground, and the WWAN card inside the shielded box is connect to the dummy WWAN card via coaxial cable to read the RSSI data again. The right figure shows the dummy WWAN module with chassis ground and WWAN card to read the RSSI data. The RSSI data read from the same RF receiver with three different conditions described above will help engineers easily identify the platform noise.

4.2 Analysis of near-field coupling from antenna to embedded devices

Since the antenna is usually implemenred in the proximity of TP and camera, the near-field coupling to the nearby TP and camera would sometimes result in malfuction due to the transmitted power (Figure 21). The electromagnetic field distribution on TP due to antennaradiation is shown in Figure 22 below.

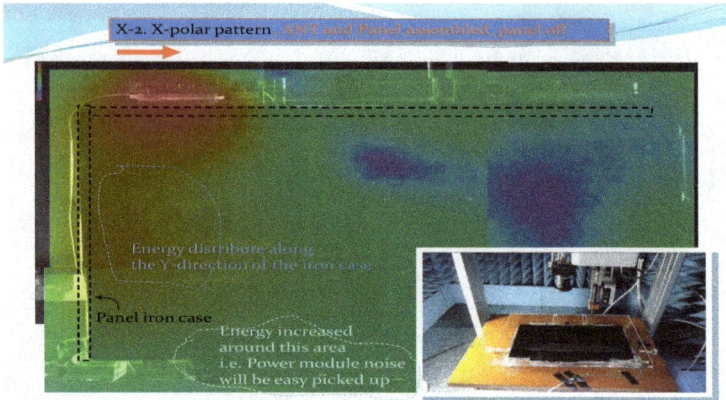

Fig. 22. Antenna radiation coupling.

4.3 Resonance issues of mechanic structure

The LCD panel of Notebook and the enclosure or PCB would usually create a resonant structure. The related resonance issue could be explained with the following measuring methodology.

1. Connect port 1 of network analyzer to LVDS cable and WEB camera cable via a coupling fixture (Balun and low-mu ferrite core as absorbing clamp) for energy transfering to the LVDS cable.
2. Connect port 2 of network analyzer to the embedded antenna of Notebook via its own mini-coaxial cable.
3. Measure the VSWR of port 1 and port 2 respectively. Both VSWR must be less than 2.5 to ensure the testing setup is good enough for efficient energy transfer.

When enclosure of LCD panel was removed, the maximal difference of S21 mentioned above can reach upto 20dB. Hence we can conclude that LVDS and Web camera cables, embedded antenna and its mini-coaxial cable, LCD panel all will lead to amplification of coupling between the interference sources and receiving module.

5. Platform noise measurement techniques

Conductive or near-field coupling platform noise will raise the level of RF receiver noise floor and thus cause performance degradation, and thus we need to investigate the isolation required as shown in Figure 23. We will describe the related measurement techniques in this section.

Platform Noise Isolation

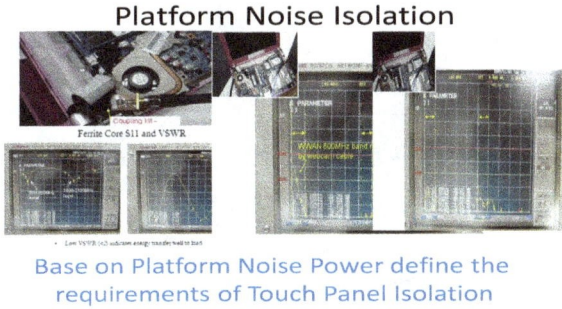

Base on Platform Noise Power define the
requirements of Touch Panel Isolation

Fig. 23. Platform noise coupling.

5.1 Noise level measurement

The result of noise level measurement represents the de-sense degree caused by IC, component, module, power circuits, PCB layout, interconnect wires, connector, and even the mechanical construction of the product. This section will describe the different methodology for the measuring procedures.

5.1.1 System noise measured with spectrum analyzer

- Measured by frequency domain sweep for multi-bands (GSM 850, 900, PCS1800 etc.) as shown in Figure 24.
- The Noise Distribution Function at area A is close to Gaussian distribution (Maximum Hold – Average = 10dB) , area B is close to CW (little amplitude variation with time)
- Noise limit level means the probability density function (PDF) of noise power is equivalent to Gaussian. Limit line for area A applies for the calculation described above, but PDF correlation is needed for area B.

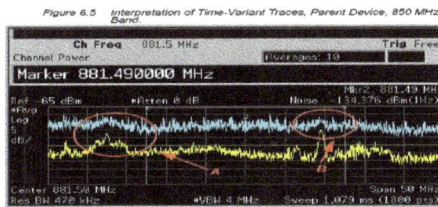

Fig. 24. System noise measured with spectrum analyzer.

5.1.2 System noise measured with WWAN card

In practice, it is more reasonable to measure system noise with a WWAN card because the RSSI is ready to appear at I/Q demodulator output, since it is the same sampling clock and module size for WWAN communication. The noise power measured from RSSI of WWAN add-on card can therefore in compliance with the definition of N0. The testing configuration and measurement are both shown in Figure 25.

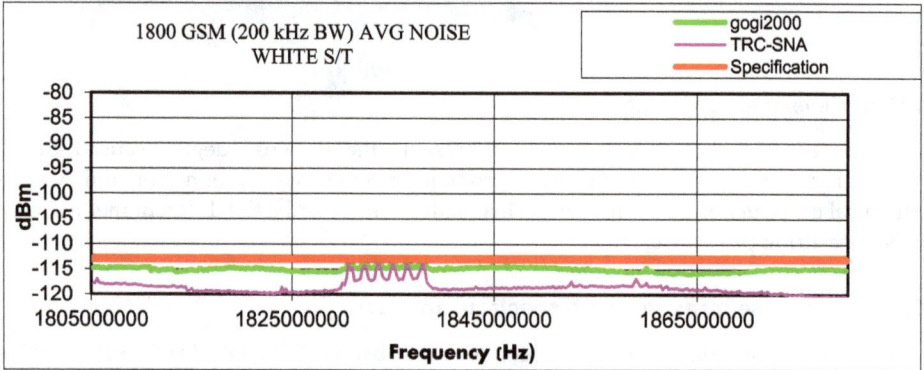

Fig. 25. Test configuration and system noise measured with WWAN card.

The SNA option shown in Figure 26 can also provide the De-Sense Measurement function. The software provides the Base-Station simulator, and provides VSG for WWAN and GPS measurement. The hardware provides switch and combiner for testing signal condition selection.

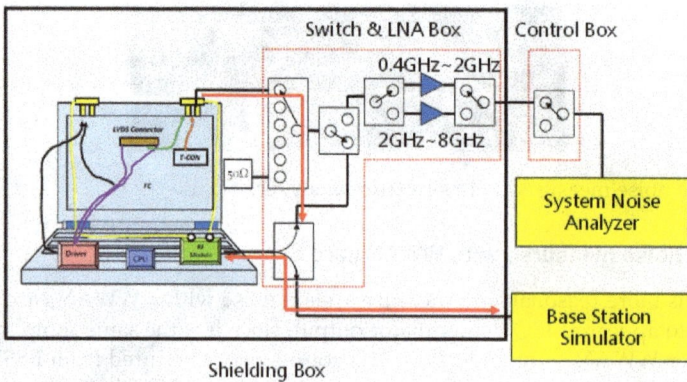

Fig. 26. The SNA test configuration.

The benifit and applications of SNA can be listed as following:

- Provide measurement adapter for fixing the antenna cable routing inside NB and Tablet PC.
- Provide testing fixture for different size LCD display panel.
- Provide testing fixture for SSD noise budget measurement
- Provide measurement adapter for ground isolation between smart phone and SNA (two Balun back-to-Back in series)
- Provide debugging plot to view each band at the same time.

5.1.3 SNA calibration at limit level

Since the WWAN or LTE cards are not originally designed for the noise measurement, and these cards need to provide about 60dB dynamic range for receiver. Therefore the variable-gain amplifier must be utilized for AGC purpose at the same time when the noise floor of receiver reaches up to 6dB. SNA is here designed to measure the platform noise. The dynamic range of receiver is from 4dB under limit line level up to 16dB above limit line level. With front end LNA implemented, the noise floor of SNA system is around 2dB and the average noise floor level is 4dB lower than that of WWAN or LTE card.

The noise floor level for WWAN is -115dBm. However, the limit line level for some particular applications like identifying the main noise source (system noise or panel noise) will be -114dBm. We can use the receiver with -115dBm noise floor level to receive the noise with level of -114dBm, and there is 1dB uncertainty for signal to noise level result. Since the average noise floor level can be obtained for SNA is -119dBm, it can provide more accurate signal to noise ratio for -114dBm measurement in general cases as in Figure 27.

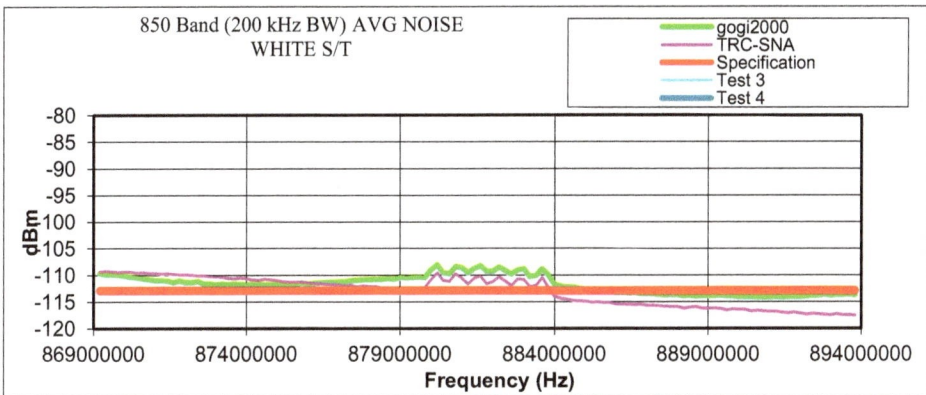

Fig. 27. Noise floor level obtained for SNA.

5.2 Throughput measurement

The throughput standard is widely adopted in different applications like video stream, online game, video conference. For example, you may need 1Mbps or more for YouTube HD video. Throughput is the most common for user experience regarding wireless RF performance, therefore throughput test need to consider how the user interacts with the RF device. For example, 90dB path loss for laptop 802.11g WLAN may represent that AP is 200 meters away in the free space environment, however the hand and head may cause the range for 90dB path loss much less than 20 meters for smart phone VOIP application. Some real-life products throughput test result has been illustrated in Figure 12 and the Root Cause Analysis (RCA) procedure will be addressed later in section 6.

5.3 Antenna surface current and near-field surface scanning

The current distribution on antenna surface represents the different sensitivity on antenna near field boundary, because the physical geometry of antenna will cause different field intensity coupled via the uniform magnetic flux. Antenna surface current also represents the immunity level for TP and Camera. As to digital noisy components, we can utilize the noise level results to locate the noisy components and identify their noise radiation pattern. We can utilize the near-field surface scanning method to observe the surface current distribution of antenna and locate the noise sources for platform noise analysis.

5.4 De-Sense measurement

De-Sense is self-referred to same subjective device operating in different condition and compared the platform noise impact. For example, a GPS module is first performed conducted test in a shielded box and obtains the C/N=40dB at -138dBm receiving level. While the same module is then bundled in a wireless platform and performed the conducted test again, the receiving level become -130dBm for keeping C/N=40dB. In this example, we found that the conducted platform noise causes GPS module de-sensed by 8dB. When the antenna terminal and GPS signal is fed to GPS receiver through a combiner, we now need -123 dBm to keep C/N =40dB and thus a 7dB de-sense caused by platform noise picking up from antenna. Furthermore, when the internal device, like Bluetooth, is active or hand-held device resting on desk, the interaction between platform noise and antenna resulting in different de-sense effect can be easily observed in all those test cases. The noise current at different locations can also be represented with the de-sense Data. The methodology and measuring technique illustrated below in Figure 28 can be applied for investigating GPS degradation of sensitivity caused by platform noise.

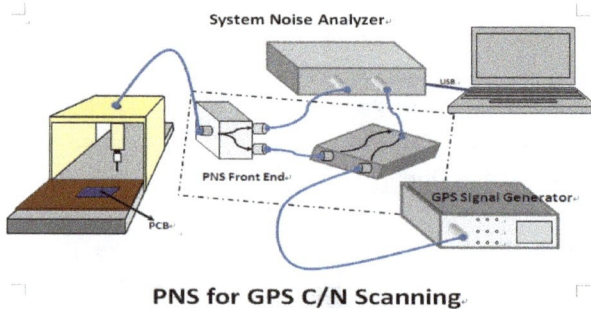

PNS for GPS C/N Scanning

Fig. 28. GPS C/N Scanning.

5.5 Noise measurement at antenna terminal

Since the antennas are the most important port for wireless communications as electromagnetic energy receiving component, they are also susceptible to nearby platform noise. Hence the analysis and measurement of noise level at antenna port is critical for RCA of wireless device to improve link performance. Figures 29-31 show the measurement techniques and configuration for noise measurement at antenna terminal.

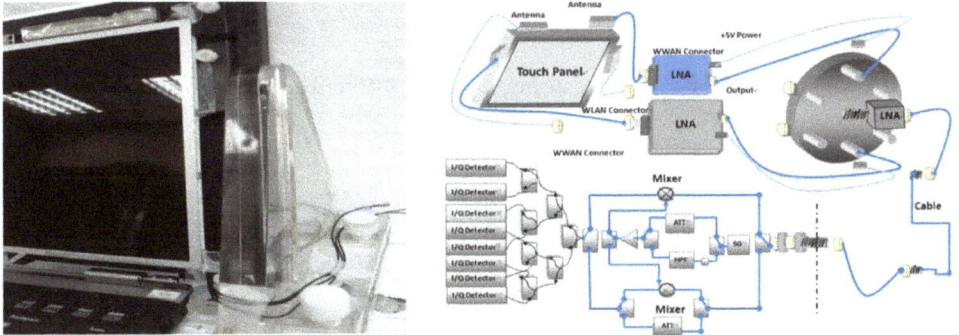

Fig. 29. LCD panel test fixture and the measurement circuits.

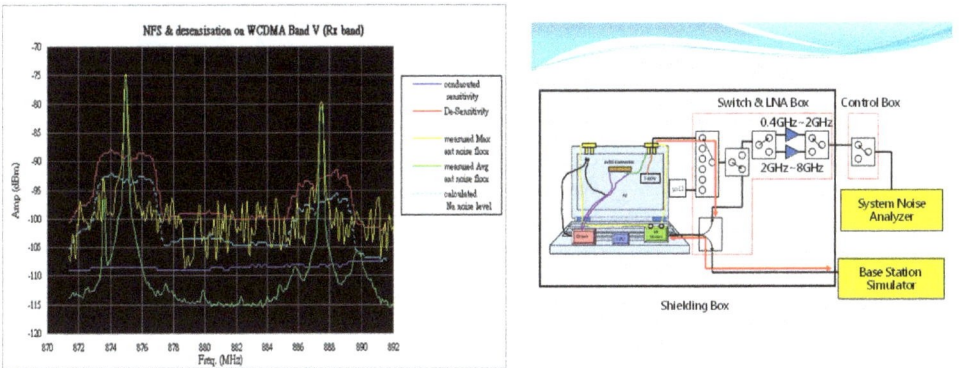

Fig. 30. Platform noise and de-sense measurement at antenna terminal.

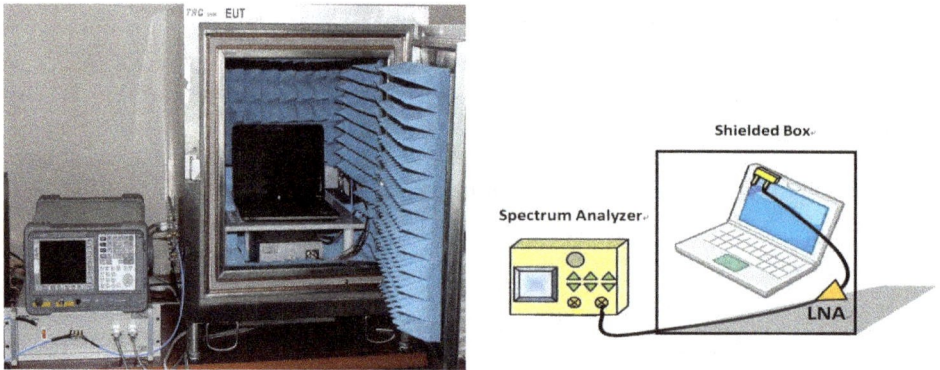

Fig. 31. NFS system and measurement setup.

From the various platform noise measuring techniques described above, we can summarize the comparison as Table 3.

	Measurement Type	Measurement Speed	End to End Calibration	Software User Interface	Correlation
SNA Same LNA and Switch as TRC NFS	Channel	10 times faster than WWAN card	Gaussian Noise Level Cal. At limit line level.	Same as the NFS & Provide the debug mode "View Plot" display test result	Within Limit Line + - 3dB range within 1dB error with WWAN & LTE Card
WWAN Card LTE Card	Channel	Very Slow	N.A .	Depend on Card Vendor of NPT utility	Same WWAN card cab be 2dB error
TRC NFS	Sweep Freq.	Very Faster	LNA & Switch	Well Accepted	Big Difference in Broadband

	Receiver Noise Floor Level	Antenna VSWR Check Before Noise Measurement	Test Fixture, Test Kit and Measurement adapter	De-Sense versus Platform Noise Plot
SNA	Average At 200KHz Channel -119dBm	SNA provide the option for the internal VSWR bridge, Each time debug and close the enclosure should be check the antenna, ensure the antenna is keep the same as original.	Measure the Adapter Insertion Loss for Correction. LCD Test Fixture Can be integrated with LNA and the antenna protect by dielectric material.	Software support the for different brand of base station simulator and VSG, Measure the De-Sense corresponding to the Noise Level.
WWAN Card LTE Card	Average -115dBm	N.A.	N.A.	N.A.
TRC NFS	Depend on Spectrum.	N.A.	N.A.	N.A.

Table 3. Measurement techniques comparison.

6. Design techniques for platform noise improvement

This section will present the Tablet PC as case study to describe the problem-resolving methodology for throughput degradation.

6.1 Identify the main noise source

The first step to solve the interference problem is to identify the main noise source, and then we can further to implement resolving techniques like filtering, shielding, and re-layout, etc. The procedure can be demonstrated below in Figure 32.

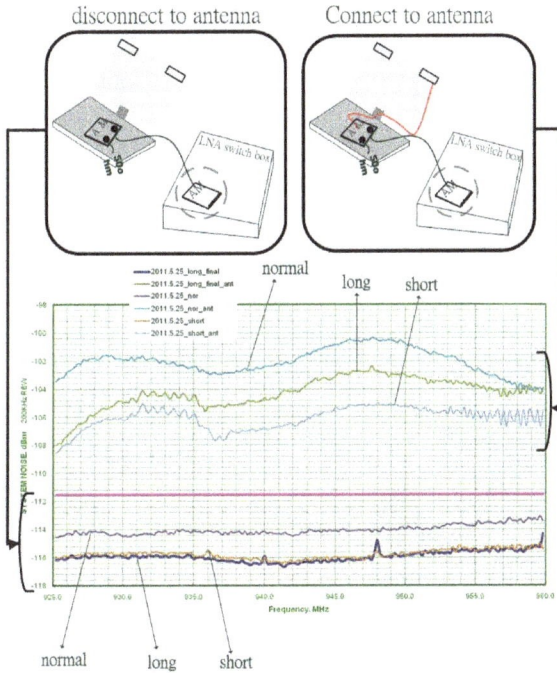

Fig. 32. Use built-in antenna to identify noise source from measured spectrum.

Those engineers who are responsible for resolving noise problem may add the copper foil or put an absorber to cover the IC, routing the antenna cable, re-installing the panel, plugging antenna connector to test port, and they may affect the antenna or noise test results after those activities. Therefore, the antenna VSWR characteristic must be checked before noise measurement. The VSWR measurement of antenna is a quick way to check if the antenna is still kept the same configuration as originally implemented, because 3D radiation pattern and efficiency measurements of antenna usually take 2~3 hours. However, the SNA option described early can provide an embedded VSWR bridge and therefore could automatically check VSWR of each measuring port before noise measurement as shown in Figure 33 for LTE band. There are two kinds of LCD panel testing fixture can be designed for different panel size as shown in Figure 34 and 35.

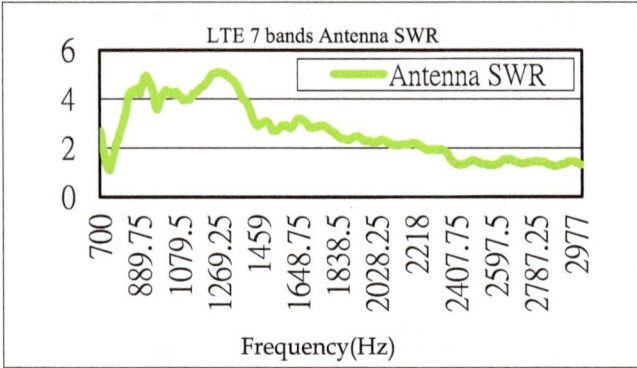

Fig. 33. VSWR of each measuring port before noise measurement for LTE band.

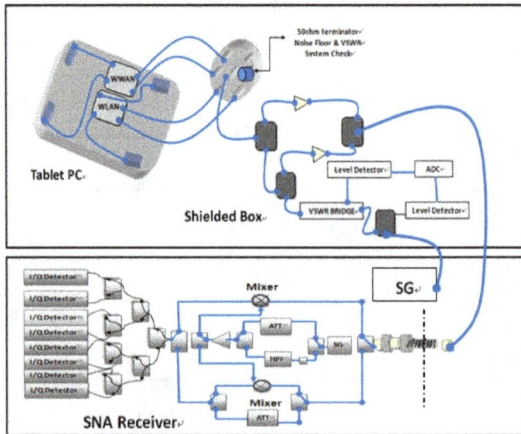

Fig. 34. LCD panel testing fixture integrated with LNA and SNA result for LTE.

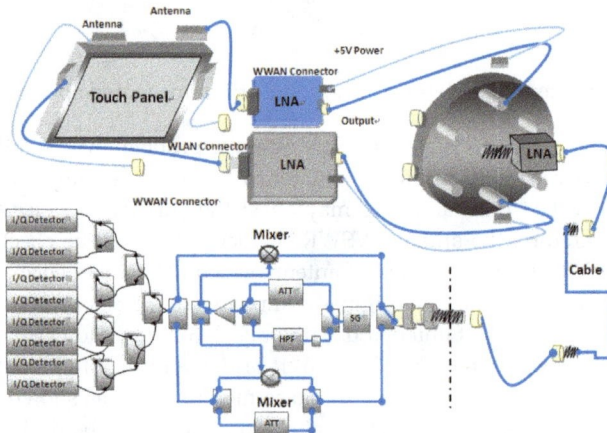

Fig. 35. LCD panel testing fixture with LNA and antenna protected by dielectric material.

6.2 Platform noise isolation

TP and camera are not the only modules resulting in the platform noise problem, but the product's assembly construction such as antenna and its mini-coaxial cable routing, TP and its LVDS cable, camera and its USB cable will also bring up noise problem. We can utilize some design techniques, like component placement and orientation, cable routing, shielding etc. to improve platform noise isolation.

6.3 Antennas isolation [8]

Noise current distribution and antenna surface current are most important message for solving the sensitivity dedradation problem. We can use a network analyzer deliver the energy to TP's LVDS or camera's USB lines and measure the insertion loss at antenna port, and that is the most commonly utilized technique to obtain the isolation situation of platform noise.

With the highly integration of powerful computing and multi-radio communications devices in a single product nowadays, multiple antennas are usually implemented to achieve the seamless and convenient communication services. However, the closely placed antennas have resulted in intra-system coupling interference and therefore severely degraded the performance of various kinds of wireless communications. The isolation technique for antenna systems must be implemented to reduce the mutual coupling between coexistent various RF systems. In this section, we will show the optimal isolation achieved from antennas separation, orientation, and utilization of periodic structure to reduce the mutual coupling interference.

The isolation requirement between coexistent RF systems is shown in Table 4. The placement of two chip antennas under investigation for Bluetooth and 802.11b/g WiFi systems inside the mold notebook computer is shown in Figure 36. The chip antennas are fabricated on FR4 with dimension of 1.6 mm thickness and 35mm × 30mm area, and it is fed by microstrip to achieve 50 Ω impedance-matching. The configurations for different spacing and orientation between coexistent antennas are shown in Figure 37 to analyze the mutual coupling effect.

Transmitter	Minimum Isolation Recommendations			
	Receiver			
	Bluetooth	802.11b/g	802.11a	GSM
Bluetooth	n/a	40dB	20dB	20dB
802.11b/g	40dB	n/a	n/a	20dB
802.11a	20dB	n/a	n/a	20dB
GSM	20dB	20dB	20dB	n/a

Table 4. Isolation requirement between coexistent systems.

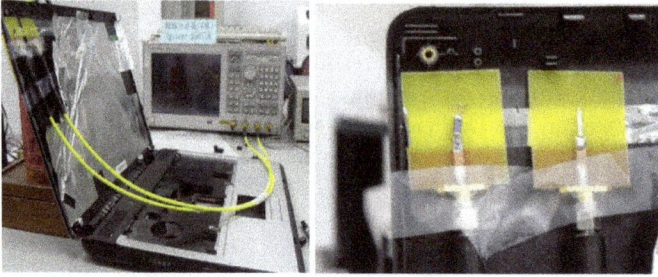

Fig. 36. Placement of two chip antennas insidemold.

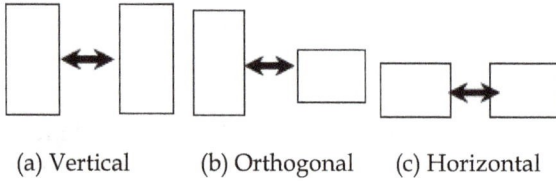

(a) Vertical   (b) Orthogonal   (c) Horizontal

Fig. 37. Three orientation for antennas placement.

The reflection coefficients (S11 and S22 are reflection coefficients for Bluetooth and 802.11b/g respectively) and isolation (S21) between each other were then measured and shown in Figure 38. The results clearly show that the isolation between antennas is much worse when they are both placed in vertical polarization with main lobes coupling. We then oriented the antennas in orthogonal direction to each other and separated antennas by moving 0mm, 10mm, and 20mm respectively. The measured results of reflection coefficient covering 2.4GHz-2.483GHz in Figure 39 show that the isolation is much better due to orthogonal polarization to each other when they are oriented in orthogonal direction. Finally, we placed both antennas in horizontal direction and adjusted the separation between them, the results in Figure 40 also show better isolation than vertical placement due to main lobes decoupling. Table 5 compares the isolation performance between various orientation and separation for both antennas, and it shows that the orthogonal and horizontal orientations gain almost 6-9dB improvement in isolation except 4 dB difference for 0mm separation.

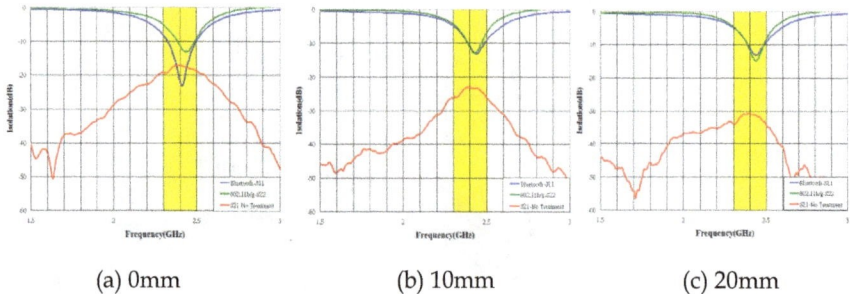

(a) 0mm   (b) 10mm   (c) 20mm

Fig. 38. Measured results for various antenna spacing with both antennas oriented in vertical direction.

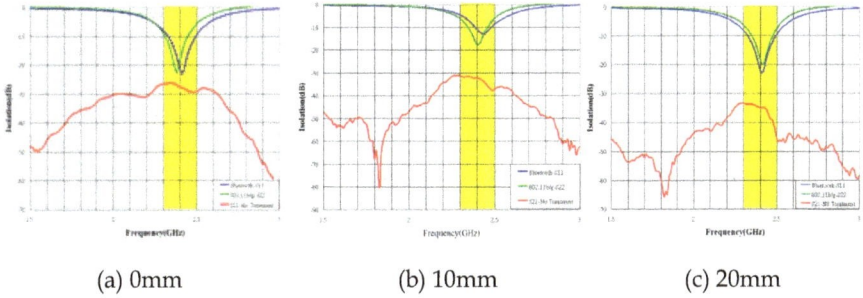

(a) 0mm   (b) 10mm   (c) 20mm

Fig. 39. Measured results for various antenna spacing with both antennas oriented in orthogonal direction.

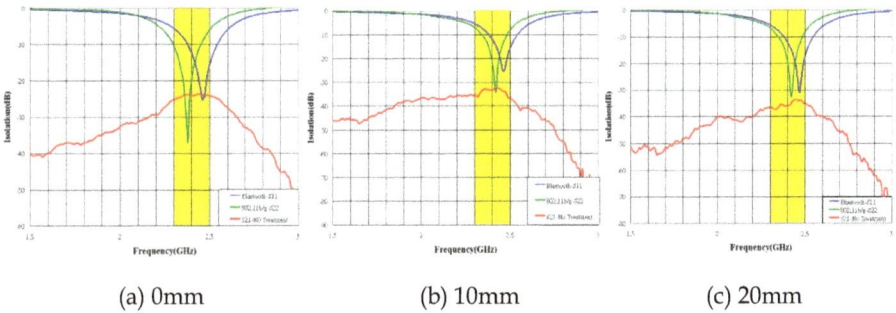

(a) 0mm   (b) 10mm   (c) 20mm

Fig. 40. Measured results for various antenna spacing with both antennas oriented in horizontal direction.

Orientation	Vertical	Orthogonal	Horizontal
Separation	Isolation		
0mm	16.9dB	28dB	24.1dB
10mm	23.1dB	32.6dB	33.1dB
20mm	30.8dB	36.2dB	36dB

Table 5. Measured results for various antennas placement configuration.

6.3.1 Suppression of mutual coupling interference between coexistent antennas [8]

The applications of EBG structure in antenna not only could improve gain and radiation efficiency, it could also help suppress side lobes and reduce coupling effect. Since isolation requirement could not be met by orientation and separation arrangement between antennas from the above measurement, we therefore chose the best placement configuration with orthogonal orientation and 20 mm separation for further investigation utilizing EBG structure. We placed the EBG structure beneath the antennas with 7.5mm (less that $\lambda/4$) and 15mm (about $\lambda/4$) distances as shown in Figure 41 and investigated the mutual coupling characteristics. Figure 42 shows the results with 7.5 mm separation between antennas and EBG structure for various antenna orientations. Because of high impedance surface from EBG structure, the 40 dB isolation requirement between antennas could be achieved and parallel-plate guided wave coupling could also be suppressed. Figure 43 shows the results with 15 mm separation between antennas and EBG structure for various antenna orientations.

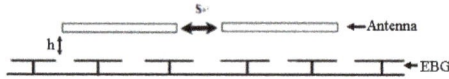

Fig. 41. Configuration of antennas and EBG structure placement.

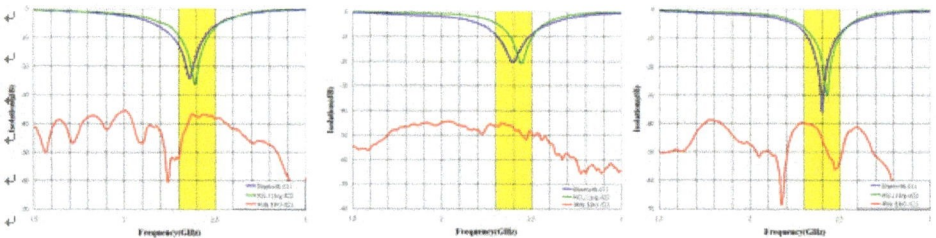

| (a) Vertical | (b) Orthogonal | (c) Horizontal |

Fig. 42. Isolation and S11 of antennas for various orientation with antennas separation 20mm and EBG-antennas spacing 7.5mm.

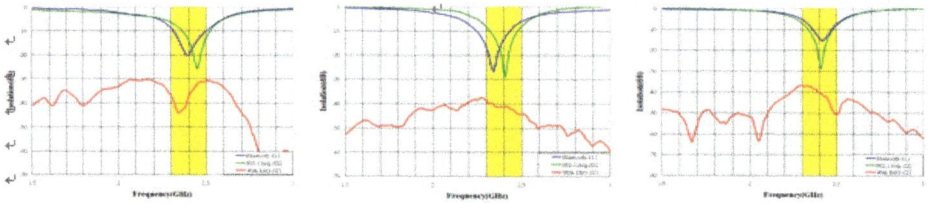

| (a) Vertical | (b) Orthogonal | (c) Horizontal |

Fig. 43. Isolation and S11 of antennas for various orientation with antennas separation 20mm and EBG-antennas spacing 15mm.

Table 6 compares the isolation performance with and without EBG structure for various antenna orientations, and it shows that 37.7 ~ 48dB and 35.7 ~ 40.9dB isolation between antennas can be achieved with EBG structure placed beneath antennas 7.5mm and 15mm respectively. When EBG structure moves closer to antennas, we can obtain better isolation between antennas.

Orientation	Vertical	Orthogonal	Horizontal
Separation Between Antennas and EBG / Isolation			
Without EBG	30.8dB	36.2dB	36dB
7.5 mm	37.7dB	48dB	47.4dB
15 mm	35.7dB	40.9dB	40.5dB

Table 6. Comparison of isolation improvement from EBG structure with antennas separation 20mm.

6.4 Keyboard grounding requirements

The large metal plate of a keyboard can act as an integrated shield to prevent noise radiating from the base chassis. However, the mechanical grounding structure needs to be taken serious care of its EMI effect on radio performance, because it is not uncommon that EMI noise radiated from the ground-base to antennas mounted around LCD panel when the keyboard is removed as shown in Figure 44. This means that we are not getting any shield effect from the keyboard, and in fact the keyboard helps EMI ground noise radiating from the base like a large antenna. To make the keyboard a useful shield for all WWAN frequencies, it is necessary to conductively contact ground structure every 1/20th wavelength or so as shown in Figure 45.

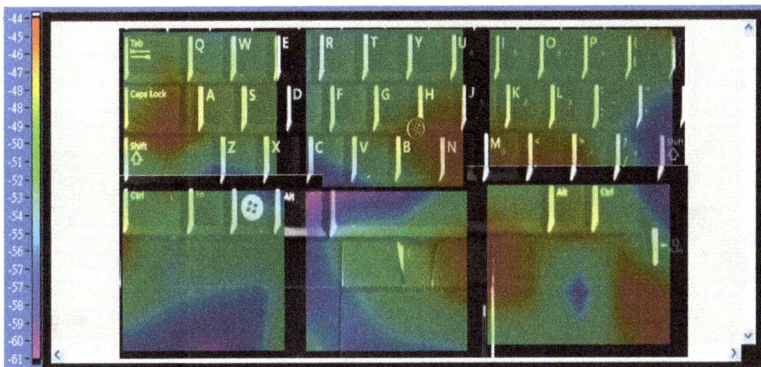

Fig. 44. Field distribution on metal shield of LCD panel and nearby antenna.

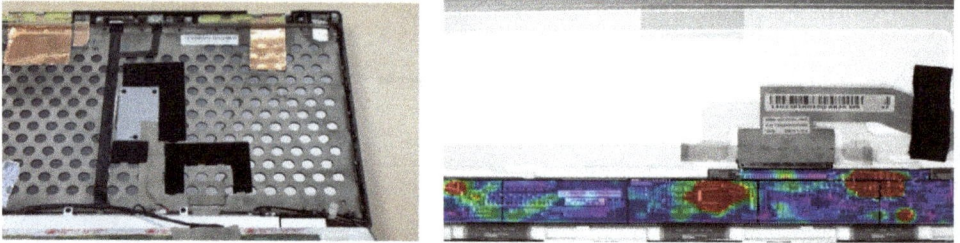

Fig. 45. Metal plate for LCD panel shielding purpose.

Grounding the motherboard to the chassis wherever possible (screw holes, connectors, etc.) will help reduce EMI radiation from the motherboard. The grounding contacts should follow EMC design guidelines for length to width ratio requirement to prevent from adding a radiation structure rather than providing a noise-reduction ground point. Every effort should be made for each screw and contact point of the motherboard to the chassis for a good chassis ground point.

The grounding of the heat sink also raises a problem for ESD, EMI, and platform noise of wireless devices, therefore improvement should be made in the grounding of the heat sink (cooler). The improvement should include: enabling better DC grounding at "spring contact" points, and no non-grounded arms longer than 7mm are allowed.

6.5 Component and signaling cable considerations

Two components (local oscillator and clock chip circuit) on motherboard probably will need an extra local shield placed on it to pass regulatory testing if the components radiate over-limit spurious noise. Hence the mechanical chassis design should also provide sufficient space with height clearance to accommodate this kind of SMT shield/can.

The camera module is also a potential RF noise source of the system, especially when it is located in the proximity of antennas. A mechanical shielding solution – sheet metal, EMI paint, foil, and/or magnesium walls should be considered to isolate the EMI noise from the antennas as shown in Figure 46. Although most of the camera modules equipped with metal shielded enclosure, but the glass which cover the CCD gate, transfer gate may leak or coupling the magnetic field to the nearby traces through the glass aperture.

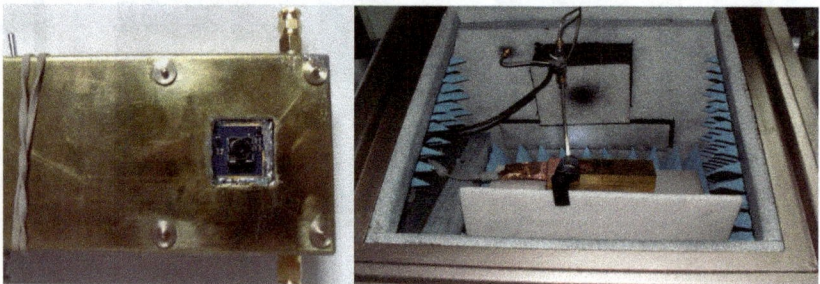

Fig. 46. Shielded enclosure and loop probe for noise measurement of camera module.

LVDS is now a popular signaling system that can deliver information at very high speed over twisted pair copper cables. LVDS technology uses the voltage difference between two wires to carry signal information for high-speed data transfer through the panel hinge to minimize EMI related problems. In order to minimize the aforementioned EMI problems, we can utilize LVDS cables with the following mechanical recommendations:

1. Make sure that the twisted pair cabling, twin-axial cabling, or flex circuit with closely coupled differential lines is used by grouping members of each pair together.
2. Differential impedance of the cables should be 100 ohms
3. Make sure that the cables used are well shielded
4. Place ground pins between pairs wherever it is possible
5. Connect shielding directly to the connectors of driver and receiver enclosure respectively.

The use of LVDS system must be cautious that most of the LVDS cables have good degree of balance for the fundamental frequency, but the balanced pair becomes unbalanced for those frequencies higher than 10 harmonics when the harmonic signal across the LVDS cable. Therefore it will usually generate the common mode voltage resulting in common mode radiation, and the shield of the LVDS cable become ground return.

The impedance mismatching at DDR2 or DDR3 memory socket usually causes higher noise current distribution around the socket area. These magnetic filed may couple to the WWAN module when the shielding effectiveness of WWAN module's shielded enclosure is not adequate as shown in Figures 47-49. Lower impedance will tmake dI/dt increase and dramatically increase the current drawn from supply (not good for the design of power distribution system), while higher impedance will emit more EMI and also become more susceptible to external interference.

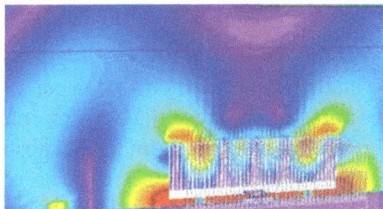

Fig. 47. RF module location and bearby field distribution.

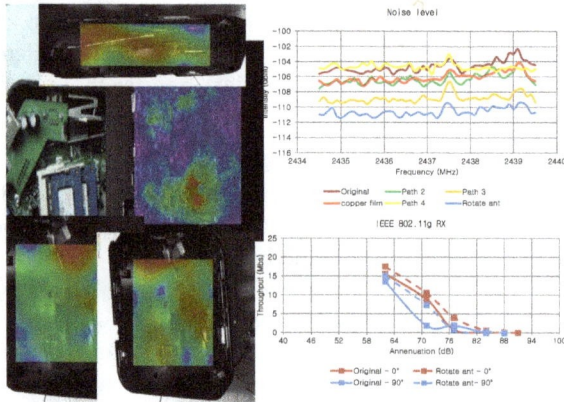

Fig. 48. Embedded antenna mini-coaxial cable and noise distribution.

Fig. 49. Circuit ground and chassis ground.

7. Application of noise budget concept

Finally, we will propose the Noise Budget concept for platform noise suppression. The noise budget for the wireless communication device can be considered as near-field EMC limit. Noise budget is a powerful tool to apply for the wireless product from initial design stage, QA, QC, and all the way to final production testing. For communications community and RF device manufactures, the link budget is established for system planning, therefore most of RF engineers understand that the characteristics of antenna, LNA, mixer play important role for range and coverage extending. However the parameters discussed in link budget are all about signal transmission alone, another important component parameter related to noise level, the Noise Budget, is the missing puzzle for the system integration.

7.1 Introduction to noise budget concept

Since the electronics of the notebook or labtop are the interference source for RF wireless device as discussed earlier. This final section covers some design guidelines and EMI measuring techniques of components, because EMI from internal ICs is also a major contribution to impact the RF performance. The primary purpose here is to address the idea of "Component Noise Budget for Wireless Integration". The concept of noise budget for devices on wireless communication product stems from the link-budget for RF Tx/Rx performance. It also borrows the idea from EMC testing requirements for automobile industry to identify the potential interference sources that might cause safety problem and thus to provide design guideline for compliance. The noise budget concept helps system designers to manage the EMC issues, as early as possible, such as coupling mechanisms, module placement, grounding, and routing for EMC test. The methodology is intended to develop modular architecture of analysis to accelerate system design and also provide the solutions for potential problems to improve performance in all aspects.

The preliminary goal of this research is to establish the noise budget for components and devices on laptop computer for further RF sensitivity analysis. To utilize the near-field EM scanner to detect the EMI sources on laptop, we can locate the major noise sources in 2D hot-spot distribution graph. From the emission levels and locations of the noisy components, we can figure out their impact on throughput and receiving sensitivity of wireless communication and find the solution to improve performance. The final goal of noise budget, however, is to establish the EMI limits for each digital components related to layout location, it would therefore help designers to choose the appropriate components for optimal placement and cost consideration to meet product requirement. The application of noise budget accompanied with near-field surface scanner not only can locate the EMI source and further to solve the problem, but also can utilize the EM analysis to improve the design efficiency and the performance of wireless communications.

The factors that would affect the receiving performance of a wireless receiver can be illustrate in Figure 50. Table 7 shows the relationship between link budget and noise budget for wireless communication system implementation.

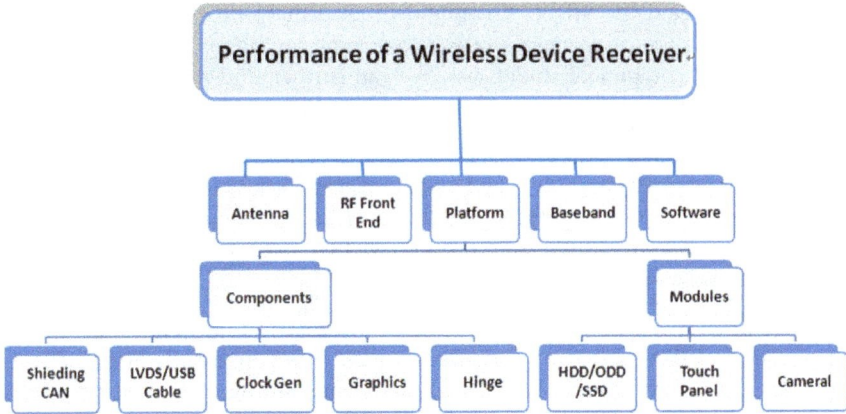

Fig. 50. Factors affecting the receiving performance of a wireless receiver.

PATH LOSS		NOISE LEVEL
ANTENNA GAIN		CPU
DIPLEXER INSERTION LOSS	-2dB	NORTH BRIDGE(LVDS DIRVER....)
MISMATCHING LOSS	-0.5dB	MEMORY
LNA GAIN	+20dB	LVDS SOKET
LNA NOISE FIGURE	2dB	LVDS FLAT FLEX CABLE
SAW FILTER LOSS	-2dB	LCD PANEL (T-CON, BACKLIGHT)
MIXER CONVERSTION LOSS	-0.5dB	
		TOUCH ME
SAW FILTER LOSS	-2dB	WEB CAM
LNA GAIN	+20dB	MINI COAXIAL CABLE FOR RF
I/Q CONVERSTION LOSS	-8dB	
		SOUTH BRIDGE

Table 7. LINK budget (left) vs. NOISE budget (right).

The common interference noise sources on integrated high-speed digital wireless communication product nowadays include: CPU, LCD panel, Memory, digital components, high-speed I/O interconnect, wires and cables, etc. The modules' placement of the test setup is illustrated in Figure 51. The above mentioned noises are usually coupled to nearby sensitive devices through radiation, conduction, or crosstalk. The resulted EMI problem will further degrade the system sensitivity and performance for wireless communications.

After the emitted noise level and corresponding location of each component has been identified, we investigate the effect of component placement on in-band noise level at antenna port and thus the performance of wireless communications by changing distance between antenna and component under test. We can further find the optimal orientation and location of component to improve overall communications performance[9,10].

Because a variety of digital components exist inside laptop computer, we focus on LCD Panel that is equipped in all computers and usually placed in the proximity of antennas. To investigate the effect of various operation modes, (such as off, standby, and key-in alphbat H pattern mode) on noise level at the antenna port, we first arranged the test setup as laptop normally working and scanned the ambient noise.

To clarify the influence of LCD panel noise on antenna port and thus receiving sensitivity, we first fixed the function setting on computer to avoid effect from software's inconsistent running mode. After activating the LCD Panel for various testing mode, we measured the noise spectrum at antenna port to find out the interference frequencies and then use near-field probes to scan the EMI noise from LVDS cables, connectors, and driver ICs of the LCD panel control circuits. Finally, we can investigate the impact of different LCD operation mode on the frequency bands of wireless communications by analyzing the measured throughput results.

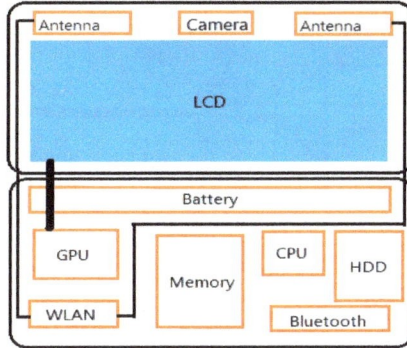

Fig. 51. Internal layout of laptop computer.

7.2 Analysis of platform noise effect from built-in CAMERA module

7.2.1 Test setup for noise level measurement

The system platform noise under investigation is first analyzed by noise floor measurement system. The complete PNS (platform noise measuring system) is composed of shielded box, pre-amplifier, spectrum analyzer, and EUT (Laptop computer). The noise level measuring system and setup for frequency domain is shown in Figure 52.

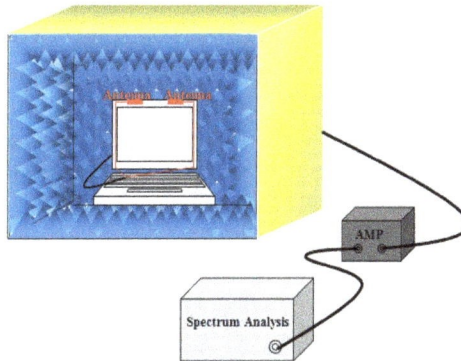

Fig. 52. Setup for antenna port noise level measurement.

Since the CAMERA or CMOS camera module is most adopted to the popular mobile devices like cellular phone or Netbook, we hence focus on EMI analysis of the built-in camera module by application of IEC 61967-2[1].

7.2.2 Test setup for TEM cell measurement[11]

The test setup for TEM cell method in this study is shown in Figure 53. One end of the TEM cell is terminated with a 50 Ω resistance terminator, and the other end is connected to spectrum analyzer via pre-amplifier.

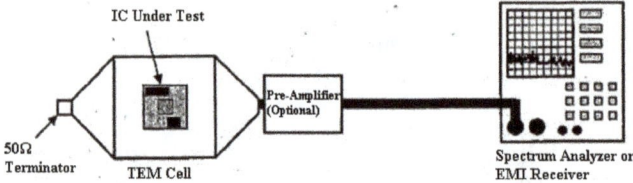

Fig. 53. TEM cell setup for IC or module EMI test.

There are two PCBs with module under test on it and are marked as 1 and 2 shown in Figure 54 and Figure 55. The tested boards for Webcam DUT are driven by USB with clock frequency 48 MHz, and the grounding connection between camera chip and PCB is utilized with wire bonding. The function of module and inner PCB routine for both boards under test are identical except for different number of bonding wires connecting to ground, there are more grounding wires for No. 2 PCB than No.1. The purpose of this measurement is to investigate the effect of multi-point grounding scheme to EMI level. Since the bonding wire is equivalent to inductance, we expect to reduce ground bounce and hence the EMI emission by parallel connection of multiple grounding wires. The connection between camera module and testing board is shown in Figure 56.

The experimental procedure for EMI test using TEM cell is following:

1. Connect the pre-amplifier (if needed) in front of spectrum analyzer at one end, and connect a 50Ω terminator at the other end.
2. Define or identify the four side of TEM cell to place the DUT oriented along all four directions and measure EMI one for each time as shown in Figure 57.
3. Set the measurement frequency range of spectrum analyzer from 150 kHz to 1 GHz.
4. Set the resolution bandwidth of spectrum analyzer around 9 to 10 kHz, and video bandwidth as more than three times of resolution bandwidth to meet the IEC standard specification[12].

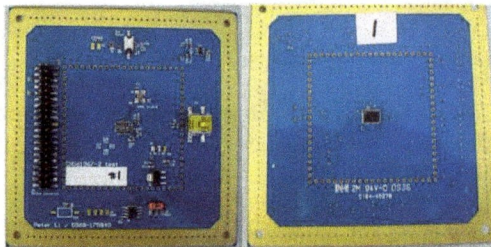

Fig. 54. Physical PCB with CAMERA module marked as NO.1.

Fig. 55. Physical PCB with CAMERA module marked as NO.2.

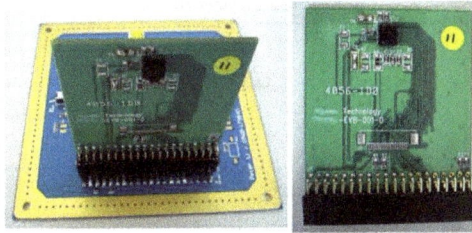

Fig. 56. Physical camera connected to testing board.

There were two operation modes for camera module to be analyzed on EMI measurement. The first mode simulates the cellular phone activating the camera module for video communication. In the case of first mode, the camera is simply turned on for full function but does not execute the video file transferring from capturing camera to store on hard disc or memory card. However, the second mode simulates the cellular phone activating the camera module for video recording. In the case of the second mode, the camera is not only activated for full function but also execute the video file transferring from capturing camera to store on hard disc or memory card. The measured results for both operation modes are shown in Table 8 and 9 respectively. Compare the measured results for both operational mode, we can observe the occurring EMI phenomena during video file transferring from capturing camera to storage device. It can be used to find that if the more functions IC executes, would the severe EMI noise be generated or not.

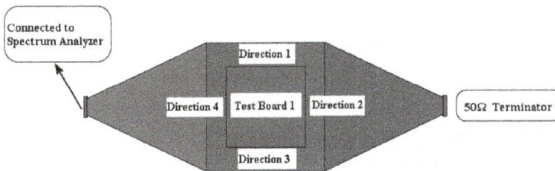

Fig. 57. Definition of 4 directions for TEM cell test orientation.

Board \ Direction	1	2	3	4
1	9.41 dBuV at 480 MHz	7.37 dBuV at 72 MHz	10.45 dBuV at 480 MHz	6.39 dBuV at 120 MHz
2	8.50 dBuV at 312 MHz	6.44 dBuV at 120 MHz	9.06 dBuV at 312 MHz	7.54 dBuV at 72 MHz

Table 8. Maximum EMI level with corresponding PCB orientation and frequency for video communications mode (mode 1).

Board \ Direction	1	2	3	4
1	10.63 dBuV at 480 MHz	6.61 dBuV at 720 MHz	11.15 dBuV at 480 MHz	6.39 dBuV at 120 MHz
2	11.14 dBuV at 72 MHz	7.74 dBuV at 960 MHz	9.25 dBuV at 480 MHz	7.32 dBuV at 815 MHz

Table 9. Maximum EMI level with corresponding PCB orientation and frequency for video file transfer mode (mode 2).

7.2.3 Test setup for near-field surface scanning[2,13]

The setup as shown in Figure 58 is to detect the EMI noise from LVDS cables, connectors, and driver ICs of the LCD panel control circuits. From the measured results, we can identify the locations of the significant EMI noise sources.

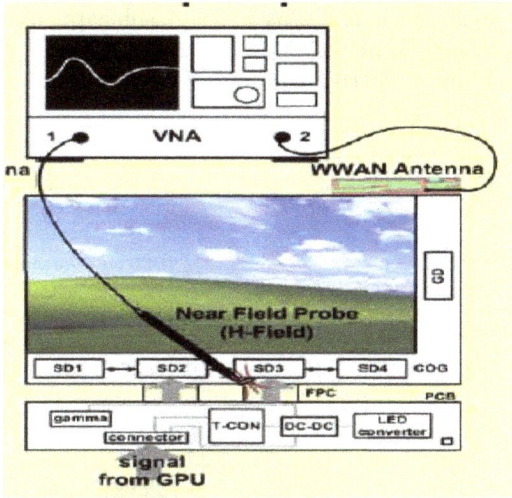

Fig. 58. Setup for LCD panel noise measurement.

7.2.4 Test setup for throughput

The setup of the throughput measurement for the analysis of communications performance in this study is shown in Figure 59. The throughput measurement system consists of WLAN AP (access point), device under test (DUT), attenuator, and Chariot software for data rate control.

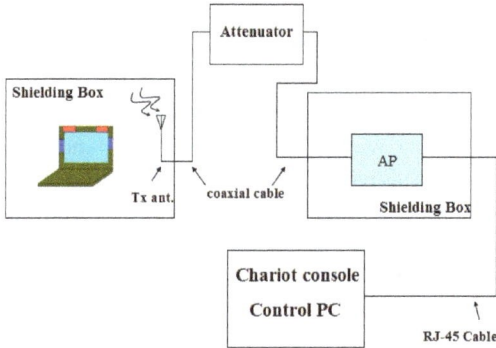

Fig. 59. Setup of throughput measurement system.

7.3 Analysis of measurement results[10-14]

7.3.1 Results of noise level measurement for different LCD panel

Variation of noise level for Camera module at different operation mode is shown in Figure 60. We can observe the significant variation of noise level in 2586~2600MHz frequency range when Camera is activated and operated at Record mode. Since the crystal oscillation of Camera is 48 MHz, we can conclude that its 50th and 54th harmonics just fall at 2400MHz and 2592MHz, the most significant noise level frequencies, respectively. Therefore, the receiving sensitivity and thus the communications performance in 2.4 GHz band are degraded by the activation of Camera functions.

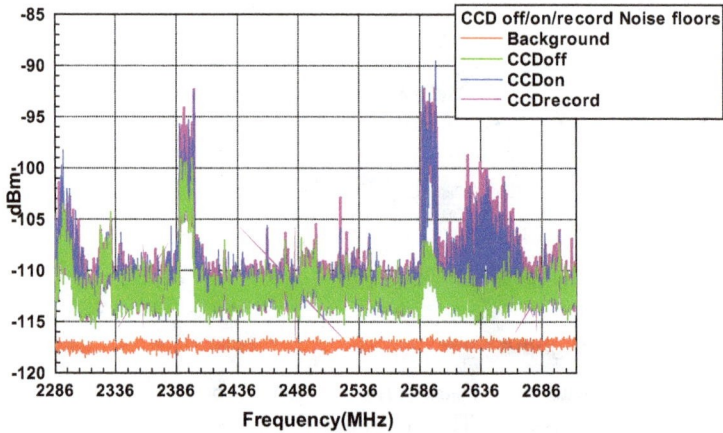

Fig. 60. Noise level for different Camera operation mode.

7.3.2 Results of surface-scanning measurement

From the results of noise level measurement, we first obtained the interference frequencies generated from LCD panel. We then used the magnetic near-field probe to observe the noise influence on wireless communication bands via antenna ports from LVDS cables, connectors, and driver ICs of the LCD panel control circuits when LCD panel was set to various operation modes, (such as off, standby, and key-in alphbat H pattern mode). Figures 61-63 show the change of transmission coefficients between antenna port and LVDS cables of the LCD panel control circuits. The measured result on left is for horizontal orientation of near-field probe placement, and the one on right is for vertical orientation. When magnetic probe is placed in vertical orientation, it is in parallel with routing traces of LVDS connector and thus results in higher sensitivity. We also observed that the coupling level are much higher for LCD panel operated in standby or key-in alphbat H pattern mode than shut off.

Figures 64 and 65 show the change of transmission coefficients between antenna port and driver IC of the LCD panel control circuits. The measured result on left is for horizontal orientation of near-field probe placement, and the one on right is for vertical orientation. We observed that the noises coupled to antenna port are much higher for LCD panel is turned on or displays H pattern mode than shut off, because the control IC is activated.

(a) Measurement position

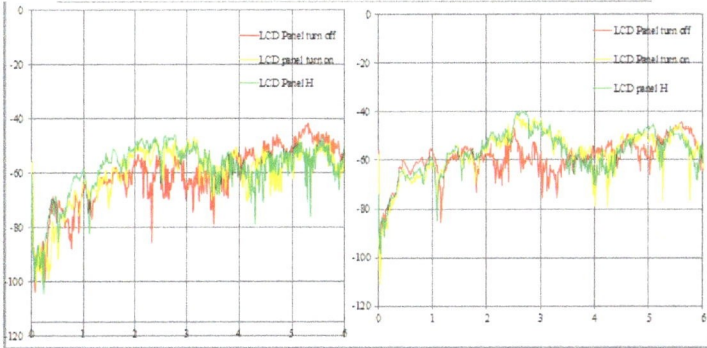

(b) Measured result of S21

Fig. 61. Measurement position (a) and (b) result of LCD Panel control circuit.

(a) Measurement position

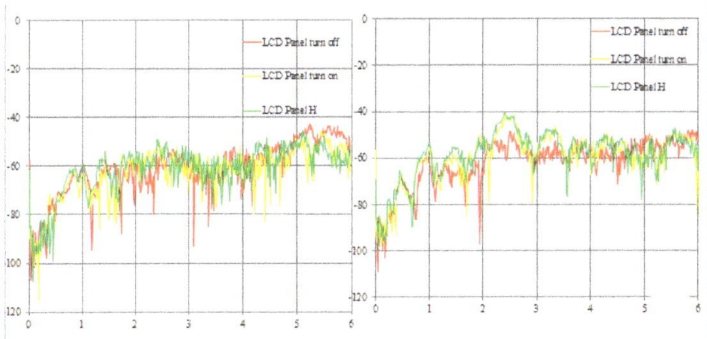

(b) Measured result of S21

Fig. 62. Measurement position (a) and (b) result of LCD Panel control IC.

(a) Measurement position

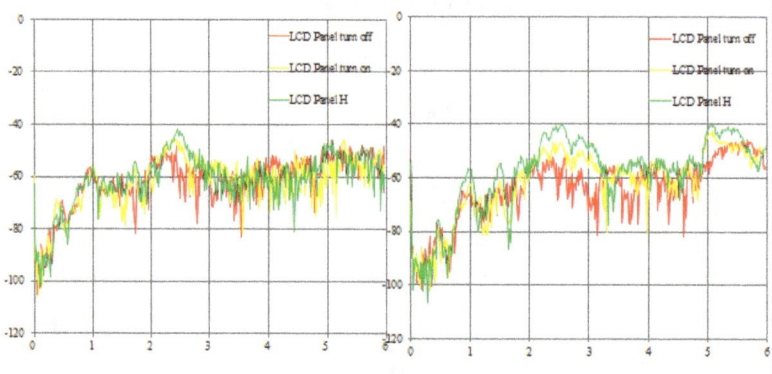

(b) Measured result of S21

Fig. 63. Measurement position (a) and (b) result of LCD Panel control circuit.

(a) Measurement position

(b) Measured result of S21

Fig. 64. Measurement position (a) and (b) result of LCD Panel control circuit.

(a) Measurement position

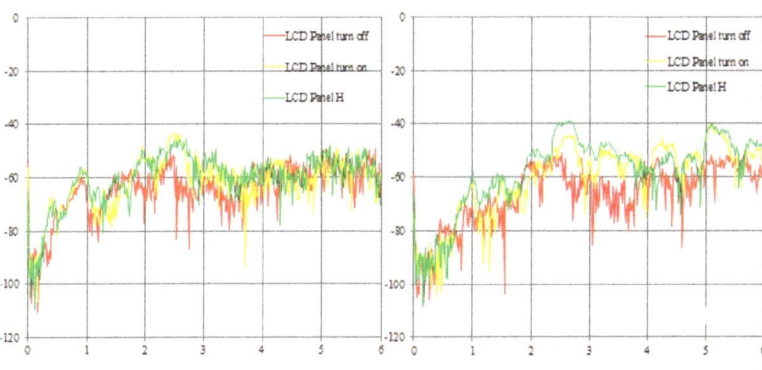

(b) Measured result of S21

Fig. 65. Measurement position (a) and (b) result of LCD Panel control circuit.

7.4 Summary

Since the development of IC technologies advancing toward nm processing technology and higher operating frequencies in recent years, the systems of highly integrated high-speed digital circuits and multi-radio modules are now facing the challenge from performance degradation by more complicated electromagnetic noisy environment. With the development of the analyzing and measuring methodologies for this wireless platform noise problem and establishment of noise budget for digital component in the near future, we can provide the EMI coupling mechanism and noise level for each component to help system engineer analyze and design the EMC compliant wireless product in the first beginning as shown in Figure 66 and 67.

TP Noise Budget Fishbone Diagram

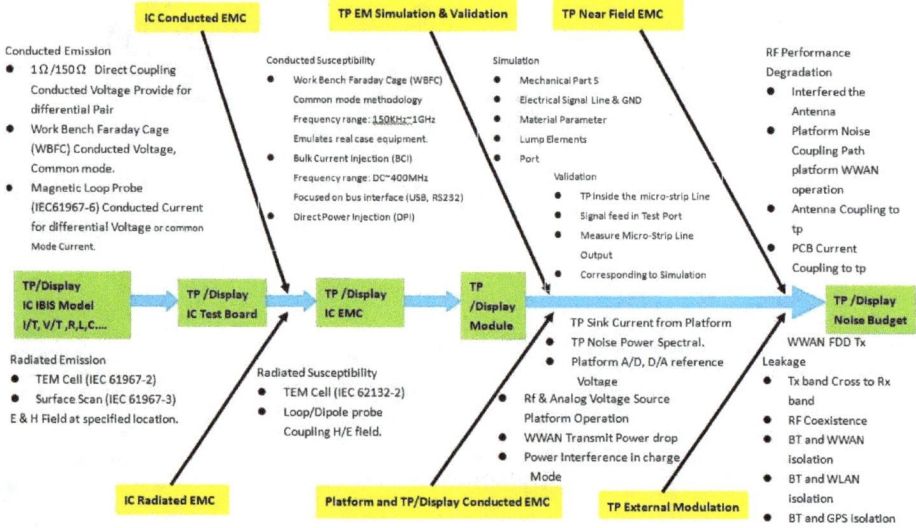

Fig. 66. Noise budget consideration for touch panel.

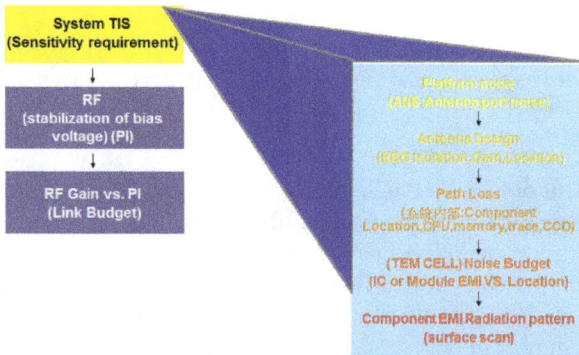

Fig. 67. System view for noise budget.

8. Acknowledgments

The author would like to thank the old friend and research partner Frank Tsai from TRC (Training and Research Company, Taiwan) for his inspiration, technical support and measurement assistance. The author would also like to thank the funding support from BSMI (Bureau of Standards, Metrology and Inspection) and NSC (National Science Council) Taiwan.

9. Reference

IEC 61967-2: Integrated circuits - Measurement of electromagnetic emissions, 150 kHz to 1 GHz - Part 2: TEM cell method, International Electrotechnical Commission (IEC), Geneva, Switzerland Int. Std., July 2005.

IEC 61967-3 Edition 1.0 (2005-06) Integrated circuits - Measurement of electromagnetic emissions, 150 kHz to 1 GHz - Part 3: Measurement of radiated emissions – Surface scan method.

Test Plan for Mobile Station Over the Air Performance: Method of Measurement for Radiated RF Power and Receiver Performance. Ver. 3.1, CTIA - The Wireless Association, January 2011

Han-Nien Lin,Ching-Hsien Lin, Tai-Jung Cheng, Min-Chih Liao,*Antenna Effect Analysis of Laptop Platform Noise on WLAN Performance*, PIERS 2009 in Beijing, Session 2A8, March 21-25, 2009

Han-Nien Lin, Ming-Cheng Chang, Jia-Li Chang, Yung-Chi Tang, Jay-San Chen, *Influence Analysis of LCD Modules Noise on Performance of 802.11b*, 2011 APEMC in Korea, Poster I-6 P55,May 16-19,2011

Han-Nien Lin, Ching-Hsien Lin, Ming-Cheng Chang, Yu-Yang Shih, *nalysis of Platform Noise Effect on WWAN*, AMEMC 2010 in Beijing, TH-PM-A1-2: SS-13, April 12-16, 2010

Nada Golmie, *Coexistence in Wireless Networks: Challenges and System-Level Solutions in the Unlicensed Bands*, Cambridge, 2006

Han-Nien Lin, Ching-Hsien Lin,Chun-Chi Tang, and Ming-Cheng Chang,*Application of Periodic Structure on the Isolation and Suppression for Notebook Multi-antennas Coupling* PIERS 2010,Xi'an, Proceeding, p.160-164, March 22-26, 2010

Han-Nien Lin, Chung-Wei Kuo , Jhih-Min Liao , Jian-Li Dong , *Design of TEM cell and high sensitive probe for EMI analysis of built-in Webcam module*, 2010 EDAPS in Singapore, THPMTS84, December 07-09,2010

Han-Nien Lin, Jing-Ting Cheng, Jian-Li Dong, Jay-San Chen, *Radiated EMI analysis for CMOS Camera module with TEM Cell and Far-field testing*, 2011 APEMC in Korea, Poster I-1 P47,May 16-19,2011

Han-Nien Lin, Ming-Feng Cheng, Han-Chang Hsieh, Jay-San Chen, *Design and characteristic analysis of TEM Cell for IC and module EMC testing* , 2011 APEMC in Korea, T-Tu3-5 P103,May 16-19,2011

IEC 61967-1: Integrated circuits - Measurement of electromagnetic emissions, 150 kHz to 1 GHz - Part 1: General conditions and definitions, International Electrotechnical Commission (IEC), Geneva, Switzerland Int. Std., March 2002.

Han-Nien Lin, Chung-Shun Chang, Gang-Wei Cao, Cheng-Chang Chen, Jay-San Chen, *Design of High Sensitivity Near-Field Probe and Application on IC EMI Detection* , 2011 APEMC in Korea, Poster I-4 P53,May 16-19,2011

Han-Nien Lin, Tai-jung Cheng, Chih-Min Liao, *Radiated EMI Prediction and Mechanism Modeling from Measured Noise of Microcontroller*, AMEMC 2010 in Beijing, TH-PM-A1-4: SS-13, April 12-16, 2010

7

Gallium Nitride-Based Power Amplifiers for Future Wireless Communication Infrastructure

Suramate Chalermwisutkul
The Sirindhorn International Thai-German Graduate School of Engineering
King Mongkut's University of Technology North Bangkok
Thailand

1. Introduction

Progress in wireless communication technology has enabled applications which were unthinkable as the first digital mobile phone came into the market. Integration of digital camera into a mobile phone was an important step of the convergence between telecommunication and information technology as users started to require transfer of digital pictures besides conventional voice and text information. In addition, fast progress in digital technology has been an immense driving force of the needs for high data rates in telecommunications. Digital multimedia contents e.g. pictures, music, video clips are expected to be available anytime and anywhere which results into tremendous requirements in research and development in wireless technology.

Even though the industry tends to be majorly driven by software applications as well as "look and feel" of mobile devices, enabling hardware technologies in the background also deserve appropriate attention from R&D engineers. As soon as the performance of mobile communication systems cannot fulfil the expectation of users in terms of data rate and error robustness, the importance of the enabling hardware technology becomes obvious.

In order to cope with the rapid growth of the needs in wireless data transmission with constantly increasing data rates, new technical challenges arise perpetually on every layers of the OSI reference model. Whereas new modulation and multiple access techniques e.g. OFDM and OFDMA are introduced to support higher data rates and intelligent network configuration deals with the optimization of routing to increase the capacity and to improve load distribution, progress in hardware components in mobile devices and mobile base stations on the physical layer is also required to serve the needs of the higher OSI layers. Such progress on the physical layer includes techniques and hardware architectures which can enhance power efficiency of the system components while still complying with other specifications regarding linearity, noise, interference, etc.. Also, novel semiconductor device technology provides improved power handling capability resulting in smaller hardware size and high impedance which simplifies the design of matching networks. Moreover, large bandwidth and high impedance offer the possibility to create multiband components by designing the matching networks to be reconfigurable (Fischer, 2004).

This chapter aims to review state-of-the-art research in power amplifiers for wireless communication infrastructure featuring advantages of Gallium Nitride (GaN)-based power devices including large bandwidth capability, high power density and high output impedance. Regarding the issues of power amplifier design, state-of-the-art power amplifier architectures will be discussed with various prospects. For wireless communication standards with high data rates e.g. WCDMA, WiMAX and LTE, their modulation schemes and multiple access techniques lead to non-constant signal envelope with high peak to average power ratio. As a consequence, power amplifiers in wireless communication infrastructure are required to operate in a wide dynamic range making it difficult to maintain high average efficiency over time. This chapter will discuss widespread techniques for average efficiency enhancement including Doherty power amplifier concept and envelope tracking (ET) with state-of-the-art results. Another possibility for power efficiency improvement is the switched-mode power amplifier where the waveforms of the voltage and current are optimized to achieve low power dissipation at the power transistor. GaN-based power transistors have demonstrated in numerous research works to be suitable power devices for the switched-mode architecture as well as for average efficiency enhancement techniques e.g. Doherty power amplifier and envelope tracking. As examples, results of 2.45 GHz GaN class AB power amplifier and GaN VHF class E power amplifier will be presented in this chapter. The wide band capability of GaN-based devices also supports design of reconfigurable and wideband power amplifiers. With all advantages of GaN-based devices, they are still not a mature technology in terms of reliability and memory effects. Results from investigation on memory effects and parasitics of GaN-based devices will also be discussed in the chapter showing promising improvements in these regards which make GaN-based devices interesting and promising power devices for future wireless communication infrastructure.

2. Power amplifiers in the wireless communication infrastructure

In a mobile communication system, power amplifier is an important component which boosts the transmitted signal power before it is sent via the antenna to the receiving device through wireless channels (see Fig. 1.). In a base station for mobile communication standards e.g. GSM, UMTS or LTE, power amplifier is the part which consumes the largest portion of power. Thus, the efficiency of power amplifier has the greatest influence on the entire system's efficiency. In addition, cooling requirement of a base station is also dominated by its power amplifier. In terms of cost, power amplifier is also the most expensive part of a base station. For the first generation of UMTS base stations, the costs of power amplifier and cooling are about 30%-35% of the cost of an entire base station (Chalermwisutkul, 2007). Besides the efficiency, linearity is also an important specification of power amplifiers which ensures that the transmitted signal is not distorted by the nonlinearity to an unacceptable level causing excessive bit errors.

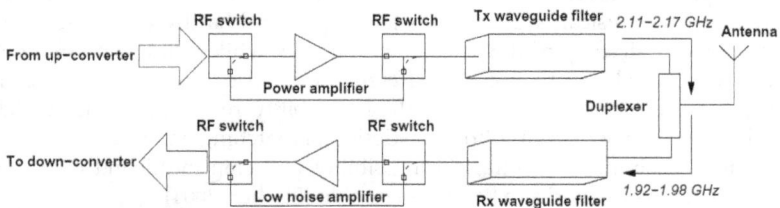

Fig. 1. Block diagram of a UMTS base station transceiver showing power amplifier and other system components.

2.1 Typical architecture and power device for base station power amplifier

In general, power amplifiers in mobile base stations are class AB amplifiers which offer both acceptable power efficiency and linearity. The operating point for the power device of this amplifier class is a compromise between those of highly efficient class B and highly linear class A. The conduction angles, output drain current waveforms, active load-lines and operating points of class A, AB, B, C, E and F amplifiers are depicted below in Fig. 2..

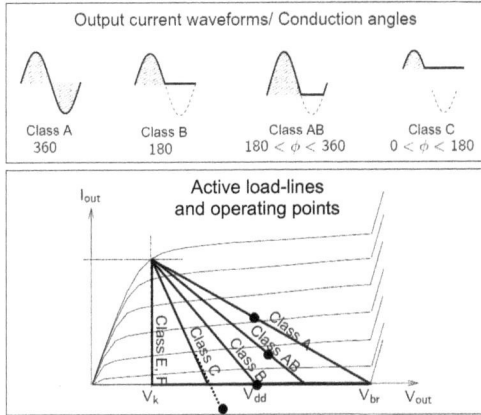

Fig. 2. Conduction angles, output drain current waveforms, active loadlines and operating points of class A, AB, B, C, E and F amplifiers. V_{out}, I_{out}, V_k, V_{dd} and V_{br} are drain output voltage, drain current, knee voltage, drain voltage supply and drain breakdown voltage, respectively (source (Chalermwisutkul, 2007)).

Typically, lateral diffused metal oxide semiconductor (LDMOS) field effect transistors based on Silicon are used as power devices for base station power amplifiers. Silicon LDMOS is considered a mature power device technology for mobile base station amplifiers due to its high efficiency, high power density and high thermal conductivity. However, main reasons which make LDMOS standard device technology for base station amplifiers are its low cost and high reliability. Although it is known that the operating frequency of LDMOS devices is limited to a few GHz, progress in LDMOS technology is still ongoing and new LDMOS devices are continuously introduced into the market with higher operating frequency and other progresses in terms of power efficiency, linearity, etc. (Ma et al, 2005). Due to this fact, the dominance of LDMOS devices in low GHz high power applications has been ensured since the first devices came into the market. However, new challenges in power device technology keep emerging as modern wireless communications are required to cope not only with higher data rates at limited frequency resource, but also with energy saving issues. In other words, there are increasing demands in high power efficiency besides spectrum efficiency for the wireless communication infrastructure. In this regard, there are several cases where it is worth to look for alternative power device to overcome limitation of existing device technologies.

Despite of all advantages of LDMOS, the main drawback of this device is the bandwidth capability. Due to high output capacitance of LDMOS device, the Q factor tends to be high and the bandwidth is small. Also, the operating frequency limit hinders this device from being used in high frequency applications which are served with other device technology e.g. GaAs MESFET and HEMT. The research interest has been then attracted by wide-bandgap semiconductor materials for high frequency power devices. Silicon Carbide (SiC) is superior in thermal conductivity compared to other wide-bandgap semiconductors. However, the cost of SiC is relatively high. Moreover, this material is not appropriate for applications with very high operating frequencies. For Indium Phosphide (InP), another wide-bandgap compound semiconductor, the focus of research is on extremely high-speed digital applications where high power is not required.

The most prominent wide-bandgap semiconductor is Gallium Nitride (GaN). Comparing with Silicon device technology which is mainly driven by microprocessor and computer industries, GaN found its applications in screen industries enabled by GaN OLED (organic light emitting diode) technology and data storage industries utilizing blue laser produced by GaN laser diode to read out the data from a Blue-ray Disc™. In automotive applications and power electronics, GaN devices are attractive due to high operating temperature and high breakdown field for switching power supply. For RF power amplifiers, GaN-based power devices offer extremely large bandwidth, high power density, high operating frequency and high output impedance. The advantages of GaN-based power devices for wireless communications will be discussed more thoroughly in the next section.

2.2 Techniques for enhancement of average power efficiency

Modulation schemes and multiple access techniques allowing high data rates in wireless communication standards lead to non-constant signal envelope with high crest factor or peak to average power ratio (PAPR). Since a typical class AB power amplifier in mobile base station offers highest power added efficiency (PAE) about at one dB compression area in the power sweep plot, high peak to average power ratio leads to power back-off from the peak efficiency point which leads to efficiency reduction (see Fig. 3.). As a result, average efficiency over time is much lower than the peak efficiency. From the system point of view, reduction of peak to average power ratio can be done with different techniques at the cost of

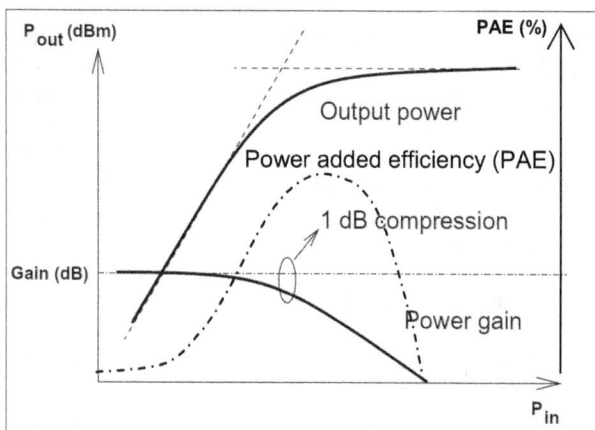

Fig. 3. Typical power sweep plot of a class AB power amplifier showing efficiency degradation when the power is backed-off from 1 dB compression point.

reduced data rate, transmit signal power increase, BER performance degradation, computational complexity increase, and so on (Jiang and Wu, 2008). Independent of the reduction techniques, rest of the peak to average power ratio still exists, so that for further average efficiency improvement, power amplifier architecture which can keep power efficiency high also when the transmitted power is backed-off must be considered.

Envelope elimination and restoration (EER) or Kahn technique

This average efficiency enhancement technique is based on the idea to separate the amplitude modulated envelope from the constant envelope, phase modulated carrier signal. The envelope is amplified with high efficiency envelope amplifier, whereas the carrier is amplified with nonlinear but highly efficient power amplifier. The output of the envelope amplifier is supplied to the carrier amplifier which reconstructs the typical signal with non-constant envelope of modern wireless communication standards (Diet et al, 2004).

Envelope Tracking (ET)

Similar to Kahn technique, supply voltage level of the RF amplifier is dynamically modified depending on the level of the signal envelope. A slight difference is that the input of the RF amplifier is still amplitude and phase modulated. Only with excessive signal power, the supply voltage of the RF amplifier is modified. The RF amplifier of this technique operates also in a linear mode unlike the Kahn technique, where the RF amplifier operates solely in a nonlinear mode.

Outphasing or Chiriex technique

Also known as linear amplification using nonlinear components (abbr. LINC), this technique uses two nonlinear high efficiency power amplifier to boost up two signals with differently controllable phases. The two amplified signals are then combined with vector addition and the phase difference between the two signals defines the power level of the resulting signal. Compared to EER and ET, phase is the dynamically changing quantity and not the supply voltage of the RF amplifier (Helaoui et al, 2007).

Doherty technique

The concept of Doherty power amplifier utilizes two power devices which are operated as main and auxiliary amplifiers. As soon as a certain level of input power is reached, main amplifier — normally class B — is running into saturation providing its maximum efficiency. As the main amplifier starts to saturate, the auxiliary amplifier starts to conduct current. The saturation condition of the main amplifier is maintained by load modulation caused by the current from the auxiliary amplifier, so that the main power device acts like a voltage source. At peak output power, the auxiliary amplifier just begins to saturate and high efficiency is ensured for both amplifiers. Block diagram of a Doherty power amplifier is depicted in Fig. 4.. Details about Doherty amplifier can be found in the literature (Raab, 1987).

Compared to other efficiency enhancement techniques, Doherty concept has gained its popularity due to simple architecture which deals with RF circuit design issues only, whereas other techniques make use of digital signal processing to improve average efficiency. Thus, it is more straightforward to design a Doherty power amplifier to cope with new peak to average power ratio value where high average efficiency is desirable. This can be simply achieved by modifying the input power division ratio between main and auxiliary amplifier. If necessary, three amplifiers can also be used in order to maintain high average efficiency over a high dynamic range.

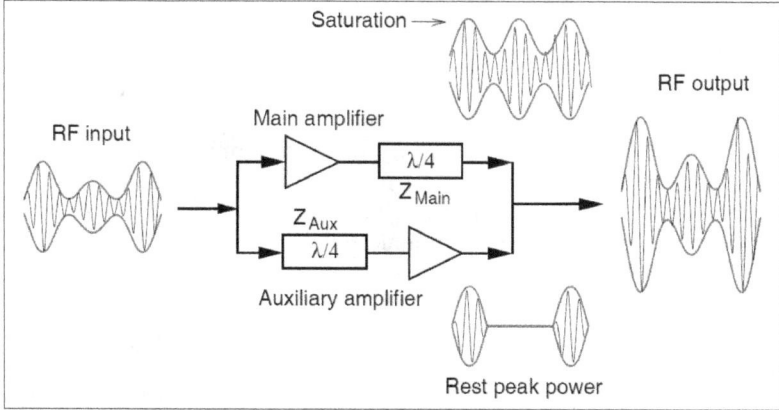

Fig. 4. Block diagram of a Doherty amplifier.

2.3 Switched-mode power amplifiers

In subsection 2.2, average efficiency enhancement techniques with the goal of maintaining high efficiency over a wide range of input power have been described. Considering peak efficiency at peak output power, switched-mode power amplifiers can achieve higher efficiency than widespread class AB power amplifiers. In case of switched-mode, the power transistor operates as a switch so that output voltage and current of the device (drain of FETs and HEMTs or collector for BJTs and HBTs) do not have high values at the same time. For the "off" state, the current is near to zero and the voltage is high and vice versa for the "on" state resulting in theoretical efficiency of 100%. In the following, switched-mode class E, F and D will be briefly described.

Class E

The first class E amplifier has been proposed by Sokal in 1975 (Sokal, 1975). Thereafter, other variations of class E amplifiers have been constantly presented with higher operating frequency where not only class E operation is ensured, but also, practical issues such as small circuit size and simple matching have been taken into account. A good example of such progress in class E amplifier design was represented by the class E amplifier with parallel circuit proposed by Grebennikov (Gebrennikov, 2002). Class E offers high efficiency by avoiding simultaneous existence of high drain voltage and high drain current and thus, avoiding power dissipation of the power transistor. Control of the output current and voltage waveforms at drain or collector node of the device is achieved using an output load network. Theoretically, as the transistor turns on, the voltage drops to zero and the current starts to flow so that the output capacitance is gradually charged. As soon as the control voltage of the switch is lower than the switching voltage threshold, the transistor is turned off and the current drops to zero while the output voltage of the device starts to increase. The ideal class E voltage and current waveforms are depicted in Fig. 5. Variations of class E amplifiers are reported to offer high power as 1 kW for switching applications at low frequency, whereas for RF applications, operating frequency of 10 GHz was already presented (Weiss, 1999). Class E is a promising switched-mode amplifier concept due to its simple architecture and flexibility compared to other switched-mode classes. Combination of a class E amplifier with average efficiency enhancement techniques e.g. EER or Doherty has been reported in the literature (Diet et al, 2004 and Kim et al 2010).

Class F

High efficiency of class F amplifiers is achieved by shaping the wave forms of output current and voltage of the power transistor which operates as a switch. Compared to class E, where load network is required to ensure the ideal switching condition (on state with high current, zero voltage and off state with high voltage and zero current), load network of class F has additional function which attempts to shape the output voltage and current waveforms at the device's drain or collector node. For conventional class F, odd harmonic peaking of the device's output voltage is realized by providing high impedance (open circuit condition) at the odd harmonic frequencies. As a result, the voltage waveform approximates a square wave. For the drain current, even harmonics are provided in addition to the fundamental by offering the device a short circuit condition at even harmonic frequencies. As a result, the current waveform approximates a half wave signal. Ideal current and voltage waveforms of a class F amplifier are shown in Fig. 5. Another alternative variation of class F is the inverse class F where the current waveform approximates a square wave, whereas the voltage waveform approximates a half wave signal. Efficiency of class F amplifiers can be increased by offering appropriate termination (open or short) at higher harmonics. However, this occurs at the cost of circuit's complexity. Similar to class E, class F and inverse class F amplifiers can be combined with Doherty technique to obtain high average efficiency for wireless communication signals with high peak to average power ratio. By using class F or inverse class F in a Doherty transmitter, peak efficiency is increased compared to the variation with class B main amplifier (Goto et al, 2004).

Class D

Unlike other switched-mode amplifier classes, class D uses at least two transistors as switches. In case of current mode class D (CMCD), the transistor's output current has a form of a square wave whereas the voltage mode class D (VMCD) shows a square output voltage of the transistor (see Fig. 5.). For both CMCD and VMCD, a tank filter is required to obtain the sinusoidal signal at the load. For CMCD, additional BALUN is also required, whereas for VMCD, two supply voltage sources are needed (see Fig. 6.). When one of the switches is turned on, the other one is turned off, so that high current and high voltage cannot exist at the same time. Theoretically, 100% efficiency can be achieved. In practice, the efficiency is compromised by limited switching speed and device's output capacitance. Due to these reasons, frequency of operation is limited for class D amplifiers. Experimental, state-of-the-art RF class D power amplifiers can operate at frequencies in the region near to 1 GHz (Aflaki et al, 2010).

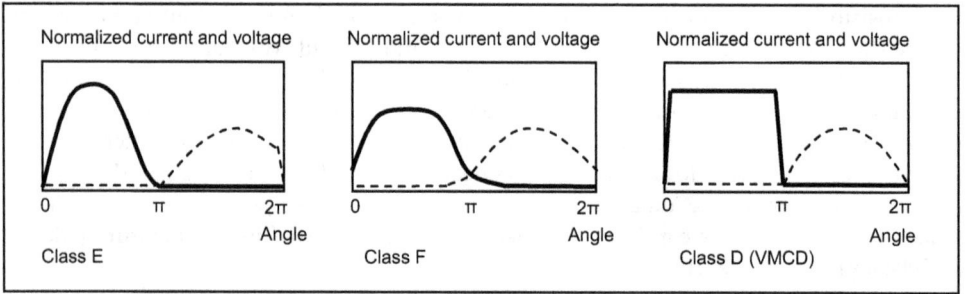

Fig. 5. Ideal current and voltage waveforms of class E, F and D switched-mode amplifiers (Raab et al., 2002). Broken lines represent the device's output current and the solid lines represent the device's output voltage.

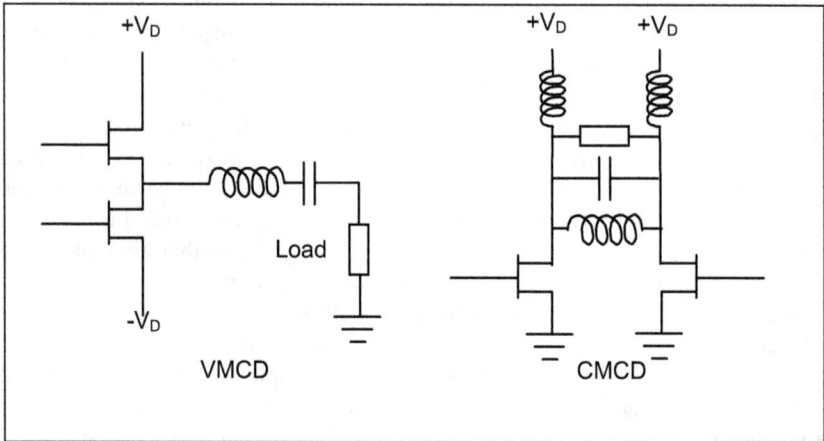

Fig. 6. Configurations of voltage mode class D (VMCD) and current mode class D (CMCD) amplifiers.

2.4 Linearization techniques

In general, a trade-off exists between efficiency and linearity of power amplifiers. For conventional transconducdance amplifier classes e.g. class A, AB and B, it is obvious that high efficiency classes are nonlinear. In subsection 2.2, average efficiency enhancement techniques aiming to keep the efficiency high over a wide dynamic range have been discussed. Even though efficiency is the main goal of such techniques, linearity was also taken into consideration so that none of such techniques would have severe impact to linearity. However, when the desired efficiency profile is achieved, linearity might not comply with wireless communication standards leading to unacceptable error vector magnitude and bit error rates. In such a case, linearity improvement techniques can be utilized to eliminate the excessive nonlinearity of the amplifier. Widespread linearization techniques are reviewed below.

Feedback linearization

In order to force the RF output to follow the input, feedback of the RF signal is realized using a directional coupler. The simplest variation of this technique subtracts the RF feedback from the input signal. However, the compensation of the non linearity with this technique is not very efficient as the transmitter's gain is reduced. Another variation detects the envelope of the RF feedback and the input signal and subtracts the first from the latter to realize the linearization of the amplitude. For the compensation for both phase and amplitude nonlinearities, another variation called Cartesian feedback was conceived that the feedback signal is down converted to I and Q values which are used to compensate the I and Q of the input signal. For a relatively small bandwidth, the two tone IMD can be reduced by 10 to 35 dB with this technique.

Feedforward linearization

This linearization technique is excellent in terms of bandwidth and IMD reduction. In order to generate the error signal, the power amplifier's output and the input signal are sampled using directional couplers and the first is then subtracted from the latter (see Fig. 7.). The error signal is then amplified and subtracted from the power amplifier's output to obtain the linear output signal. Since this technique utilizes an open loop concept, additional loop control is required in order to compensate the degradation of the power device over time to ensure the right settings of phase shift and gain for maximum linearity. IMD reduction of 20-40 dB can be achieved for bandwidth up to 100 MHz. The drawback of this technique is the complexity of the system.

Digital predistortion

In order to obtain undistorted signal at the transmitter output, the input signal can be intentionally distorted before being fed to a nonlinear power amplifier. The predistorter generates nonlinearities which operate in the opposite way to the nonlinearities generated by the power amplifier, so that the overall response at the PA-output is linear (see Fig. 8). The linearization is done in the digital regime using FPGA which makes the system very flexible and adaptive for changes in power device over time to ensure linear output. As computational power of FPGA is continuously increasing, linearization over larger bandwidth can be realized with this technique. In the literature, linearization with digital predistortion technique which can cope with dynamic nonlinearity caused by electrical memory effects has also been reported (Lee et al, 2009).

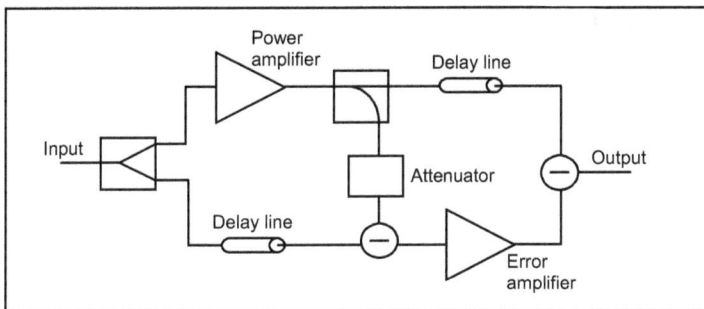

Fig. 7. Block diagram of a feedforward transmitter.

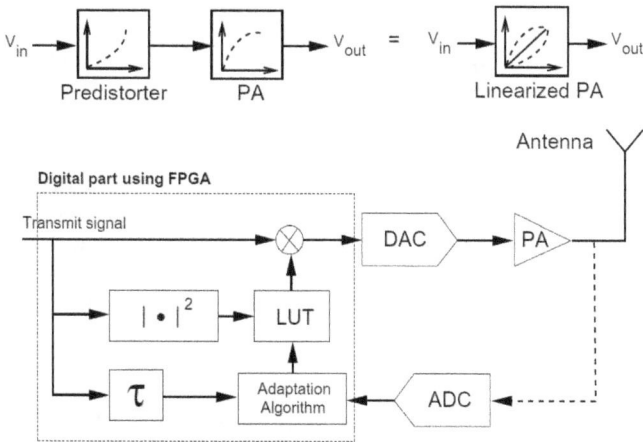

Fig. 8. Principle and block diagram of digital predistortion for linearization of power amplifiers.

In comparison to other techniques, digital predistortion offers higher efficiency and greater flexibility at low cost and represents a mature linearization technique for mobile base stations. Due to the mentioned flexibility and simple architecture, digital predistortion has gained it popularities in the power amplifier design community. In most of the cases where no extremely large bandwidth is required, high efficiency amplifiers e.g. Doherty and switched-mode amplifiers are combined with digital predistortion to improve the linearity.

3. GaN-based power amplifiers

As mentioned in section 2.1, GaN is a promising semiconductor material for high power and high frequency power transistors which are used as power devices in mobile base station power amplifiers. The advantages of GaN originate from physical properties of this wide-bandgap semiconductor. Table 1 shows physical properties of various semiconductor materials including GaN.

Material/Properties	Si	GaAs	InP	SiC	GaN
Bandgap (eV)	1.1	1.4	1.3	3.2	3.4
Saturation Velocity ($*10^7$ cm/s)	1.0	2.1	2.3	2.0	2.7
Thermal Conductivity (W/cmK)	1.3	0.46	0.7	4.9	1.7
Breakdown Field ($*10^6$ V/cm)	0.3	0.4	0.7	2.0	2.7
Electron Mobility (cm2/Vs)	1350	8500	5400	800	1500

Table 1. Physical properties of semiconductor materials for RF and microwave applications.

From table 1, advantages of GaN compared to other semiconductor materials for RF and microwave applications are obvious. GaN offers very high saturation velocity leading to high operating frequency up to 100 GHz or higher. High breakdown field allows GaN-based devices to operate with high supply voltage which is advantageous for the off state of switched mode amplifiers and for obtaining high output power with high output impedance. Due to higher supply voltage, efficiency is also improved due to the reduction of the need for voltage conversion. For extreme operating environment e.g. for automotive applications, GaN offers wide bandgap and high thermal conductivity leading to the capability to operate at high temperature.

The most prominent GaN-based device for RF and microwave applications is GaN-based high electron mobility transistor (GaN HEMT). This kind of device offers extremely high operating frequency due to high electron mobility in the so-called 2DEG channel (Smorchkova, 2001). Moreover, one of the most impressive features of this device is the extremely high power density meaning that the device's size can be much smaller compared to other device technology for the same output power. With size reduction, output impedance becomes larger and parasitic capacitances smaller leading to large bandwidth and uncomplicated matching to 50 Ohm. It was also mentioned in the literature that GaN HEMT can offer better noise performance than that of MESFET's (Mishra et al, 2007).

For wireless communication infrastructure, GaN HEMT has proven itself to be an attractive alternative power device besides LDMOS FET for base station power amplifiers. For WCDMA base station, a GaN HEMT-based transmitter with output power higher than 200 W and supply voltage of 50 V was published in 2004 (Kikkawa et al, 2004). Reliability--one of the biggest concerns regarding GaN HEMT compared to LDMOS--was also presented in that work. However, at this point, it is not possible to foresee when GaN HEMT's will take the place of LDMOS FET's in base station power amplifiers. Even if the frequency of operation is limited to a few GHz for LDMOS, this device technology is continuously developed regarding power, reliability, linearity, etc.. Moreover, LDMOS is considered a cost-effective and mature power device technology with a large LDMOS amplifier designer community. Consequently, knowhow and design experience for this device is available to a great extent. Regarding this consideration, GaN HEMT will find its importance first in applications where large bandwidth is required or high power is desirable at high frequency. Besides reliability, charge carrier trapping in GaN HEMT has been a big issue for device technology improvement. Numerous investigations have been done regarding trapping effects of GaN devices. Charge carrier traps can cause dependency of the pulse-measured I-V characteristic on the quiescent point. This is a phenomenon of the so-called electrical memory effect (Chalermwisutkul, 2008). Other phenomena of memory effects are gate lag and drain lag in time domain where the drain current reaches its final value after some delay as the bias voltages are abruptly changed. In frequency domain, dispersion of output impedance is the consequence of electrical memory effect leading to dynamic nonlinearity with a large bandwidth of spectral regrowth (Fischer, 2004). Improvement of GaN device technology regarding charge carrier trapping and reliability has been reported occasionally e.g. SiN passivation or use of the field plate for traps reduction (Mishra, 2007). In this section, results from the works regarding GaN device modeling and GaN power amplifier design in which the author has been involved will be presented.

3.1 GaN device modeling

Computer simulation of the performance belongs to a typical design flow of power amplifiers. As many as possible components in the amplifier circuit should be characterized and described by models in order to obtain accurate prediction of circuit's performance from the simulations. As the main component of a power amplifier, quality of power transistor model plays a significant role in the accuracy of circuit simulation. Especially for power amplifier design, nonlinearities of the device must also be described by the device model unlike for small signal amplifiers, where it is sufficient to have the device's S-parameter sets of a few bias points of interest.

Even for one device technology, it is not practical to create a universal model which can describe the device's behavior under all operating conditions. In order to describe more effects and dependencies of the device's behavior on dynamic thermal and electrical conditions, more and more model parameters and nonlinear equations are required. In that case, the model would become very complex and long simulation time is needed. Though computational resource can be increased, complex device models suffer from poor robustness, that the simulation would be often terminated without convergence and reasonable results. For switched-mode power amplifiers e.g. class E, F, inverse F or D, a concept of using switch model in combination with the "on" state resistance R_{on} and output capacitance C_{ds} instead of empirical transistor model exists (Negra et al, 2007). This simple model is capable of providing good trend of power and efficiency and of verifying switched-mode operating conditions. At this point, there exist some discussions regarding the accuracy of such switch model for switched-mode power amplifier applications. Especially for power devices with charge carrier trapping and thus, memory effects, the switch model is not able to describe such effects which can have influence in efficiency and output power of switched-mode amplifiers (Chalermwisutkul, 2008).

Electrical memory effects

Even when electrical memory effects of GaN HEMT are still not negligible compared to those of GaAs HEMT, but the benefit of high power density, high output impedance, high frequency, etc. of GaN HEMT can be used, when the device is accurately described including the memory effects by the device model. First of all, the extraction of model parameters should be done using multibias pulsed measurement data. In such a measurement process, the bias voltages of the transistor is pulsed starting from the so-called quiescent point to other bias points in the I-V characteristics and drain current I_{ds} as well as S-parameters of that bias point are measured. Pulsed measurement has a significant advantage which is the isothermal measurement condition. The measured I-V characteristic of a pulsed measurement does not contain the self-heating of the transistor at high V_{ds} and I_{ds} as seen in DC measurement which is more familiar to the realistic operating condition. Moreover, quiescent point of pulsed measurement can be chosen equal to the operating point of the amplifier class of interest in order to create a device model which corresponds to the behavior of the device under realistic operating condition. In particular, the quiescent point dependent device model is necessary for a power device with significant trapping effects (see Fig. 9.). Theoretically, the dependence on quiescent point could be included into the model making the device model a general purpose one. However, as described above, this would increase the complexity and decrease the robustness of the model. Promising results of high power GaN HEMT have been published in 2004 showing the progress in

GaN device technology in term of reduction of trapping effects where the DC measurement of I-V curves shows no significant difference in the level of drain current compared to a pulsed measurement with a quiescent point at high drain voltage region (Kikkawa et al, 2004). In such a case, the quiescent point dependence of the device model would not be so critical. For power transistor manufacturers, normally, only one device model is provided to the circuit designer. As a result, the model of a mature power device regarding trapping will offer more accurate results for arbitrary classes of amplifiers.

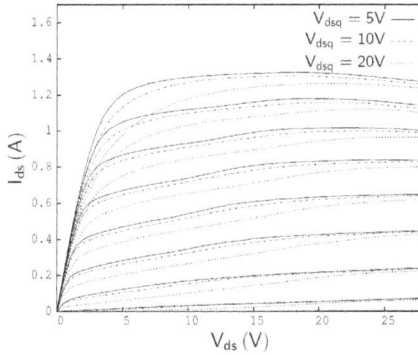

Fig. 9. Dependence of I-V characteristic of a GaN HEMT on quiescent point. The quiescent voltage was constant at a pinch-off value (no quiescent current) whereas the drain quiescent voltage V_{dsq} was varied.

Knee walkout

In contrast to GaAs HEMT and MESFET, the knee voltage of a GaN HEMT depends on the gate voltage and the drain quiescent voltage. With high gate voltage, the knee of the I-V curve becomes more round than at lower gate voltage where the knee is relatively angular. In addition, the knee voltage is shifted to the right toward higher drain voltage when gate voltage is high. This so-called *knee walkout* effect observed only with GaN HEMT and not with GaAs HEMT or MESFET cannot be modeled with standard EEHEMT model. By adding dependency of the knee voltage on the gate voltage and the drain quiescent voltage, the knee region of the I-V curve with high gate voltage can be better described (see Fig. 10.). As a result more accurate power and efficiency simulation can be done (see Fig. 11.) (Chalermwisutkul, 2007).

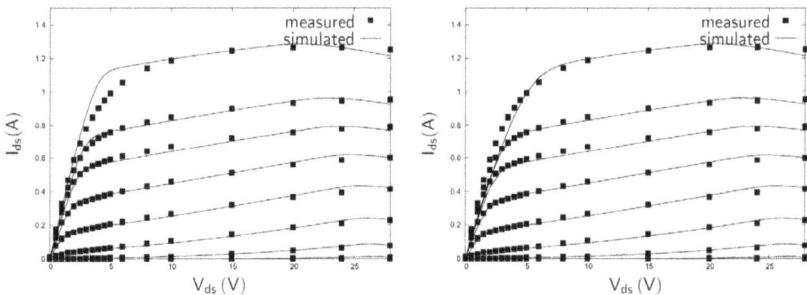

Fig. 10. I-V curves fitting results without (left) and with (right) the description of the knee-walkout.

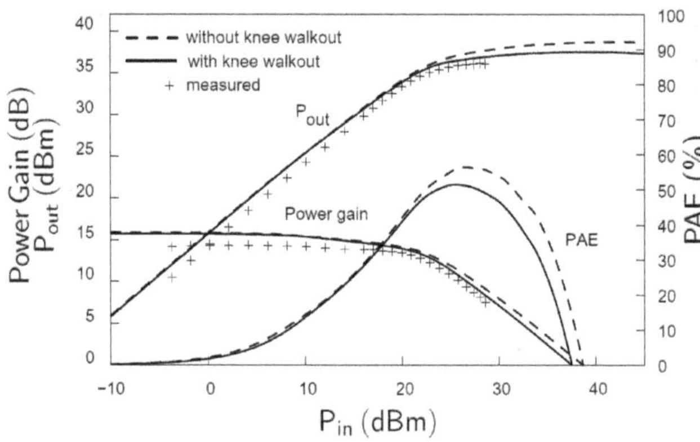

Fig. 11. Power sweep simulation without and with the description of the knee-walkout compared to measured values.

Large signal behavioral model

As discussed before, large signal model is required in order to describe nonlinearities of the power device. However, device modeling is a complex task which requires extensive experience of modeling engineers and special modeling software, so that power amplifier design engineers are mostly forced to rely on the large signal model provided by device's manufacturer. Due to progress in RF measurement techniques, a measurement system has been developed which allows measurement of the so-called X-parameters (Betts et al, 2011). Unlike with S-parameters, not only small signal behavior of the device can be described, but also nonlinearities arising under large signal conditions. In general, the input signal power is swept and the output response at the fundamental as well as at higher harmonics is measured. The measured information is then concluded into the X-parameter set which can be directly used in the circuit simulation software as the device's behavioral model. This kind of device modeling is very convenient and can be combined with source and load tuners to obtain load dependence of the X-parameters. In addition, the extracted behavioral model is accurate, robust and does not require large computational resource. However, the behavioral model cannot provide insights into physical properties of the device and the measurement setup is relatively expensive for small companies and educational institutions with low budget.

Package modeling

Packaged transistors comprise also parasitic components of the package and bond wires. These typical parasitic inductance and capacitance can compromise the performance of the amplifier circuit especially at high frequencies. For example, for class F amplifiers where short or open circuit must be provided at the drain node of the transistor at harmonic frequencies in order to shape the output current and voltage waveforms for high efficiency. Optimization for efficiency can be done best, if the package model of the transistor is known. The current and voltage waveforms which are optimized for minimum overlap should be presented at the internal drain node of the device inside the package and not at the external drain port (Schmelzer and Long, 2007).

Design examples of GaN HEMT power amplifiers

As examples, two GaN power amplifiers are presented. The first one is a 2.45 GHz GaN HEMT class AB power amplifier (Monprasert et al, 2009). This power amplifier is intended for the use in a WLAN system. The power transistor used in this amplifier is NPTS00004 GaN HEMT from Nitronex Corporation. The performance of the 2.45 GHz power amplifier is shown in Table 2. The drain supply voltage was varied with V_{dsq}=20V and 28V. For the drain supply voltage of 28V, the output power is not as high as in the case with V_{dsq}=20V since the drain current was increased as the device started to be saturated. The DC power exceeded the limit of 7 Watts given in the datasheet and the device was damaged. Fig. 12 shows a photograph of the fabricated class AB amplifier.

Drain quiescent voltage	V_{dsq} = 20 V	V_{dsq} = 28 V
Maximum output power	34.68 dBm	30.93 dBm
Maximum Power Added Efficiency	42.5 %	20.8%
Small signal gain	12.27 dB	13.69 dB

Table 2. Measured performance of 2.45 GHz GaN HEMT class AB power amplifier.

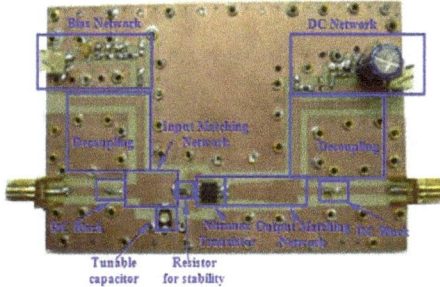

Fig. 12. Fabricated 2.45 GHz GaN HEMT class AB power amplifier.

Another design example is the VHF class E power amplifier (Khansalee et al, 2010). Using the same GaN power device Nitronex NPTB00004, a class E power amplifier for the operating frequency from 140 MHz to 170 MHz has been designed and fabricated. The values of load network L, C, L_0 and C_0 (see Fig. 13.) were determined using equations in the work published by Gebrennikov (Gebrennikov, 2002).

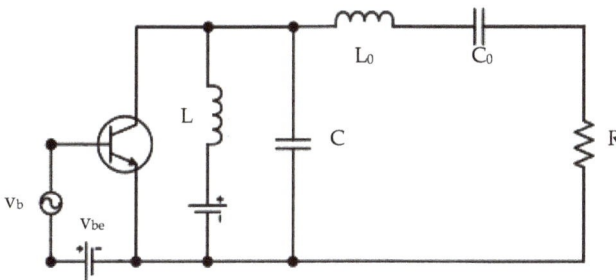

Fig. 13. Schematic of class E power amplifier with parallel circuit.

The optimal load impedance was determined using load pull simulation in Advance Design System (ADS). Simulated drain voltage and current waveforms show that class E operation is achieved (see Fig. 14.).

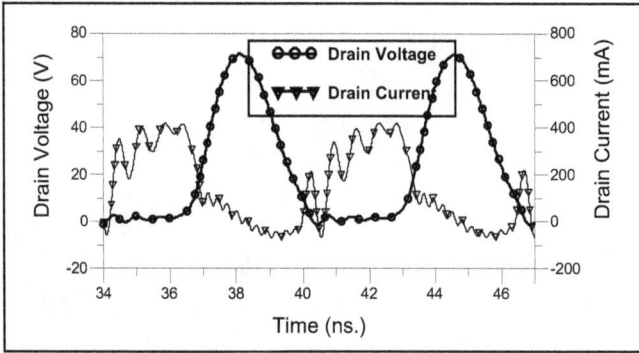

Fig. 14. Simulated drain current and voltage waveforms of the class E amplifier.

The fabricated class E power amplifier delivers maximum output 33.9 dBm, peak Power-Added Efficiency (PAE) of 72.5% and power gain of 16.4 dB at the center frequency of 155 MHz. Fig. 15. shows output power, efficiency and gain over the required operating frequency from 140 MHz to 170 MHz. A photograph of the fabricated GaN class E amplifier is depicted in Fig. 16.

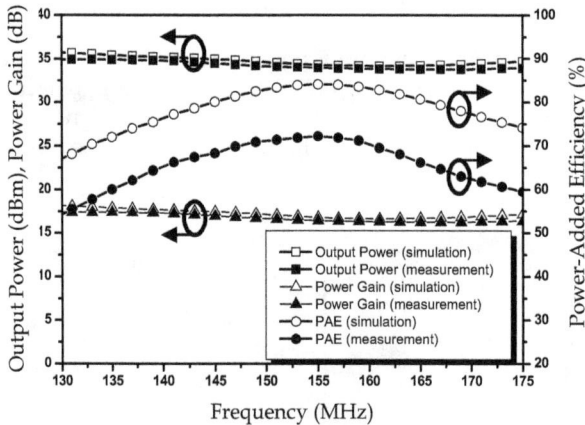

Fig. 15. Simulation and measurement results of power gain, output power, and PAE over the frequency 140 MHz to 170 MHz at input power of 18 dBm with the drain supply voltage of 24 V and gate supply voltage of -1.4 V over frequency.

Fig. 16. Fabricated class E GaN VHF power amplifier.

4. Future research in power amplifiers for wireless communications

Needs for high data rates anywhere and anytime while the spectrum resource is limited will be a great challenge for future mobile and wireless communications. In order to utilize the bandwidth efficiently, new approaches on the network layers are being standardized and conceived including opportunistic, software defined and white space radio. The challenge of such frequency agile concepts will be not only be on the network and system layers e.g. spectrum sensing for vacant frequency slots, but also on the physical layer regarding the need of transmitters which can cope with extremely wide band or can be reconfigured for dynamic band migration. Regarding efficiency and power management, issues on every layer must be taken into consideration which would lead to interlayer optimization from network over system to physical layers. Active antenna and multiple inputs, multiple outputs (MIMO) concept will also be important topics which will require co-design and integration of amplifiers and antennas.

Energy saving is and will be a big issue not only in automotive and electrical power areas but also in wireless communications. To fulfil the intention for the "green transmission", high efficiency must be provided by all infrastructure components e.g. base stations. Also, the trend of modern wireless communication standards is going in the direction of low power and small base stations will small cell size. This means that not only the mobile devices e.g. smart phone or tablets require aesthetic design but also the infrastructure components which should be well integrated into the environment. High efficiency will contribute to this requirement by offering small size of base stations. Regarding efficiency, research and development efforts will be spent in high efficiency signal transmission including design of switched-mode high efficiency power amplifiers with modulated input for improved efficiency e.g. class S amplifiers for delta sigma modulated signal (Pivit et al, 2008). Considering the demand of wide bandwidth and the capability to deliver high switching speed at high power, GaN-based devices are promising device technology for future wireless communications.

5. Conclusion

In this chapter, GaN-based power amplifiers for wireless communication infrastructure have been discussed. GaN HEMT's offer superior performance compared to state-of-the-art power devices for base station power amplifiers e.g. LDMOS. Especially high power density and high supply voltage of GaN HEMT's leads to smaller size of the device and thus, to lower parasitic capacitance, higher output impedance and large bandwidth which are advantageous for switched-mode and reconfigurable power amplifiers. In addition, wide range of operating frequency can be covered by GaN-based power devices. The concerns of GaN transistors regarding charge carrier trapping and reliability is gradually extenuated by the progress in GaN device technology.

Device modelling is another important issue which ensures the power amplifier design community fast design process and accurate simulations. As examples, VHF class E amplifier and 2.45 GHz class AB amplifier have been presented.

6. Acknowledgment

The author would like to thank his family for the support and understanding during the preparation of the manuscript. Also, the author would like to express his appreciation to the research assistants, staffs and students of the RF and Microwave Laboratory, the Sirindhorn International Thai-German Graduate School of Engineering, King Mongkut's University of Technology North Bangkok for their interest in RF and microwave topics as well as for their support.

7. References

Fischer, G. (2004). Architectural benefits of wide bandgap RF power transistors for frequency agile basestation systems, *Proceedings of the IEEE/MTT Wireless and Microwave Technology Conference*, Clearwater Beach, Florida , April 16, 2004.

Chalermwisutkul, S. (2007). *Large Signal Modeling of GaN HEMTs for UMTS Base Station Power Amplifier Design Taking into Account Memory Effects*, PhD. Thesis, Faculty of Electrical Engineering and Information Technology, RWTH Aachen University, Germany, April 2007.

Chalermwisutkul, S. (2008). Phenomena of Electrical Memory Effects on the Device Level and Their Relations, *Proceedings of the 5th International Conference on Electrical Engineering/Electronics, Computer, Telecommunications and Information Technology, 2008. ECTI-CON 2008*, pp. 229 - 232, ISBN 978-1-4244-2101-5, Krabi, Thailand, May 14-17 2008

Ma, G.; Qiang Chen; Tornblad, O.; Tao Wei; Ahrens, C. and Gerlach, R. (2005). High Frequency Power LDMOS Technologies for Base Station Applications Status, Potential, and Benchmarking, *Electron Devices Meeting, 2005. IEDM Technical Digest. IEEE International*, pp. 361 – 364, ISBN 0-7803-9268-X, Washington DC, USA, 5 Dec. 2005

Jiang, T. and Wu, Y. (2008). An Overview: Peak-to-Average Power Ratio Reduction Techniques for OFDM Signals, *IEEE Transactions on Broadcasting, Vol. 54, No. 2, JUNE 2008*, pp. 257 – 268, ISSN 0018-9316

Diet, A.; Berland, C.; Villegas, M. and Baudoin, G. (2004). EER architecture specifications for OFDM transmitter using a class E amplifier, *Microwave and Wireless Components Letters, IEEE, Volume: 14 Issue:8, Aug. 2004* , pp. 389 – 391, ISSN 1531-1309

Helaoui, M.; Boumaiza, S.; Ghannouchi, F. M.; Kouki, A. B. and Ghazel, A. (2007). A New Mode-Multiplexing LINC Architecture to Boost the Efficiency of WiMAX Up-Link Transmitters, *Transactions on Microwave Theory and Techniques, IEEE, Vol.: 55 Issue:2, Feb. 2007*, pp. 248 – 253, ISSN 0018-9480

Raab, F.H. (1987). Efficiency of Doherty RF Power-Amplifier Systems, *IEEE Transactions on Broadcasting, Vol.: BC33 Issue:3, Feb. 2007*, pp. 77 – 83, ISSN 0018-9316

Sokal, N. O. and Sokal, A. D. (1975). Class E-a new class of high efficiency tuned single-ended switching power amplifiers, *IEEE Journal of Solid-State Circuits, vol. SC-10, no. 3, June 1975*, pp. 168-176, ISSN 0018-9200

Grebennikov, A. and Jaeger, H. (2002). Class E with parallel circuit – A new challenge for high-efficiency RF and microwave power amplifiers, *IEEE MTT-S Int. Micro. Symp. Dig., vol. 3*, pp. 1627-1630, June 2002, ISBN 0-7803-7239-5, Seattle, WA , USA, 02 Jun 2002 - 07 Jun 2002

Kim, I.; Moon, J., Jee, S. and Kim, B. (2010). Optimized Design of a Highly Efficient Three-Stage Doherty PA Using Gate Adaptation, *IEEE Transactions on Microwave Theory and Techniques, Vol. 58, No. 10, October 2010*, pp. 2562 – 2574, ISSN 0018-9480

Goto, S.; Kunii, T.; Inoue, A.; Izawa, K.; Ishikawa, T. and Matsuda, Y.; Efficiency enhancement of Doherty amplifier with combination of class-F and inverse class-F schemes for S-band base station application, *2004 IEEE MTT-S International Microwave Symposium Digest, Vol.2*, pp. 839 – 842, ISSN 0149-645X

Aflaki, P.; Negra, R. and Ghannouchi, F. M. (2009). Enhanced Architecture for Microwave Current Mode Class-D Amplifiers Applied to the Design of an S-Band GaN-Based PA, *IET Microwave Antenna & Propagation, Vol.3, No. 6, pp.997–1006, Sep. 2009*, doi:10.1049/iet-map.2008.0282

Raab, F.H.; Asbeck, P.; Cripps, S.; Kenington, P.B.; Popovic, Z.B.; Pothecary, N.; Sevic, J.F.; Sokal, N.O. (2002). Power amplifiers and transmitters for RF and microwave, *IEEE Transactions on Microwave Theory and Techniques, Vol. 50, No. 3, March 2002*, pp. 814 - 826, ISSN 0018-9480

Lee, M. W; Lee, Y. S.; Kam, S. H.; Jeong, Y. H. (2010), A wideband digital predistortion for highly linear and efficient GaN HEMT Doherty Power Amplifier, *Microwave and Optical Technology Letter, Volume 52, Issue 2, February 2010*, pp. 484–487, DOI: 10.1002/mop.24951

Smorchkova, I. P.; Chen, L. ; Mates, T.; Shen, L.; Heikman, S.; Moran, B.; Keller, S.; DenBaars, S. P.; Speck, J. S. and Mishra, U. K. (2001). AlN/GaN and (Al,Ga)N/AlN/GaN two-dimensional electron gas structures grown by plasma-assisted molecular-beam epitaxy, *Journal of Applied Physics, vol. 90, no. 10*, pp. 5196–5201, Nov. 15, 2001. doi:10.1063/1.1412273

Mishra, U.K.; Shen Likun; Kazior, T.E.; Yi-Feng Wu (2008). GaN-Based RF Power Devices and Amplifiers, *Proceedings of the IEEE, Volume: 96 Issue:2, Feb. 2008* pp. 287 – 305, ISSN 0018-9219

Kikkawa, T.; Maniwa, T.; Hayashi, H.; Kanamura, M.; Yokokawa, S.; Nishi, M.; Adachi, N.; Yokoyama, M.; Tateno, Y.; Joshin, K. (2004). An Over 200-W Output Power GaN HEMT Push-Pull Amplifier with High Reliability, *2004 IEEE MTT-S International Microwave Symposium Digest, Vol.3*, pp. 1347 – 1350, 6-11 June 2004, ISSN 0149-645X

Negra, R.; Chu, T.D.; Helaoui, M.; Boumaiza, S.; Hegazi, G.M.; Ghannouchi, K. (2007). Switch-based GaN HEMT model suitable for highly-efficient RF power amplifier design, *IEEE/MTT-S International Microwave Symposium, 2007*, pp. 795 – 798, 3-8 June 2007, Honolulu, HI, USA, ISSN 0149-645X

David Schmelzer and Stephen I. Long (2007). A GaN HEMT Class F Amplifier at 2 GHz with > 80 % PAE, *Compound Semiconductor Integrated Circuit Symposium, 2006. CSIC 2006. IEEE*, pp. 96 - 99 San Antonio, TX, Nov. 2006, ISBN 1-4244-0126-7

Loren Betts; Dylan T. Bespalko and Slim Boumaiza (2011). Application of Agilent's PNA-X Nonlinear Vector Network Analyzer and X-Parameters in Power Amplifi er Design, Agilent Technologies White Paper, May 12, 2011

Monprasert, G.; Suebsombut, P.; Pongthavornkamol, T. and Chalermwisutkul, S. (2009). 2.45 GHz GaN HEMT Class-AB RF power amplifier design for wireless communication systems, *Proceedings of the 2010 International Conference on Electrical Engineering/Electronics Computer Telecommunications and Information Technology (ECTI-CON)*, pp. 566 – 569, Chiangmai, Thailand, 19-21 May 2010, ISBN 978-1-4244-5606-2

Khansalee, E.; Puangngernmak, N. and Chalermwisutkul, S. (2010). A high efficiency VHF GaN HEMT class E power amplifier for public and homeland security applications, 2010 *Asia-Pacific Microwave Conference Proceedings (APMC)*, pp. 437 - 440 Yokohama, Japan ,7-10 Dec. 2010 ISBN 978-1-4244-7590-2

Florian Pivit; Jan Hesselbarth; Georg Fischer and Suramate Chalermwisutkul (2008), Radio Frequency Transmitter, Pub. No.: WO/2009/062847 International Application No.: PCT/EP2008/064659, International Filing Date: 29.10.2008

Part 3

Channel Estimation and Capacity

8

Indoor Channel Measurement for Wireless Communication

Hui Yu and Xi Chen
Shanghai Jiao Tong University
China

1. Introduction

In the past few years, Multiple Input Multiple Output (MIMO) system received lots of attentions, since it is capable in providing higher spectrum efficiency, as well as better transmission reliability. This improvement is brought by the multiple antennas at both sides of transmission. Since there are additional parallel sub-channels in spatial domain, system can not only increase the reliability by spatial diversity technology, but also provide higher data rate utilizing spatial multiplexing (see [1],[2], etc.). Orthogonal Frequency Division Multiplexing (OFDM) can also accommodate high data rate requirement by providing frequency multiplexing gain. To fully utilize the benefits of both technologies, the combination of the above two, MIMO-OFDM, has been employed in many wireless communication systems and protocols, such as WLAN [3] and LTE systems [4].

The introduction of MIMO-OFDM raises plenty challenges in channel estimation and measurement. Transmitted signals are reflected and scattered, resulting in a multipath spread in the received signals. Moreover, the transmitters, receivers, and reflecting or scattering objects are moving, which means that channels change rapidly over time [5]. Inter-Channel interference may also bring a destructive effect in transmission, which should be cancelled by accurate channel measurement or estimation.

As an important application of MIMO-OFDM technology, WLAN is suitable in providing high data rate service in hotspot area, such as office buildings, airports, libraries, stations, hospitals and restaurants. This means lots of MIMO-OFDM applications (such as WLAN) take place in indoor situations, where both transmitters and receivers are surrounded by mobile and static scatters. Different from outdoor scenario, there are some unique characteristics of indoor scenario. More scatters result in larger influence by multipath effect; higher density of users and overlap between different access points bring larger interference. Because of these differences in channel parameters, it is crucial to obtain a better understanding of indoor channels. Statistics such as delay spread, Doppler spread, angle spread and path loss must be estimated by detail channel measurement. This requirement rises the interests in indoor channel measurement in the past few years.

Channel measurement or estimation schemes can be divided into two major categories, blind and non-blind. Blind channel estimation method requires large data and can only exploit the statistical behavior of channel, hence, suffers a lot in fast fading channels.

On contrast, non-blind method utilizes pre-determined information in both transmitters and receivers, to measure the channel. One of the most frequently used methods is the pilot/data aided channel measurement/estimation. The method can be further divided into two sub-methods considering the resources occupied by pilots with each resource unit. In the first sub-method, pilots occupy the whole resource unit, for example, an OFDM training symbol. This type of pilot arrangement is largely used for pure purpose of channel measurement without the request of communication. In real-time communication, it is only suitable for slow channel variation and for burst type data transmission schemes. In the second sub-method, pilots only occupy part of the resource unit; the other part of the unit is allocated to data. This pilot arrangement is capable to provide real-time communication which takes place in time and frequency varying channels. However, a linear interpolation or higher order polynomial fitting should be applied to recover the whole channel, which will certainly cause some errors.

Recently, there are plenty works considering indoor channel measurement and estimations, with non-blind pilot/data aided method, each of which focuses on different aspects of the issue. In [6], authors introduced a detail design of a MIMO channel sounder. In their measurement process, they used a PN sequence as the probing signal (pilot), which occupy the whole frequency and time resources. Their measurement took place in Seoul railway station, and they used the results to illustrate the characteristics of delay and Doppler spread of indoor channel. In [7], authors provide a PC-FPGA design in solving a similar problem for WLAN system. Instead of occupying the whole channel resource, the PN sequences only insert in certain parts of the resource unit. In this way, channel measurement and transmission can be carried out simultaneously. The effect of polarized antennas has also been considered in [8]. Wireless situations include non-line-of-sight, propagation along the corridor and propagation over a metallic ceiling.

Other important applications include several scenarios such as: near-ground indoor channel aiming military or emergency usage [9], and indoor channel model for wireless sensor network and internet of things [10]. Pilot signal design is flexible. Instead of a PN sequence, other pseudolite signals are available, too. Also, the kinds of carrier signals are variable, such as an OFDM signal [7] or a GPS-based signal [11].

In addition, some rough estimations of channel parameters are provided. Most of these works based on the assumption that indoor channels follow the Ricean distribution. The most important parameter of Ricean distribution for indoor channels is the K-factor, which represents the ratio between the average power of deterministic and random components of the channel. In [12], a two-moment method of the Ricean K-factor is provided theoretically. Experimental results can be found in [13], which gives an application for the two-moment estimation of the Ricean K-factor in wideband indoor channels at 3.7 GHz. Although the Ricean distribution can only provide an unclear view of the channel, it is convenient and low-complexity estimation; thus can be applied in scenarios that require only partial channel state information.

2. Measurement schemes

As is introduced before, channel measurement schemes can be divided into two major categories, blind and non-blind. Since blind measurement and estimation is much less reliable, we only discuss non-blind pilot/data aided schemes. Two of the most frequently used schemes are discussed: measurement based on PN sequence which occupies a whole resource unit, and measurement based on OFDM pilot who occupies only part of an OFDM symbol.

2.1 Indoor channel model

The low-pass time-variant channel impulse response (CIR) is denoted as $h(\tau;t)$, which represents the response of the channel at time t due to an impulse applied at time $t-\tau$. Then transmission can be expressed as:

$$y(t,\tau) = \sum_{n=0}^{N_{CIR}-1} h(n,\tau)x(t-n,\tau)+w(t,\tau)_k \qquad (1)$$

Where x is the transmit signal, and y is the transmit signal, N_{CIR} is the duration of the CIR, w is the noise sequence.

By taking the Fourier transform of $h(\tau;t)$, we can obtain the time-variant channel transfer function

$$H(f;t) = \int_{-\infty}^{+\infty} h(\tau;t)e^{-j2\pi f\tau}d\tau \qquad (2)$$

On the assumption that the scattering of the channel at two different delays is uncorrelated, the autocorrelation function of $h(\tau;t)$ can be defined as:

$$\frac{1}{2}E\left[h^*(\tau_1;t)h(\tau_2;t+\Delta t)\right] = \varphi_h(\tau_1;\Delta t)\delta(\tau_1-\tau_2) \qquad (3)$$

If we let $\Delta t = 0$, the resulting autocorrelation function $\varphi_h(\tau) \equiv \varphi_h(\tau;0)$ is called delay power profile of the channel. The range of values of τ over which $\varphi_h(\tau)$ is essentially nonzero is called the multipath spread of the channel. Then the scattering function of the channel is defined as:

$$S(\tau;\lambda) = \int_{-\infty}^{+\infty} \varphi_h(\tau;\Delta t)e^{-j2\pi\lambda\Delta t}d\Delta t \qquad (4)$$

By taking the integration of $S(\tau;\lambda)$, we obtain the Doppler power spectrum of the channel as:

$$S(\lambda) = \int_{-\infty}^{+\infty} S(\tau;\lambda)d\tau \qquad (5)$$

The range of values of λ over which $S(\lambda)$ is essentially nonzero is called the Doppler spread of the channel.

Indoor channel conditions are much more complex than that of outdoor channel. There are plenty kinds of scattering figures, such as walls, tables, etc. People indoor can also act as scattering figures, and the movements of cell phones caused by this bring about worse channel conditions. As a result, angle of arrival, multipath spread and scattering factor of indoor channels are different from those of outdoor channels. Consequently, measurement schemes should be redesign carefully according to the above distingue characteristics, so as to to fulfill the needs of indoor channel measurements.

It is worth noticing that channels of phone calls made in indoor conditions are mixtures of both indoor and outdoor channels. Large scale fading, small scale fading and shadow fading should be equally considered in such channels. Each of these factors can cause a large channel capacity decrease.

2.2 Channel measurement using PN sequence

In this scheme, Pseudo-Noise (PN) Sequence is used as a probing signal. To authors' best knowledge, the most widely used binary PN sequence is the Maximum-Length-Shift-Register (MLSR) sequence. The length of MLSR sequence is $N_{PN} = 2^m - 1$ bits. And one of the possible generators of this sequence is an m-stage linear feedback shift register (see [14]). As a result, MLSR sequence is periodic with period n. Within each period, there are 2^{m-1} ones and 2^{m-1} zeros.

One of the most important characteristics of the periodic PN sequence is its sharp auto-correlation. Consider an PN sequence x_k, we have :

$$\sum_{k=0}^{N_{PN}-1} x_{k+m}x_k = \begin{cases} N_{PN}, & m = 0, \pm N_{PN}, \pm 2N_{PN}, \ldots \\ 0, & \text{Others} \end{cases} \tag{6}$$

If $N_{PN} \gg 1$, it is approximate that:

$$\frac{1}{N_{PN}} \sum_{k=0}^{N_{PN}-1} x_{k+m}x_k \approx \sum_{i=-\infty}^{+\infty} \delta_{m-iN_{PN}} \tag{7}$$

We can represent the received signal y_k as the convolution of transmitted signal x_k and the CIR h_k as:

$$y_k = \sum_{n=0}^{N_{CIR}-1} h_n x_{k-n} + w_k \tag{8}$$

The transmitter use PN sequence as the transmitting data x_k. If the generators of PN sequence in both transmitter and receiver are synchronous, at each time slot, receiver is aware of the transmitting PN sequence. Hence, receiver can obtain the CIR by doing a cross-correlating between y_k and x_k :

$$\hat{h}_k = N_{\text{PN}} \sum_{m=0}^{N_{\text{PN}}-1} y_{m+k} x_m$$

$$= \frac{1}{N_{\text{PN}}} \sum_{m=0}^{N_{\text{PN}}-1} \left(\sum_{n=0}^{N_{\text{CIR}}-1} h_n x_{m+k-n} + w_{m+k} \right) x_m$$

$$= \sum_{n=0}^{N_{\text{CIR}}-1} h_n \frac{1}{N_{\text{PN}}} \sum_{m=0}^{N_{\text{PN}}-1} x_{m+k-n} x_m + \frac{1}{N_{\text{PN}}} \sum_{k=0}^{N_{\text{PN}}-1} w_{m+k} x_m \qquad (9)$$

$$= \sum_{n=0}^{N_{\text{CIR}}-1} h_n \sum_{i=-\infty}^{+\infty} \delta_{k-n-iN_{\text{PN}}} + \frac{1}{N_{\text{PN}}} \sum_{k=0}^{N_{\text{PN}}-1} w_{m+k} x_m$$

$$= \sum_{i=-\infty}^{+\infty} h_{k-iN_{\text{PN}}} + w'_m$$

It can be seen that the result of cross-correlating is a summation of noise w' and the periodic extension of h_k. If $N_{\text{CIR}} \le N_{\text{PN}}$, a period of $\hat{h}_k$ can be used as the estimate of h_k [7].

There are two drawbacks in PN sequence measurement. First, the PN sequence takes up a great amount of time and frequency resources (most of the time, all the transmitting resources). This results in a great loss of throughput, as well as a channel mismatch caused by the delay between channel measurement and data transmission. Second, the accuracy of synchronizers in both sides should be in a high level, which raises the cost of equipments for channel measurement.

2.3 Channel measurement using OFDM pilot

Unlike measurement based on PN sequence, the pre-determined data of measurement based on OFDM pilot only occupied a relatively small percentage of time and frequency resources. Channel segments located on the pilots can be measured directly and correctly. However, other channel segments can only be estimated indirectly with some interpolations. Although the pilot-based measurement can only give an imperfect result, it provides a possibility of transmitting data and measuring channel simultaneously. This characteristic is crucial for real systems with limited feedback, such as WLAN, WiMax, LTE and LTE-A systems. There are two major problems in this scheme: how to design the pilot pattern and how to interpolate with discrete channel value on both time and frequency domain.

2.3.1 OFDM pilot pattern

OFDM pilots may be inserted in both time and frequency resources. A pilot pattern refers to the places where pilots are inserted in every OFDM symbol. An effective pilot pattern needs to be designed carefully in both frequency and time domains.

In frequency domain, according to the Nyquist sampling theorem, if we want to capture the variation of channel, the frequency space D_f between pilots should be small enough:

$$D_f \le \frac{1}{\tau_{\max} \Delta f} \qquad (10)$$

where τ_{max} represents the maximum delay of channel, and Δf denotes the frequency space between subcarriers.

If the above condition is not satisfied, the channel estimation cannot sample the accurate channel, since channel fading in frequency cannot be detected fast enough.

In time domain, spacing between pilots inserted in the same frequency is determined by the coherence time. In order to capture the variation of channel, the time space of pilots D_t must be correlated with coherence time, and must be small enough:

$$D_t \leq \frac{1}{2 f_{doppler} T_f} \tag{11}$$

Where $f_{doppler}$ represents the maximum Doppler spread of channel, and T_f denotes the duration of each OFDM symbol.

However, it is worth pointing out that pilot allocation is a tradeoff of many factors in real systems. These include channel estimation accuracy, spectral efficiency of the system, wasted energy in unnecessary pilot symbols, and fading process not being sampled sufficiently. As a result, there is no optimal pilot pattern for all the channels, as fading process are varied.

Another important element of pilot pattern is the power allocation. Power is equally allocated to pilots and data symbols in regular cases. However, the accuracy of channel estimation can increase greatly with the power allocated to pilots. Considering the total power constraint, this will result in a decline of data symbols' SNR. Hence, another tradeoff between channel estimation and transmission capacity has to be evaluated.

There is a lack of pilots at the edges of OFDM symbols, which leads to a much higher error rate in such places. One simple but less effective way is to place more pilots at the edge. The drawback of this scheme is obvious: it reduces the frequency efficiency of systems. Some other scheme utilizes periodic behavior of the Fourier Transform, and establishes certain correlations between the beginning and the end of OFDM symbols. Simulations are reported to verify the effectiveness of this scheme.

2.3.2 Measurements on pilots

Channel segments locating on the pilots can be measured directly by some well-known algorithm, such as Least Square (LS) and Linear Minimum Mean Square Error (LMMSE).

Both LS and LMMSE algorithm aim to minimize a parameter: $\min\{(y_k - x_k h_k)^H (y_k - x_k h_k)\}$.

Using LS algorithm, we have:

$$\tilde{h}_k^{LS} = x_k^{-1} y_k = h_k + x_k^{-1} w_k = [\frac{y_k^1}{x_k^1}, \frac{y_k^2}{x_k^2}, \cdots, \frac{y_k^{M_k}}{x_k^{M_k}}]^T \tag{12}$$

where M_k is the length of transmit and receive signal.

Using LMMSE algorithm, we have:

$$\tilde{h}_k^{LMMSE} = \mathbf{R}_{H_k H_k} (\mathbf{R}_{H_k H_k} + \frac{\beta}{SNR} \mathbf{I})^{-1} \tilde{H}_k^{LS} \tag{13}$$

where $\mathbf{R}_{H_k H_k} = E\{H_k H_k^*\}$ autocorrelation of channel. and $\beta = E\left\{|x_k|^2\right\} E\left\{1/|x_k|^2\right\}$.

Comparing both LS algorithm and LMMSE algorithm, we can draw the following conclusion. LS algorithm is much easier to realize and apply. LS algorithm only needs one discrete divider to estimate the channels on all the pilots. Moreover, statistical information about channel and noise are not necessary while employing LS algorithm. However, the accuracy of LS algorithm is very sensitive to the noise and synchronization errors.

On the other hand, LMMSE algorithm can be seen as a filtering on the estimation result on LS algorithm. And this filtering is based on the MMSE criteria. It can be proven that, under the same MSE conditions, estimations results of LMMSE algorithm provide a larger gain than that of LS algorithm. Drawbacks of LMMSE algorithm are also obvious. Due to the inversion operation of matrices, complexity of LMMSE algorithm is relatively high. Furthermore, LMMSE algorithm requires knowing the statistical information of channel and noise in prior, which is unrealistic in applications.

While taking the errors of estimated $\mathbf{R}_{H_k H_k}$ into account, the MSE and SNR gains provided by LMMSE algorithm are marginally larger than that of LS algorithm. Considering the tradeoff between complexity and performance, LS algorithm may be a better solution than LMMSE algorithm.

2.3.3 Interpolations

By applying LS or LMMSE algorithm, one can easily obtain the CIR of piloted channel segments. However, we still have no idea of channel segments not occupied with pilots. In order to obtain CIR of these segments, interpolations should be used.

The best interpolation algorithm may be 2-D Wiener filtering, since it can cancel noise as much as possible, in both frequency and time domain. However, in order to decide the weights of Wiener filtering, channel statistics must be known. Moreover, the complexity brought by matrix inversion increases gigantically with data in pilots. All of the above prevent the usage of 2-D Wiener filtering in real systems.

Some achievable interpolations include: cascade 1-D Wiener filtering, Lagrange interpolation, and transform domain interpolation.

Cascade 1-D Wiener filtering tries to realize a 2-D Wiener filtering by cascading two 1-D Wiener filtering. The complexity of cascade 1-D Wiener filtering is much less than 2-D Wiener filtering, while the performance only decreases a little bit. There two kinds of cascade 1-D Wiener filtering, in respects of interpolation order of frequency and time domain.

In frequency domain, the major interpolations include Lagrange interpolation, LMMSE interpolation, transform domain interpolation, DFT based interpolation, and low-pass filtering interpolation. In time domain, the available schemes are LMMSE interpolation, Lagrange interpolation.

2.3.3.1 Two dimensions LMMSE interpolations

Assume that the estimated channel matrix of all the piloted subcarriers is $\tilde{H}_{n',i'}, \forall \{n',i'\} \in P$, where P is the set of positions of all the pilots, n' is the index in frequency domain, i' is the index in time domain. We have:

$$\tilde{H}_{n',i'} = \frac{Y_{n',i'}}{X_{n',i'}} = H_{n',i'} + \frac{N_{n',i'}}{X_{n',i'}}, \forall \{n',i'\} \in P \tag{14}$$

Then, estimate the channel parameters by two dimensions interpolation filtering:

$$\hat{H}_{n,i} = \sum_{\{n',i'\} \in \Gamma_{n,i}} w_{n',i',n,i} \tilde{H}_{n',i'} \quad, \Gamma_{n,i} \subseteq P \tag{15}$$

Where $w_{n',i',n,i}$ are the weights of interpolation filter, $\hat{H}_{n,i}$ is the estimated channel, $\Gamma_{n,i}$ is the number of used pilots. The number of weights in the filter is $N_{tap} = \|\Gamma_{n,i}\|$.

Applying MSE criteria, the MSE $J_{n,i}$ in subcarrier (n,i) is:

$$\varepsilon_{n,i} = H_{n,i} - \hat{H}_{n,i}$$
$$J_{n,i} = E\{|\varepsilon_{n,i}|^2\} \tag{16}$$

A filter following the MMSE criteria is a two dimensions Wiener filter. According to the orthogonality of such a filter, we have:

$$E\{\varepsilon_{n,i} \tilde{H}_{n'',i''}^*\} = 0, \forall \{n'',i''\} \in \Gamma_{n,i} \tag{17}$$

Where (n'',i'') represents the positions of pilots while channel estimation is conducted.

The Wiener-Hopf equation can be derived from (17), which follows:

$$E\{H_{n,i} \tilde{H}_{n'',i''}^*\} = \sum_{\{n',i'\} \in \Gamma_{n,i}} w_{n',i',n,i} E\{\tilde{H}_{n',i'} \tilde{H}_{n'',i''}^*\}, \forall \{n'',i''\} \in \Gamma_{n,i} \tag{18}$$

Define the correlation function as:

$$\theta_{n-n'',i-i''} = E\{H_{n,i} \tilde{H}_{n'',i''}^*\} \tag{19}$$

If the mean value of noise is zero, and is independent to the transmission data, the correlation can be further expressed as:

$$\theta_{n-n'',i-i''} = E\{H_{n,i} \tilde{H}_{n'',i''}^*\} \tag{20}$$

Define the right part of (18) as the autocorrelation of channels at pilots:

$$\varphi_{n'-n'',i'-i''} = E\{\tilde{H}_{n',i'} \tilde{H}_{n'',i''}^*\} \tag{21}$$

According to (14), it follows:

$$\varphi_{n'-n'',i'-i''} = \theta_{n'-n'',i'-i''} + \frac{\sigma^2}{E\left\{\left|X_{n',i'}\right|^2\right\}}\delta_{n'-n'',i'-i''}$$

$$= \theta_{n'-n'',i'-i''} + \frac{1}{SNR}\delta_{n'-n'',i'-i''}$$

(22)

Where $E\left\{\left|X_{n',i'}\right|^2\right\}$ is the average power of pilot symbols.

Equation (22) shows that the correlation function depends on the distance between the position of channel being estimated (n,i) and the positions of pilots employed in the estimation process (n'',i''). And the autocorrelation function depends on the distances between pilots.

Substituting (20) and (21) into (18), we have:

$$\boldsymbol{\theta}_{n,i}^T = \mathbf{w}_{n,i}^T \boldsymbol{\Phi}$$

(23)

Where $\boldsymbol{\Phi}$ is the $N_{tap} \times N_{tap}$ autocorrelation matrix, $\boldsymbol{\theta}_{n,i}$ is the correlation vector with length N_{tap}, and $\mathbf{w}_{n,i}$ is the parameter vector of filter with length N_{tap}. Therefore, the parameter of the 2-D Wiener filter is:

$$\mathbf{w}_{n,i}^T = \boldsymbol{\theta}_{n,i}^T \boldsymbol{\Phi}^{-1}$$

(24)

The full estimated channel matrix can be expressed as:

$$\hat{H}_{n,i} = \mathbf{w}_{n,i}^T \tilde{H}_{n,i}$$

(25)

In conclusion, the design of such a filter is to decide its parameters $\mathbf{w}_{n,i}$, which can be derived by the correlation function $\theta_{n-n'',i-i''}$ and average SNR. Unfortunately, the correlation of channel cannot be achieved accurately in real systems. Hence, approximate models with typical multipath delay profile $\rho(\tau)$ and Doppler power spectrum $S_{f_D}(f_D)$ are employed.

2-D Wiener filtering is the optimal interpolation scheme in respect of MMSE; it can obtain optimal performance theoretically. However, its requirement of prior statistic knowledge of channel matrix, as well as the complexity of matrix inversion, makes it almost impossible to apply in real systems.

2.3.3.2 Interpolations in frequency domain

A 2-D interpolation can be form by a cascade of two 1-D interpolations. By appropriate designs, such a cascade can largely reduce the complexity while maintaining a good performance. The order of interpolation should be taken into consideration. We propose to interpolate firstly in frequency domain, then to conduct the time domain interpolation. The reason is that frequency-time scheme can start once the piloted OFDM symbols receive, while time-frequency scheme has to wait for the arrivals of all the symbols in one frame or subframe before the interpolation begins. As a result, frequency-time scheme can decrease the delay of channel measurement, hence provides more effective interpolation.

Interpolations in frequency domain aim to obtain all the channel function respond (CFR) $\hat{H}_C$, according to the measured CFR $\tilde{H}_p$ in each piloted subcarrier:

$$\hat{H}_C = \mathbf{w} \cdot \tilde{H}_p \tag{26}$$

where $\mathbf{w}$ is the frequency domain interpolation matrix, and is the channel vector of piloted subcarriers obtained by (12) or (13). We have:

$$\tilde{H}_p = \mathbf{X}_{pp}^{-1} Y_p = H_p + \mathbf{X}_{pp}^{-1} N_p = H_p + \tilde{n} \tag{27}$$

2.3.3.2.1 LMMSE interpolation

Projecting equation (23) on frequency domain, we obtain the optimal interpolation parameter vector $\mathbf{w}$ as:

$$\mathbf{R}_{\tilde{H}_p \tilde{H}_p} \mathbf{w}^* = \mathbf{R}_{\tilde{H}_p H_C} \tag{28}$$

where $\mathbf{R}_{\tilde{H}_p \tilde{H}_p} = E\{\tilde{H}_p \tilde{H}_p^*\}$ represents the autocorrelation matrix of estimation channel segments with pilots $\tilde{H}_p$, $\mathbf{R}_{\tilde{H}_p H_C} = E\{\tilde{H}_p H_C^*\}$ denotes the correlation matrix of estimation channel segments with pilots $\tilde{H}_p$ and the real channel being interpolated H_C.

If $\mathbf{R}_{\tilde{H}_p \tilde{H}_p}$ is invertible, $\mathbf{w}$ can be expressed as:

$$\mathbf{w} = \mathbf{R}_{H_C \tilde{H}_p} \mathbf{R}_{\tilde{H}_p \tilde{H}_p}^{-1} \tag{29}$$

Combining (12) and (27), it follows:

$$\begin{aligned} \mathbf{R}_{\tilde{n}\tilde{n}} &= E\{\tilde{n}\tilde{n}^*\} = E\{\mathbf{X}_p^{-1} N_p \cdot N_p^* (\mathbf{X}_p^{-1})^*\} \\ &= \sigma_n^2 E\{\mathbf{X}_p^{-1} \cdot \mathbf{I} \cdot (\mathbf{X}_p^{-1})^*\} = \sigma_n^2 E\{(\mathbf{X}_p^* \mathbf{X}_p)^{-1}\} \\ &= \mathbf{I}_{N_p} E\{1 / |X_k^p|^2\} \sigma_n^2 \end{aligned} \tag{30}$$

where X_k^p is the constellation point of piloted channel, and σ_n^2 is the power of noise.

Substituting (12) into (29), we have:

$$\mathbf{w} = \mathbf{R}_{H_C H_p} (\mathbf{R}_{H_p H_p} + \mathbf{I}_{N_p} E\{1 / |X_k^p|^2\} \sigma_n^2)^{-1} \tag{31}$$

where $\mathbf{R}_{H_C H_p}$ and $\mathbf{R}_{H_p H_p}$ are the ideal correlation and autocorrelation matrices. Further representing the CFRs of channels with their CIRs, it is obvious that $H_C = \mathbf{F}_{CL} h_L$ and $H_P = \mathbf{F}_{PL} h_L$ (where h_L is the discrete CIR, $\mathbf{F}_{CL}$ and $\mathbf{F}_{PL}$ are the corresponding DFT transform matrices). This converts (31) as followed:

$$\mathbf{w} = \mathbf{F}_{CL}\mathbf{R}_{h_Lh_L}\mathbf{F}_{PL}^*(\mathbf{F}_{PL}\mathbf{R}_{h_Lh_L}\mathbf{F}_{PL}^* + \mathbf{I}_{N_P}E\{1/\left|X_k^p\right|^2\}\sigma_n^2)^{-1} \tag{32}$$

Applying Parseval Theorem, which certifies that the power in frequency domain equals that of time domain, we draw the following conclusion:

$$\mathbf{w} = \mathbf{F}_{CL}\mathbf{\bar{R}}_{h_Lh_L}\mathbf{F}_{PL}^*(\mathbf{F}_{PL}\mathbf{\bar{R}}_{h_Lh_L}\mathbf{F}_{PL}^* + \mathbf{I}_{N_P}\frac{E\{1/\left|X_k^p\right|^2\}\sigma_n^2}{trace(\mathbf{R}_{h_Lh_L})})^{-1}$$

$$= \mathbf{F}_{CL}\mathbf{\bar{R}}_{h_Lh_L}\mathbf{F}_{PL}^*(\mathbf{F}_{PL}\mathbf{\bar{R}}_{h_Lh_L}\mathbf{F}_{PL}^* + \mathbf{I}_{N_P}\frac{E\{1/\left|X_k^p\right|^2\}\sigma_n^2}{E(\mathbf{H}_C^*\mathbf{H}_C)})^{-1} \tag{33}$$

$$= \mathbf{F}_{CL}\mathbf{\bar{R}}_{h_Lh_L}\mathbf{F}_{PL}^*(\mathbf{F}_{PL}\mathbf{\bar{R}}_{h_Lh_L}\mathbf{F}_{PL}^* + \mathbf{I}_{N_P}\frac{E\{1/\left|X_k^p\right|^2\}E\{\left|X_k^p\right|^2\}\sigma_n^2}{E\{\left|X_k^p\right|^2\}E(\mathbf{H}_C^*\mathbf{H}_C)})^{-1}$$

In special cases where QPSK modulation and equal power allocation are adopted, the interpolation is:

$$\mathbf{w} = \mathbf{F}_{CL}\mathbf{\bar{R}}_{h_Lh_L}\mathbf{F}_{PL}^*(\mathbf{F}_{PL}\mathbf{\bar{R}}_{h_Lh_L}\mathbf{F}_{PL}^* + \mathbf{I}_{N_P}\frac{1}{SNR})^{-1} \tag{34}$$

where $SNR = \dfrac{P_r}{\sigma_n^2} = \dfrac{E\{\left|X_k\right|^2\}E\{\mathbf{H}_C^*\mathbf{H}_C\}}{\sigma_n^2} = \dfrac{E\{\left|X_k^p\right|^2\}E\{\mathbf{H}_C^*\mathbf{H}_C\}}{\sigma_n^2}$.

2.3.3.2.2 Lagrange interpolation

Lagrange interpolation is widely used and easy to implement. It is a group of interpolation algorithms, including linear interpolation, Gaussian interpolation, cubic interpolation, etc. Lagrange interpolation is suitable for both frequency and time domain interpolation. However, the disadvantage of Lagrange interpolation is obvious. It is unable to cancel the noise.

Linear interpolation in frequency domain utilizes each pair of adjacent piloted channel segments to obtain the channel function within them. The interpolation process follows the following equation:

$$\hat{H}(l+d) = \left(1 - \frac{d}{D}\right)\tilde{H}_p(l) + \frac{d}{D}\tilde{H}_p(l+D), 1 \le d \le D-1 \tag{35}$$

where D is the interval between two adjacent pilots, $\tilde{H}_p(l)$ and $\tilde{H}_p(l+D)$ are the corresponding channel estimation results.

Gaussian interpolation in frequency domain employs the measured channels of three adjacent pilots, which can be represented as followed:

$$
\hat{H}(x) = \begin{cases}
\tilde{H}_p(l_{j-1})\dfrac{x-l_j}{l_{j-1}-l_j}\dfrac{x-l_{j+1}}{l_{j-1}-l_{j+1}} + \tilde{H}_p(l_j)\dfrac{x-l_{j-1}}{l_j-l_{j-1}}\dfrac{x-l_{j+1}}{l_j-l_{j+1}} \\[2mm]
+\tilde{H}_p(l_{j+1})\dfrac{x-l_{j-1}}{l_{j+1}-l_{j-1}}\dfrac{x-l_j}{l_{j+1}-l_j}, (l_j \neq K_{max}) \\[2mm]
\tilde{H}_p(l_{j-2})\dfrac{x-l_{j-1}}{l_{j-2}-l_{j-1}}\dfrac{x-l_j}{l_{j-2}-l_j} + \tilde{H}_p(l_{j-1})\dfrac{x-l_{j-2}}{l_{j-1}-l_{j-2}}\dfrac{x-l_j}{l_{j-1}-l_j} \\[2mm]
+\tilde{H}_p(l_j)\dfrac{x-l_{j-2}}{l_j-l_{j-2}}\dfrac{x-l_{j-1}}{l_j-l_{j-1}}, (l_j = K_{max})
\end{cases}
\tag{36}
$$

Where K_{max} denotes the maximum position of pilots, $\tilde{H}_P(l_{j-1})$, $\tilde{H}_P(l_j)$, and $\tilde{H}_P(l_{j+1})$ are the channel measurement results of three used pilots, and $l_{j-1} < x < l_j$.

Cubic interpolation further increase the number of used pilots onto four. The expression for interpolation is showed as followed:

$$
\hat{H}(x) = \tilde{H}_p(l_{j-2})\frac{x-l_{j-1}}{l_{j-2}-l_{j-1}}\frac{x-l_j}{l_{j-2}-l_j}\frac{x-l_{j+1}}{l_{j-2}-l_{j+1}} + \tilde{H}_p(l_{j-1})\frac{x-l_{j-2}}{l_{j-1}-l_{j-2}}\frac{x-l_j}{l_{j-1}-l_j}\frac{x-l_{j+1}}{l_{j-1}-l_{j+1}}
$$
$$
+\tilde{H}_p(l_j)\frac{x-l_{j-2}}{l_j-l_{j-2}}\frac{x-l_{j-1}}{l_j-l_{j-1}}\frac{x-l_{j+1}}{l_j-l_{j+1}} + \tilde{H}_p(l_{j+1})\frac{x-l_{j-2}}{l_{j+1}-l_{j-2}}\frac{x-l_{j-1}}{l_{j+1}-l_{j-1}}\frac{x-l_j}{l_{j+1}-l_j}
$$
$$
(l_j \neq K_{max}, l_j \neq d, l_{j-1} < x < l_j)
\tag{37}
$$

All the above schemes are simple to apply in real systems. However, they all introduce certain level of noise, and yield effect of error floor. This can be eliminated by employing a low pass filter after interpolation.

2.3.3.2.3 Transform domain interpolation

The basic idea of transform domain interpolation is to reduce the complexity by conducting interpolation in various transform domains. The most widely used kind of transform domain interpolation is based on DFT.

The fundamental principle of DFT based interpolation is: in process of signal processing, zeroizing in time domain is equivalent to interpolating in frequency domain. If a sequence of N points has $N - N_p$ zeros in the end, its Fourier transform values at the positions of multiples of N_p are the same as the counterparts of Fourier transform of sequence formed by the former N_p points. On the other hand, the Fourier transform values not at the positions of multiples of N_p consist of linear combinations of the Fourier transform of truncated sequence.

After receiving the information of piloted channels, DFT based interpolation conducts IFFT of length N_p. Then the interpolation zeroizes the transformed sequence into a N pointed sequence. Finally, transform the sequence into frequency domain by a N points FFT.

The zeroizing can be conducted as followed:

$$\hat{h}_N(i) = \begin{cases} \tilde{h}_{N_p}(i) & 0 \leq i < N_p/2 \\ 0 & N_p/2 \leq i < N - N_p/2 \\ \tilde{h}_{N_p}(i-(N-N_p)) & N - N_p/2 \leq i \leq N-1 \end{cases}$$

(38)

Considering the influences of noise and inter-channel interference, introducing a low pass filtering before zeroizing can effective increase the accuracy of channel measurement. The block diagram of the above procedure is showed in Fig. 1.

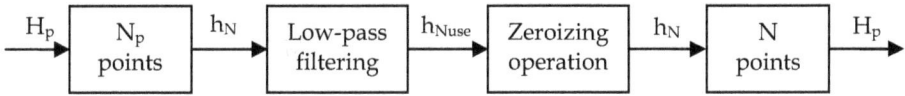

H_p → [N_p points] → h_N → [Low-pass filtering] → h_{Nuse} → [Zeroizing operation] → h_N → [N points] → H_p

Fig. 1. DFT and low-pass filter based interpolation.

Since the most complex calculations in this kind of interpolation are FFT and IFFT, the complexity of DFT based interpolation is much lower than others. However, the performance will drop largely, if the multipath spread is not a multiple of sampling period.

2.3.3.3 Interpolations in time domain

After frequency interpolation is done, we can launch the interpolation in time domain. The interpolation can also be expressed as a interpolation matrix as followed:

$$\hat{H}_{Ct} = \mathbf{w}_t \cdot \tilde{H}_{pt}$$

(39)

Where $\mathbf{w}_t$ is the time domain interpolation matrix, $\tilde{H}_{pt}$ denotes the CFR of channel segments on the pilots, $\hat{H}_{Ct}$ represents the CFR for all the channel segments. By assuming $\tilde{H}_{pt} = H_{pt} + \tilde{n}$, we consider the impact of AWGN in the interpolation.

2.3.3.3.1 LMMSE interpolation

LMMES interpolation in time domain is similar to that in frequency domain. The only difference is that we project equation (23) into time domain, so that it follows:

$$\mathbf{w}_t = \mathbf{R}_{H_{Ct}\tilde{H}_{pt}} \mathbf{R}_{\tilde{H}_{pt}\tilde{H}_{pt}}^{-1}$$

(40)

where $\mathbf{R}_{\tilde{H}_{pt}\tilde{H}_{pt}} = E\{\tilde{H}_{pt}\tilde{H}_{pt}^*\}$ represents the autocorrelation matrix of estimation channel segments with pilots $\tilde{H}_{Pt}$, $\mathbf{R}_{H_{Ct}\tilde{H}_{pt}} = E\{H_{Ct}\tilde{H}_{pt}^*\}$ denotes the correlation matrix of estimation channel segments with pilots $\tilde{H}_{Pt}$ and the real channel being interpolated H_{Ct}.

Following the steps of derivation for frequency LMMSE interpolation, the interpolation matrix of time LMMSE interpolation can be simplified as below:

$$\mathbf{w}_t = \mathbf{R}_{H_{Ct}H_{pt}} \left(\mathbf{R}_{H_{pt}H_{pt}} + \frac{\beta}{SNR} \mathbf{I}_{N_{pt}} \right)^{-1}$$

(41)

We consider a special case where QPSK modulation and average power allocation are used. Then the correlation between adjacent pilots i and i'' is:

$$R_{H_{C_t}H_{pt}}(i,i'') = J_0(2\pi f_{D_{max}}(i-i'')T_s')$$ (42)

Where $f_{D_{max}}$ is the maximum Doppler spread, T_s' is the interval between two symbols, and $J_0(x)$ represents the first zero-order Bessel function.

However, the previous frequency interpolation and the corresponding filtering cause changes in the noise power of each subcarrier. Therefore, the signal-to-noise-ratio of each channel segment no longer equals the original value. An adjustment was proposed based on the MSE after the frequency interpolation.

Assume that channel frequency response after frequency LMMSE interpolation is:

$$H_f^n = [H_0^{(n)} + \tilde{w}_0^{(n)}, \cdots, H_i^{(n)} + \tilde{w}_i^{(n)}, \cdots, H_{L-1}^{(n)} + \tilde{w}_{L-1}^{(n)}]^T$$ (43)

Where L is the number of OFDM symbols in each frame, $H_i^{(n)}$ represents the channel frequency response of ith OFDM symbol in nth subcarrier, $\tilde{w}_i^{(n)}$ denotes the corresponding residual noise. Hence, variance of $\tilde{w}_i^{(n)}$ is equivalent to the MSE after interpolation.

$$MSE_{LMMSE,n} = [R_{H_C H_C} - R_{H_C H_p}(R_{H_p H_p} + I_{N_p}E\{1/|X_k^p|^2\}\sigma_n^2)^{-1}R_{H_C H_p}^H]_{nn}, n = 0,1,2,\cdots,N-1$$ (44)

Where N is the number of subcarriers waiting for measured.

As a result, time domain LMMSE interpolation should be optimized according to the variance of noise. The interpolation matrix is then:

$$\mathbf{w}_t = R_{H_{C_t}H_{pt}}(R_{H_{pt}H_{pt}} + diag(\sigma_0^2, \cdots, \sigma_i^2, \cdots, \sigma_{L-1}^2))^{-1}$$ (45)

Where σ_i^2 represents the variance of residual noise $\tilde{w}_i^{(n)}$.

To reduce the complexity, channel segments in the same OFDM symbol can utilize their average noise variance in the interpolation, and an approximate interpolation matrix can be expressed as followed:

$$\mathbf{w}_t = R_{H_{C_t}H_{pt}}(R_{H_{pt}H_{pt}} + \frac{1}{N}\sum_{i=0}^{N-1}\sigma_0^2(i)I_L)^{-1}$$ (46)

2.3.3.3.2 Lagrange interpolations

Lagrange interpolations in time domain are almost the same as those in frequency domain. The only difference is channel response of piloted channel segments. One can refer to previous sections for details.

3. Applications

Here we show some useful and easily implemented examples to illustrate the indoor channel measurement. Measurement based on PN sequence, as well as OFDM pilot, will be discussed.

3.1 Channel measurement system using PN sequence

In this section, we present an example of 2x2 MIMO channel measurement, utilizing a semi-sequential scheme. This semi-sequential scheme uses parallel receivers and a switch at the transmitter (Fig. 2). When a measurement process starts, the probing signal is firstly transmitted from the 1st transmit antenna (TX1) and the receive signal is sampled from 1st and 2nd receive antennas (RX1 and RX2) simultaneously. Thus the channel from TX1 to RX1 and RX2 can be measured at the same time. Then, a similar process is used to measure the channel from TX2 to RX1 and RX2. Strictly speaking, the semi-sequential MIMO channel sounder measures Single Input Multiple Output (SIMO) channels directly. The MIMO channel is obtained by combine the two SIMO channels, on the assumption that the MIMO channel doesn't change significantly in a single round of sequential measurement.

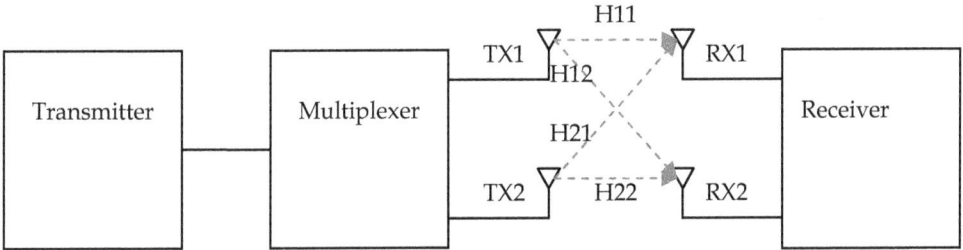

Fig. 2. Semi-sequential scheme of MIMO channel sounder [7].

Each SIMO channel can be measured by the algorithm introduced in Section 2.2 Then a combination of two SIMO channel construct the whole MIMO channel.

3.1.1 Baseband signal processing algorithm

3.1.1.1 System parameters

The link-level block diagram of the sliding correlation channel sounder for Single Input Single Output (SISO) channel is shown in Fig. 3. In the semi-sequential MIMO channel sounder (or SIMO channel sounder); there should be two parallel receivers.

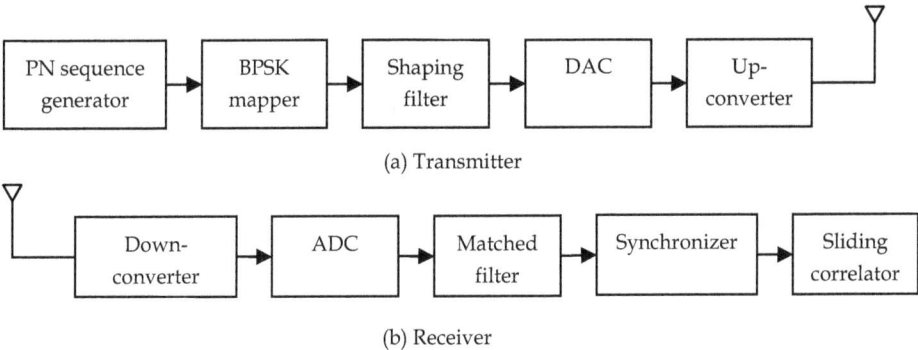

Fig. 3. Channel Sounder for SISO.

Here we only focus on the indoor wireless MIMO channel for WLAN like devices. The measurement system supports 20/40MHz bandwidth by suitable RFIC [16][17]. The system parameters of baseband are listed in Table 1.

Name	Symbol	Value
Sampling frequency	f_s	60MHz
Symbol rate of PN sequence	R_{symb}	20M Symbol/s
Period of PN sequence, express in units of T_{symb}	N_{PN}	127
Length of CIR, express in units of T_{symb}	N_{CIR}	127
Sampling interval	T_s	$1/f_s$
Interpolated sample interval	T_i	$T_{symb}/2$
Symbol interval of PN sequence	T_{symb}	$1/R_{symb}$

Table 1. System Parameters.

3.1.1.2 Symbol timing synchronizing algorithm

Symbol timing synchronizer is a critical module of the digital receiver design of the channel sounder based on sliding correlation channel measurement. Gardner's symbol timing recovery method is used in this system [18][19]. The structure of the symbol timing synchronizer is shown in Fig. 4. All the processing of this synchronizer is done in digital domain. No interaction between analog and digital part of the system is needed. This synchronizer is capable of compensating sampling phase and frequency offset and is independent of carrier phase [20].

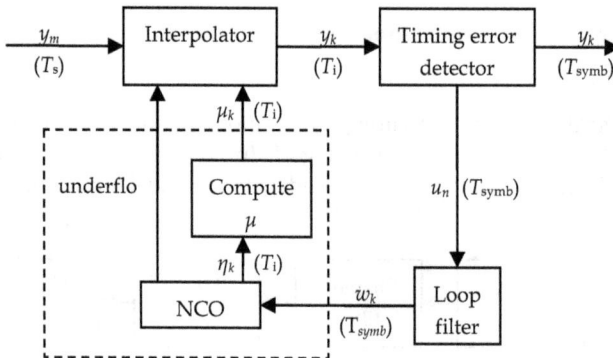

Fig. 4. Symbol timing synchronizer.

The sampled data y_m, which is filtered by matching filter, is then feed into the interpolator to compute the desired sampled strobe y_k. This is done by digital interpolation, controlled by NCO (Numerical Controlled Oscillator) and fraction interval μ_k. Ideally, the period of the NCO is $T_i = T_{symb}/K$, where K is an integer. The loop consisting of timing error detector, loop filter and NCO function just like a DPLL, where u_n, w_k and η_k represents timing error signal, NCO control word and NCO register content respectively.

In this design, the DTTL algorithm [20] is used to compute the timing error signal. This choice specifies $T_i = T_{symb} / 2$. In order to avoid up-sampling in the interpolator, T_s should be smaller than T_i. Thus, the sampling frequency f_s should be larger than two times the symbol rate R_{symb}. The interpolator performs linear interpolation, which is easy to implement. The loop filter is a proportional-plus-integral structure.

3.1.2 Hardware design

The whole measurement system hardware consists of several modules: antennas module, multi-channel AD/DA module, baseband processing FPGA board, USB access module and a computer Graphical User Interface (GUI) module. The architecture of hardware is showed in Fig. 5. The RF board is based on MAX2829, which can support MIMO operation. We choose the FFP board (IAF GmbH) as FPGA prototyping platform for baseband signal processing, RF control and interface to PC. The interface between the FFP board and the PC is an USB2.0 port. The GUI program runs on the computer for user to control the channel measurement functions and demonstrate the real time test results. Because the most effort is on the development of FPGA, here we focus on the design of baseband transceiver.

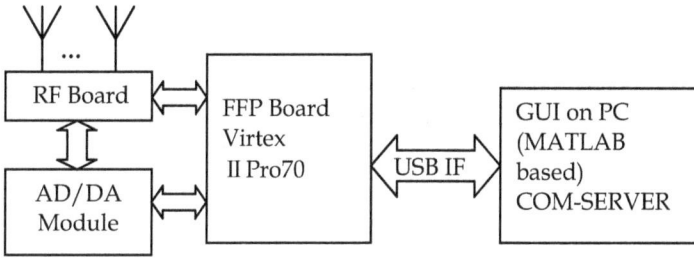

Fig. 5. Hardware architecture of the channel measurement system.

The baseband transceiver module performs the baseband signal processing of a 2x2 MIMO channel sounder. This module generates the baseband probing signal, i.e. a BPSK modulated PN sequence, and delivers the CIR extracted from the received signal to the upper-level module.

The block diagram of this module is shown in Fig. 6. The module can be divided into three parts. The first part is the transmitter, which includes signal generator and transmit multiplexer. The second part is the receiver, which includes receive buffers, signal processor, and data buffer. The last part is the control logic of the module.

The functions of the sub-modules are as follows:

- Signal generator generates the baseband probing signal, i.e. a BPSK modulated PN sequence.
- Transmit multiplexer distributes probing signal to different TX antennas.
- Receive buffers save the received signal from RX antennas.
- Receive multiplexer feed the signal stored in receive buffers into the signal processor in a sequential order.

- Signal processor performs the signal processing, i.e. filtering, symbol timing, and sliding correlation, to extract the CIR from received signal.

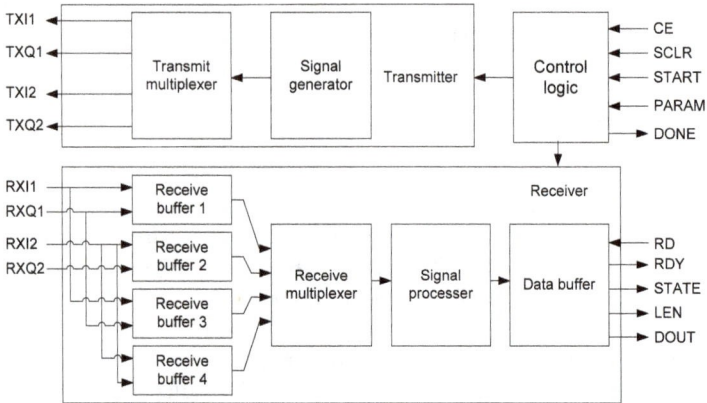

Fig. 6. Baseband transceivers.

3.1.3 GUI

To provide a user friendly human interface, we design a MATLAB based GUI. The real time data stream is accessible from the specific application software through a function call of the COM-Server from IAF. The software can provide several channel information from the original measured data. These channel information include channel impulse response, channel transfer function, delay power profile, scattering function and Doppler power spectrum. Fig. 7 is an indoor channel test result for example.

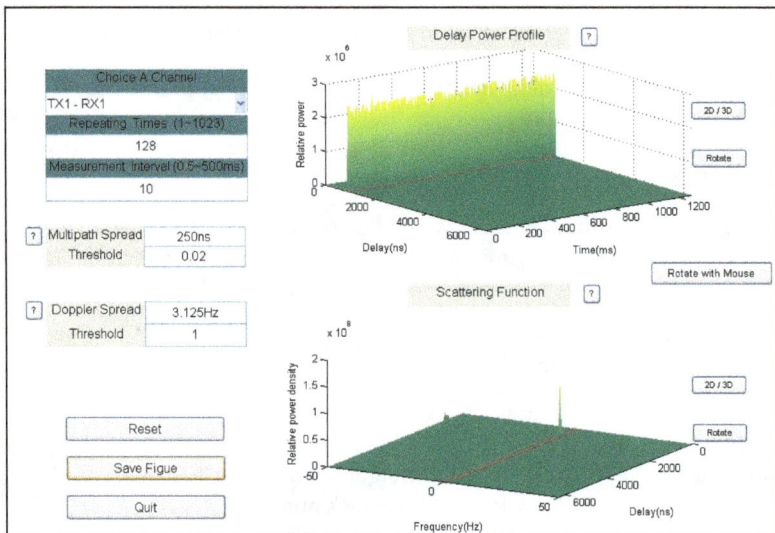

Fig. 7. Measurement result on GUI.

3.2 Channel measurement system using OFDM pilot

In the section, we present an example of OFDM-pilot-based MIMO channel measurement scheme. The measurement is conducted under LTE system. We utilize the reference signal (pilot) to carry out the 4x4 MIMO channel measurement. In this example, the measurement of channel occupied by pilots is LS algorithm, with the purpose of decreasing complexity.

One transmitter sends data according to the LTE agreement, so that each transmitted subframe consists of pilots and useful data. Receiver breaks down each subframe to obtain pilot segments and data segments, respectively. Such measurement equipment can implement the channel measurement without interrupting communications.

A cascade 1-D filtering is used for the 2-D interpolation. This cascade 1-D filtering firstly interpolates the channel in frequency domain with LMMSE interpolation, and then finishes the whole interpolation with a linear time domain interpolation.

There are several reasons why we choose a cascade of frequency LMMSE interpolation and time linear interpolation. LMMSE interpolation certainly has the best MSE performance among all the interpolation schemes. However, the complexity of LMMSE interpolation is much larger than that of linear interpolation. Thus, a tradeoff between performance and complexity has to be made. In frequency domain, LMMSE can provide a large performance increase. When achieving the same BLER or throughput performance, LMMSE interpolation can save about 2 dB SNR. On the other hand, the performance improvement in time domain by applying LMMSE interpolation is marginal, saving only 0.25 dB average. Considering the above, the usage of a cascade of frequency LMMSE interpolation and time linear interpolation is reasonable.

A block diagram of this example is showed in Fig. 8.

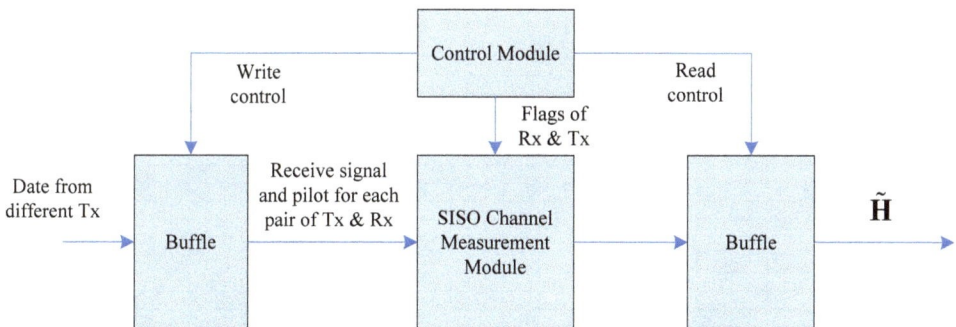

Fig. 8. A block diagram of LTE MIMO channel measurement system.

In addition, the pilot pattern of LTE system with 4 antennas can be seen in Fig. 9.

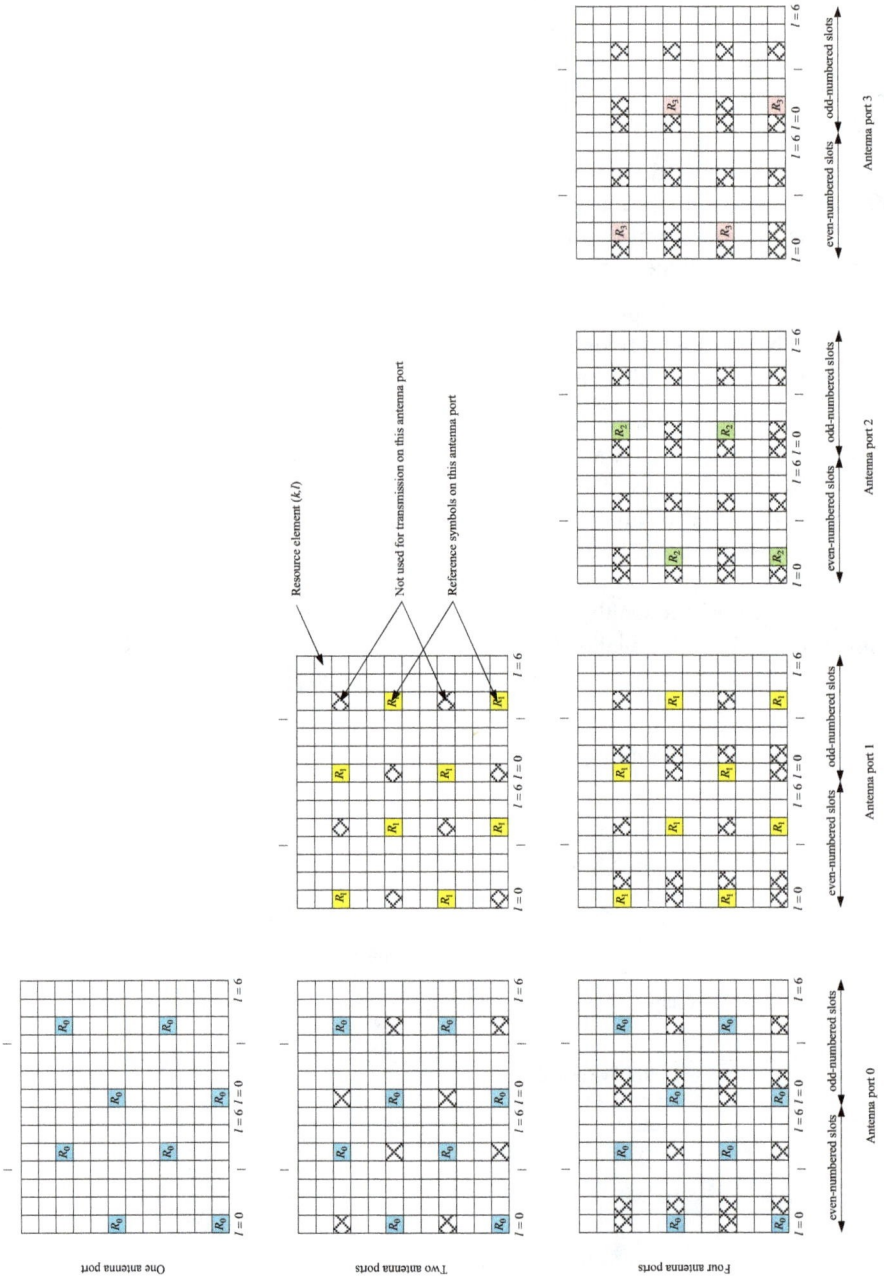

Fig. 9. Pilot pattern of LTE system [4].

3.2.1 Measurements on LTE pilots

Since LTE symbols are typical OFDM symbols, classic measurement schemes such as LS and LMMSE can be applied directly on LTE pilots. A matrix form of (12) for measured channel by LS algorithm is showed as followed:

$$\tilde{H}_p^{LS} = X_{pp}^{-1} Y_p = H_p + X_{pp}^{-1} N_p = [\frac{Y_p^1}{X_p^1}, \frac{Y_p^2}{X_p^2}, \cdots, \frac{Y_p^{M_p}}{X_p^{M_p}}]^T \tag{47}$$

Since LMMSE algorithm is very vulnerable to the speed of mobile stations, the benefit brought by LMMSE will be negligible while comparing to its large processing burden. Furthermore, LS algorithm can be helpful in cancelling the effect of noise brought by LMMSE interpolation, so that the overhead of LMMSE interpolation can be reduced.

3.2.2 Design of frequency domain interpolation

Considering equation (34), there are three major challenges in realizing frequency domain LMMSE interpolation: estimating autocorrelation matrix $\bar{R}_{h_L h_L}$, determining Signal-to-Noise-Ratio in receiver and obtaining the inversion of matrix.

3.2.2.1 Autocorrelation matrix $\bar{R}_{h_L h_L}$

Since the real channels are time-varying, it is impossible to obtain the accurate autocorrelation of channels. The most widely used scheme is to estimate the approximate autocorrelation through some known channel models. It is well-known that two of the most important factors in wireless channel models are multipath spread and Doppler spread. While in the frequency domain, we mainly consider the influence of multipath spread, and propose a simple but useful construction scheme for wireless channels as followed.

The CIR of such a multipath channel is showed as followed:

$$h(\tau) = \sum_{l=0}^{N_L-1} h_l \delta(\tau - \tau_l) \tag{48}$$

Where τ_l and h_l are the delay and amplitude of the lth path. N_L denotes the max number of taps. δ represents the impulse function.

Define $L = \{0, 1 \cdots N_L - 1\}$. Define $h \in C^{N \times 1}$, $h_l = 0, \forall l \in \{N_L \cdots N - 1\}$ as the multipath amplitude vector, N as subcarriers in each OFDM symbol.

Within digital baseband, we assume that the discrete delay as:

$$\tau_l = \frac{lT_s}{N}, l \in L \tag{49}$$

Where is the length of an OFDM symbol.

Further assume that power σ_l^2 of independent Rayleigh-distributed tap h_l is fading exponentially with time constant τ_d:

$$\sigma_l^2 \sim e^{-\frac{l}{\tau_d}}, l \in L \tag{50}$$

Then the normalized CIR autocorrelation can be expressed as:

$$\bar{R}_{h_L h_L} = \frac{R_{h_L h_L}}{\left\| R_{h_L h_L} \right\|} = \frac{diag(\sigma_0^2 \cdots \sigma_{N_L-1}^2)}{\left\| R_{h_L h_L} \right\|} = \frac{diag(\sigma_0^2 \cdots \sigma_{N_L-1}^2)}{\sum_i \sigma_i^2} \tag{51}$$

Base on the above derivation, we need to determine N_L and τ_d to obtain $\bar{R}_{h_L h_L}$. The number of available taps N_L can be same as the length of cyclic prefix (CP), with the purpose of simplification. Yet such a simplification is reasonable, since the multipath spread is less than the length of CP in most of the time. The multipath spread can be estimated with real-time scheme, so as to refine the channel model, as well as the autocorrelation $\bar{R}_{h_L h_L}$.

One of the possible schemes to estimate the multipath spread is provided as followed:

Step 1. Measure the channel matrix of piloted segments $\tilde{H}_p$ in a symbol, with LS algorithm. Take a N_p points IFFT to obtain the rough CIR $\hat{h}_L$, and set $N_L^{max} = N_p$ as the max length of multipath spread.

Step 2. Define a parameter $\hat{h}_{pow}^s$ as:

$$\hat{h}_{pow}^s = \left| \hat{h}_s \right|^2 \quad s = 1,2,\cdots N_L^{max} \tag{52}$$

Where $\hat{h}_{pow}^s$ denotes the square of amplitude for the s-th element in $\hat{h}_L$.

Then obtain a decision object K_s as followed;

$$K_s = \frac{(\sum\limits_{j=s-9}^{s} \hat{h}_{pow}^j)/(2 \times 10)}{(\sum\limits_{k=s+1}^{N_L^{max}} \hat{h}_{pow}^k)/(2 \times (N_L^{max} - s))} \tag{53}$$

Step 3. Find a value of s by the following procedure:

Decrease the value of s from $N_L^{max} - 15$ to 1 with a step of 5. Take the first value of s that satisfies $K_s > 2.55$ as the estimate multipath spread. One can refer to [15] for the reason of choosing 2.55 as the threshold.

After determining the multipath spread, one can obtain $\bar{R}_{h_L h_L}$ by following previous derivation.

3.2.2.2 Signal-to-noise-ratio

SNR value may be measured or estimated in other blocks of receiver. If it is not, the following estimation scheme can be applied.

Denote $\bar{R}_{h_L} = \bar{R}_{h_L h_L}^{1/2}$, $\mathbb{F}_{PL} = F_{PL} \bar{R}_{h_L}$. Do a singular value decomposition on $\mathbb{F}_{PL}$, so that $\mathbb{F}_{PL} = USV^*$. Project estimated channel matrix $\tilde{H}_p$ and real channel matrix H_p as

$U^*\tilde{H}_p$ and U^*H_p. The element in the project of real channel U^*H_p tends to zero when the singular value of $\mathbb{F}_{PL}$ is zero. But things are different in $U^*\tilde{H}_p$. Since we have $U^*\tilde{H}_p = U^*(H_p + X_{pp}^{-1}N_p) = U^*H_p + U^*X_{pp}^{-1}N_p$, it is clear that when the last elements of U^*H_p are zeros, the corresponding elements of $U^*\tilde{H}_p$ reflect the impact of noise. As a result, we can estimate the noise power by these elements.

Let $s = \{N_p - N_s \cdots N_p - 1\}, 1 \le N_s < N_p$ be the range of index, N_p denotes the number of pilots, N_s represents the number of zero singular value in $\mathbb{F}_{PL}$. Then the estimated noise power is $\tilde{p}_n = \dfrac{1}{N_s}\left\|U_{.s}^*\tilde{H}_p\right\|_2^2$, and the corresponding signal power is $\tilde{p}_s = \dfrac{1}{N_p}\left\|\tilde{H}_p\right\|_2^2 - \tilde{p}_n$.

Assume that the SNR is constant within adjacent k pilots, then an average SNR can be obtain as followed:

$$\overline{SNR} = \frac{\sum_{i=1}^k \tilde{p}_{n_i}}{\sum_{i=1}^k \tilde{p}_{s_i}} \tag{54}$$

Since the last element in $U_{.s}^*\tilde{H}_p$ rarely contains signal information, it is the most suitable one for SNR estimation. Therefore, we can simplify the process by setting $N_s = 1$.

3.2.2.3 Inversion of matrix

It is clear from equation (34) that in order to obtain the interpolation matrix w, a N_p order matrix inversion operation must be conducted. The overhead will be very large. Fortunately, instead of the entire matrix, we only need several discrete $\overline{R}_{h_L h_L}$ matrices. Therefore, if we apply discrete average SNR in equation (34), the parameter of interpolation matrix w will be discrete. We can pre-design the discrete range of w, and save it in a table. Then the real-time calculation is simplified as a looking up in a table, according to the measured $\overline{R}_{h_L h_L}$ and SNR.

Specifically, we can adapt a look-up table which cuts the SNR range into several intervals. Each SNR interval combines with a corresponding multipath spread $\hat{\tau}$. Each of such pairs jointly determines a pre-designed w. With this scheme, the complexity of matrix inversion in real-time process is converted to the design of look-up table. Since the look-up table is generated off-line, real-time calculation burden for LMMSE interpolation is largely reduced.

3.2.3 Design of time domain interpolation

According to LTE standardization, each transmission time interval (TTI) is of length 1ms, which is the exact length of a subframe. Consequently, mobile stations process date in units of subframe. When time domain interpolation is conducting, there are at most four pilots in each subframe. As a result, the reference of time domain interpolation of LTE system is at most four estimated channel segments. Two of the most widely used schemes in time domain interpolation are LMMSE interpolation and linear interpolation. The detailed procedures of these two interpolations are presented in previous sections, so we only provide some simulation results to illustrate the advantages and disadvantages of each scheme.

The following simulation considers a unban macro scenario, in which the bandwidth is 10MHz, center frequency is 2GHz and noise is AWGN. Fig. 10 shows the MSE performances of both LMMSE and linear interpolations under different MS speeds.

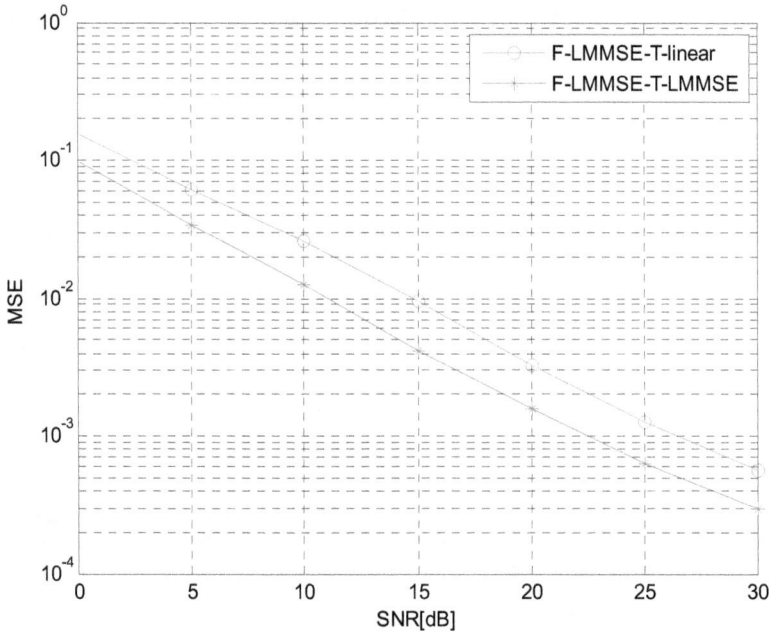

Fig. 10. (a) MSE performances of LMMSE and linear interpolations under MS speed 1m/s.

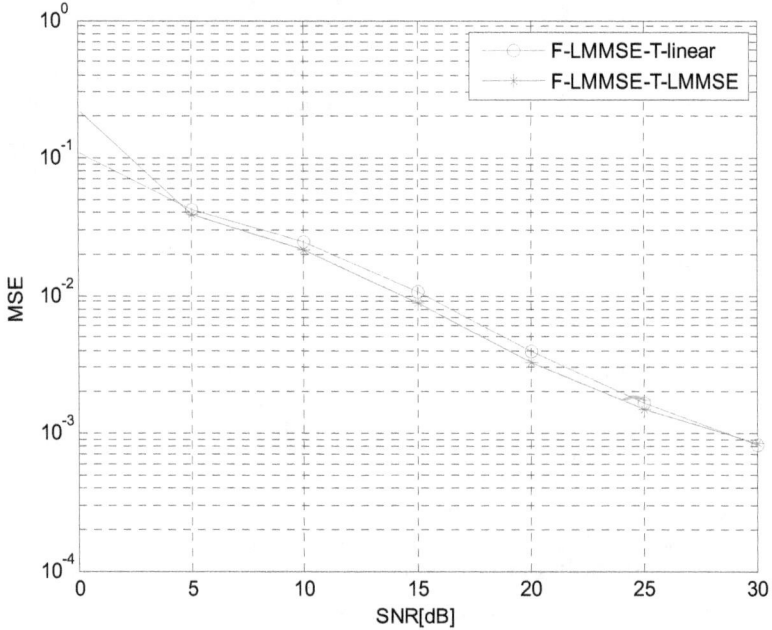

Fig. 10. (b) MSE performances of LMMSE and linear interpolations under MS speed 30m/s.

The following conclusions can be inferred from the simulation results.

When the speed of MS is small, correspondingly small Doppler spread, LMMSE interpolation can save 4 dB SNR while achieving the same MSE performance of linear interpolation. However, when the speed (as well as the Doppler spread) of MS increases to a relatively high level, performances of LMMSE and linear schemes become very close. This means that the large overhead spent on LMMSE outputs marginal gains on the performance. When the errors of Doppler spread estimations are taken into account, the MSE performance of LMMSE scheme may even be worse than that of linear interpolation. Consequently, after considering the tradeoff between performance and complexity, we propose to use a simple linear interpolation in time domain.

4. Reference

[1] I. E. Telatar, "Capacity of Multi-Antenna Gaussian channels," European Transactions on Telecommunications, vol. 10, no. 6, pp. 585–595, Nov./Dec. 1999.
[2] S. Vishwanath, N. Jindal, and A. Goldsmith, "Duality, Achievable Rates,and Sum-Rate Capacity of Gaussian MIMO Broadcast Channels," IEEE Transactions on Information Theory, vol. 49, No. 10, pp. 2658–2668, Oct. 2003.
[3] I. W. Group, "IEEE 802.11 Wireless Local Area Networks,"May 2001
[4] 3GPP TS 36.211: "Evolved Universal Terrestrial Radio Access (E-UTRA); Physical channels and modulation".
[5] M.K. Ozdemir and H. Arslan, "Channel Estimation for Wireless OFDM Systems," IEEE Communications Surveys & Tutorials, vol. 9, No. 2, pp. 18–48, 2007.
[6] Myung-Don Kim, Heon Kook Kwon, Bum Soo Park, Jae Joon Park, and Hyun Kyu Chung, "Wideband MIMO Channel Measurements in Indoor Hotspot Scenario at 3.705GHz," International Conference on Signal Processing and Communication Systems, 2010.
[7] Hui Yu, Ruikai Zhang, Xi Chen, Wentao Song, and Hailong Wang , "Design of an Indoor Channel Measurement System," International Wireless Communications and Mobile Computing Conference, 2010.
[8] Jose-Maria, Molina-García-Pardo, José-Víctor Rodríguez, and Leandro Juan-Llácer, "Polarized Indoor MIMO Channel Measurements at 2.45 GHz," IEEE Transactions on Antennas and Propagation, vol.56, no. 12, Dec., 2008.
[9] David W. Matolak, and Qian Zhang, "5 GHz Near-Ground Indoor Channel Measurements and Models," IEEE Radio and Wireless Symposium, 2009.
[10] Ye Wang, Wenjun Lu and Hongbo Zhu, "Experimental Study on Indoor Channel Model for Wireless Sensor Networks and Internet of Things," IEEE International Conference on Communication Technology, 2010.
[11] Alexandru Rusu- Casandra, Ion Marghescu and Elena Simona Lohan, "Estimators of the indoor channel for GPS-based pseudolite signal," International Symposium on Electronics and Telecommunications, 2009.
[12] L. J. Greenstein, D. G. Michelson, and V. Erceg, "Moment-Method Estimation of the Ricean K-Factor," IEEE Communications Letters, vol. 3, No. 6, pp. 175-176, 1999.
[13] Jae-Joon Park, Myung-Don Kim, and Hyun-Kyu Chung, "Characteristics of Ricean K-factor in Wideband Indoor Channels at 3.7 GHz," International Conference on Signal Processing and Communication Systems, 2010.

[14] John G. Proakis, "Digital Communications", McGraw-Hill Companies, Inc and publishing house of electronics industry, China. ,fourth edition, pp.766 , 2001.

[15] Guosong Li, "Research on channel estimation in wireless OFDM systems", Ph.D thesis, University of Electronic Science and Technology of China, 2005.

[16] IEEE P802.11n/D1.0, March 2006.

[17] MAXIM Integrated Products, Datasheet of MAX2828/2829, 19-3455, rev0, Oct. 2004

[18] F. M. Gardner, "Interpolation in digital modems – Part I: Fundamentals," IEEE Trans. Commun., vol. 41, pp. 501-507, Mar. 1993.

[19] F. M. Gardner, "Interpolation in digital modems – Part II: Implementation and performance," IEEE Trans. Commun., vol. COM-41, pp. 998-1008, Jun. 1993.

[20] F. M. Gardner, "A BPSK/QPSK timing-error detector for detector for sampled receivers," IEEE Trans. Commun., vol. COM-34, pp. 423-429, May. 1986.

9

Channel Capacity Analysis Under Various Adaptation Policies and Diversity Techniques over Fading Channels

Mihajlo Stefanović[1], Jelena Anastasov[1], Stefan Panić[2],
Petar Spalević[3] and Ćemal Dolićanin[3]
[1]Faculty of Electronic Engineering, University of Niš,
[2]Faculty of Natural Science and Mathematics, University of Priština,
[3]State University of Novi Pazar
Serbia

1. Introduction

The lack of available spectrum for expansion of wireless services requires more spectrally efficient communication in order to meet the consumer demand. Since the demand for wireless communication services have been growing in recent years at a rapid pace, conserving, sharing and using bandwidth efficiently is of primary concern in future wireless communications systems. Therefore, channel capacity is one of the most important concerns in the design of wireless systems, as it determines the maximum attainable throughput of the system [1]. It can be defined as the average transmitted data rate per unit bandwidth, for a specified average transmit power, and specified level of received outage or bit-error rate [2]. Skilful combination of bandwidth efficient coding and modulation schemes can be used to achieve higher channel capacities per unit bandwidth. However, mobile radio links are, due to the combination of randomly delayed reflected, scattered, and diffracted signal components, subjected to severe multipath fading, which leads to serious degradation in the link signal-to-noise ratio (SNR). An effective scheme that can be used to overcome fading influence is adaptive transmission. The performance of adaptation schemes is further improved by combining them with space diversity, since diversity combining is a powerful technique that can be used to combat fading in wireless systems resulting in improving link performance [3].

1.1 Channel and system model

Diversity combining is a powerful technique that can be used to combat fading in wireless systems [4]. The optimal diversity combining technique is maximum ratio combining (MRC). This combining technique involves co-phasing of the useful signal in all branches, multiplication of the received signal in each branch by a weight factor that is proportional to the estimated ratio of the envelope and the power of that particular signal and summing of the received signals from all antennas. By co-phasing, all the random phase fluctuations of the signal that emerged during the transmission are eliminated. For this process it is necessary to estimate the phase of the received signal, so this technique requires the entire amount of the channel state information (CSI) of the received signal, and separate receiver chain for each branch of the diversity system, which increases the complexity of the system [5].

One of the least complicated combining methods is selection combining (SC). While other combining techniques require all or some of the amount of the CSI of received signal and separate receiver chain for each branch of the diversity system which increase its complexity, selection combining (SC) receiver process only one of the diversity branches, and is much simpler for practical realization, in opposition to these combining techniques [4-7]. Generally, SC selects the branch with the highest SNR, that is the branch with the strongest signal, assuming that noise power is equally distributed over branches. Since receiver diversity mitigates the impact of fading, the question is whether it also increases the capacity of a fading channel.

Another effective scheme that can be used to overcome fading influence is adaptive transmission. Adaptive transmission is based on the receiver's estimation of the channel and feedback of the CSI to the transmitter. The transmitter then adapts the transmit power level, symbol/bit rate, constellation size, coding rate/scheme or any combination of these parameters in response to the changing channel conditions [8]. Adapting certain parameters of the transmitted signal to the fading channel can help better utilization of the channel capacity. These transmissions provide a much higher channel capacities per unit bandwidth by taking advantage of favorable propagation conditions: transmitting at high speeds under favorable channel conditions and responding to channel degradation through a smooth reduction of their data throughput. The source may transmit faster and/or at a higher power under good channel conditions and slower and/or at a reduced power under poor conditions. A reliable feedback path between that estimator and the transmitter and accurate channel estimation at the receiver is required for achieving good performances of adaptive transmission. Widely accepted adaptation policies include optimal power and rate adaptation (OPRA), constant power with optimal rate adaptation (ORA), channel inversion with fixed rate (CIFR), and truncated CIFR (TIFR). Results obtained for this protocols show the trade-off between capacity and complexity. The adaptive policy with transmitter and receiver side information requires more complexity in the transmitter (and it typically also requires a feedback path between the receiver and transmitter to obtain the side information). However, the decoder in the receiver is relatively simple. The non-adaptive policy has a relatively simple transmission scheme, but its code design must use the channel correlation statistics (often unknown), and the decoder complexity is proportional to, the channel decorrelation time. The channel inversion and truncated inversion policies use codes designed for additive white Gaussian noise (AWGN) channels, and are therefore the least complex to implement, but in severe fading conditions they exhibit large capacity losses relative to the other techniques.

The performance of adaptation schemes is further improved by combining them with space diversity. The hypothesis that the variation of the combiner output SNR is tracked perfectly by the receiver and that the variation in SNR is sent back to the transmitter via an error-free feedback path will be assumed in the ongoing analysis [8]. Also, it is assumed that time delay in this feedback path is negligible compared to the rate of the channel variation.

Following these assumptions, transmitter could adapt its power and/or rate relative to the actual channel state.

There are numerous published papers based on study of channel capacity evaluation. In [9], the capacity of Rayleigh fading channels under four adaptation policies and multibranch system with variable correlation is investigated. The capacity of Rayleigh fading channels under different adaptive transmission and different diversity combining techniques is also studied in [7], [10]. In [11], channel capacity of MRC over exponentially correlated Nakagami-*m* fading channels under adaptive transmission is analyzed. Channel capacity of adaptive transmission schemes using equal gain combining (EGC) receiver over Hoyt fading channels is presented in [12]. In [13], dual-branch SC receivers operating over correlative Weibull fading under three adaptation policies are analyzed.

In this chapter we will focus on more general and nonlinear fading distributions. We will perform an analytical study of the *κ-μ* fading channel capacity, e.g., under the OPRA, ORA, CIFR and TIFR adaptation policies and MRC and SC diversity techniques. To the best of authors' knowledge, such a study has not been previously considered in the open technical literature. The expressions for the proposed adaptation policies and diversity techniques will be derived. Capitalizing on them, numerically obtained results will be graphically presented, in order to show the effects of various system parameters, such as diversity order and fading severity on observed performances. In the similar manner an analytical study of the Weibull fading channel capacity, under the OPRA, ORA, CIFR and TIFR adaptation policies and MRC diversity technique will be performed.

1.1.1 *κ-μ* channel and system model

The multipath fading in wireless communications is modelled by several distributions including Nakagami-*m*, Hoyt, Rayleigh, and Rice. By considering important phenomena inherent to radio propagation, *κ-μ* fading model was recently proposed in [14] as a fading model which describes the short-term signal variation in the presence of line-of-sight (LOS) components. This distribution is more realistic than other special distributions, since its derivation is completely based on a non-homogeneous scattering environment. Also *κ-μ* as general physical fading model includes Rayleigh, Rician, and Nakagami-*m* fading models, as special cases [14]. It is written in terms of two physical parameters, *κ* and *μ*. The parameter *κ* is related to the multipath clustering and the parameter *μ* is the ratio between the total power of the dominant components and the total power of the scattered waves. In the case of *κ*=0, the *κ-μ* distribution is equivalent to the Nakagami-*m* distribution. When *μ*=1, the *κ-μ* distribution becomes the Rician distribution with *κ* as the Rice factor. Moreover, the *κ-μ* distribution fully describes the characteristics of the fading signal in terms of measurable physical parameters.

The SNR in a *κ-μ* fading channel follows the probability density function (pdf) given by [15]:

$$p_\gamma(\gamma) = \frac{\mu}{k^{(\mu-1)/2} e^{\mu k}} \left(\frac{1+k}{\bar{\gamma}}\right)^{(\mu+1)/2} \gamma^{(\mu-1)/2} e^{-\mu(1+k)\gamma/\bar{\gamma}} I_{\mu-1}\left(2\mu\sqrt{\frac{(1+k)k\gamma}{\bar{\gamma}}}\right). \qquad (1.1)$$

In the previous equation, $\bar{\gamma}$ is the corresponding average SNR, while $I_n(x)$ denotes the n-th order modified Bessel function of first kind [16], and κ and μ are well-known κ-μ fading parameters. Using the series representation of Bessel function [16, eq. 8.445]:

$$I_n(x) = \sum_{k=0}^{+\infty} \frac{x^{2k+n}}{2^{2k+n}\Gamma(k+n+1)k!},$$

(1.2)

the cumulative distribution function (cdf) of γ can be written in the form of:

$$F_\gamma(\gamma) = \sum_{p=0}^{+\infty} \frac{\mu^p \kappa^p}{e^{\mu\kappa}\Gamma(p+\mu)}\Lambda\left(p+\mu, \frac{\mu(1+\kappa)\gamma}{\bar{\gamma}}\right)$$

(1.3)

with $\Gamma(x)$ and $\Lambda(a,x)$ denoting Gamma and lower incomplete Gamma function, respectively [16, eqs. 8.310.1, 8.350.1].

It is shown in [15], that the sum of κ-μ squares is κ-μ square as well (but with different parameters), which is an ideal choice for MRC analysis. Then the expression for the pdf of the outputs of MRC diversity systems follows [15, eq.11]:

$$p_\gamma^{MRC}(\gamma) = \frac{L\mu}{k^{(L\mu-1)/2}e^{L\mu k}}\left(\frac{1+k}{L\bar{\gamma}}\right)^{(L\mu+1)/2}\gamma^{(L\mu-1)/2}\,e^{-\mu(1+k)\gamma/\Omega}I_{L\mu-1}\left(2\mu L\sqrt{\frac{(1+k)k\gamma}{L\bar{\gamma}}}\right)$$

(1.4)

with L denoting the number of diversity branches.

The expression for the pdf of the outputs of SC diversity systems can be obtained by substituting expressions (1.1) and (1.3) into:

$$p_\gamma^{SC}(\gamma) = \sum_{i=1}^{L} p_{\gamma_i}(\gamma)\prod_{\substack{j=1\\j\neq i}}^{L} F_{\gamma_j}(\gamma)$$

(1.5)

where $p_{\gamma i}(\gamma)$ and $F_{\gamma i}(\gamma)$ define pdf and cdf of SNR at input branches respectively and L denotes the number of diversity branches.

1.1.2 Weibull channel and system model

The above mentioned well-known fading distributions are derived assuming a homogeneous diffuse scattering field, resulting from randomly distributed point scatterers. The assumption of a homogeneous diffuse scattering field is certainly an approximation, because the surfaces are spatially correlated characterizing a nonlinear environment. With the aim to explore the nonlinearity of the propagation medium, a general fading distribution, the Weibull distribution, was proposed. The nonlinearity is manifested in terms of a power parameter $\beta > 0$, such that the resulting signal intensity is obtained not simply as the modulus of the multipath component, but as the modulus to a certain given power. As β increases, the fading severity decreases, while for the special case of $\beta = 2$ reduces to the well-known Rayleigh distribution. Weibull distribution seems to exhibit good fit to experimental fading channel measurements, for both indoor and outdoor environments.

The SNR in a Weibull fading channel follows the pdf given by [17, eq.14]:

$$p(\gamma) = \frac{\beta}{2a\overline{\gamma}} \left(\frac{\gamma}{a\overline{\gamma}} \right)^{\frac{\beta}{2}-1} e^{-\left(\frac{\gamma}{a\overline{\gamma}} \right)^{\beta/2}} \tag{1.6}$$

In the previous equation, $\overline{\gamma}$ is the corresponding average SNR, β is well-known Weibull fading parameter, and $a = 1/\Gamma(1+2/\beta)$.

It is shown in [18,19], that the expression for the pdf of the outputs of MRC diversity systems follows [19, eq.1]:

$$p_\gamma^{MRC}(\gamma) = \frac{\beta \gamma^{L\beta/2-1}}{2\Gamma(L)(\Xi\overline{\gamma})^{L\beta/2}} e^{-\left(\frac{\gamma}{\Xi\overline{\gamma}} \right)^{\beta/2}} ; \qquad \Xi = \frac{\Gamma(L)}{\Gamma(L+2/\beta)} \tag{1.7}$$

with L denoting the number of diversity branches.

Similary, expression for the pdf of the outputs of SC diversity systems can be obtained as (1.5)

2. Optimal power and rate adaptation

In the OPRA protocol the power level and rate parameters vary in response to the changing channel conditions. It achieves the ergodic capacity of the system, i. e. the maximum achievable average rate by use of adaptive transmission. However, OPRA is not suitable for all applications because for some of them it requires fixed rate.

During our analysis it is assumed that the variation in the combined output SNR over κ-μ fading channels γ is tracked perfectly by the receiver and that variation of γ is sent back to the transmitter via an error-fee feedback path. Comparing to the rate of channel variation, the time delay in this feedback is negligible. These assumptions allow the transmitter to adopt its power and rate correspondingly to the actual channel state. Channel capacity of the fading channel with received SNR distribution, $p_\gamma(\gamma)$, under optimal power and rate adaptation policy, for the case of constant average transmit power is given by [8]:

$$<C>_{pra} = B \int_{\gamma_0}^{\infty} \log_2 \left(\frac{\gamma}{\gamma_0} \right) p_\gamma(\gamma) d\gamma, \tag{1.8}$$

where B (Hz) denotes the channel bandwidth and γ_0 is the SNR cut-off level bellow which transmission of data is suspended. This cut-off level must satisfy the following equation:

$$\int_{\gamma_0}^{\infty} \left(\frac{1}{\gamma_0} - \frac{1}{\gamma} \right) p_\gamma(\gamma) d\gamma = 1, \tag{1.9}$$

Since no data is sent when $\gamma < \gamma_0$, the optimal policy suffers a probability of outage P_{out} equal to the probability of no transmission, given by:

$$P_{out} = \int_0^{\gamma_0} p_\gamma(\gamma)d\gamma = 1 - \int_{\gamma_0}^{\infty} p_\gamma(\gamma)d\gamma \qquad (1.10)$$

2.1 κ-μ fading channels

To achieve the capacity in (1.8), the channel fading level must be attended at the receiver as well as at the transmitter. The transmitter has to adapt its power and rate to the actual channel state; when γ is large, high power levels and rates are allocated for good channel conditions and lower power levels and rates for unfavourable channel conditions when γ is small. Substituting (1.1) into (1.9), we found that the cut-off level must satisfy:

$$\sum_{i=0}^{\infty} \frac{(kL\mu)^i}{e^{L\mu k}\Gamma(i+L\mu)i!}\left(\frac{1}{\gamma_0}\Lambda\left(L\mu+i,\frac{\mu(1+k)\gamma_0}{\bar{\gamma}}\right) - \right.$$

$$\left. \frac{\mu(1+k)}{\bar{\gamma}}\Lambda\left(L\mu+i-1,\frac{\mu(1+k)\gamma_0}{\bar{\gamma}}\right)\right)-1=0 \qquad (1.11)$$

Substituting (1.1) into (1.8), we obtain the capacity per unit bandwidth, $<C>_{opra}/B$, as:

$$\frac{\langle C\rangle_{opra}^{MRC}}{B} = \sum_{i=0}^{\infty}\frac{L\mu}{k^{(L\mu-1)/2}e^{L\mu k}}\left(\frac{1+k}{L\bar{\gamma}}\right)^{(L\mu+1)/2}\int_{\gamma_0}^{\infty}\log_2\left(\frac{\gamma}{\gamma_0}\right)\gamma^{L\mu+i-1}e^{-\mu(1+k)\gamma/\bar{\gamma}}d\gamma \qquad (1.12)$$

Now, by making change of variables, , $<C>_{opra}/B$ can be obtained as:

$$\frac{\langle C\rangle_{opra}^{MRC}}{B} = \sum_{i=0}^{\infty}\frac{(L\mu k)^i}{\Gamma(i+L\mu)i!e^{L\mu k}}\left(\int_0^{\infty}\log_2\left(\frac{t\bar{\gamma}}{\mu(1+k)\gamma_0}\right)t^{L\mu+i-1}e^{-t}dt - \right.$$

$$\left. \int_0^{\gamma_0\mu(1+k)/\bar{\gamma}}\log_2\left(\frac{t\bar{\gamma}}{\mu(1+k)\gamma_0}\right)t^{L\mu+i-1}e^{-t}dt\right) = \sum_{i=0}^{\infty}\frac{(L\mu k)^i}{\Gamma(i+L\mu)i!e^{L\mu k}}(I_1-I_2) \qquad (1.13)$$

Integral I_1 can be solved by applying Gauss-Laguerre quadrature formulae:

$$I_1 = \int_0^{\infty}f_1(t)e^{-t}dt \cong \sum_{k=1}^{R}A_k f_1(t_k); \quad f_1(t) = \log_2\left(\frac{t\bar{\gamma}}{\mu(1+k)\gamma_0}\right)t^{L\mu+i-1} \qquad (1.14)$$

In the previous equation A_k and t_k, $k=1,2,...,R$, are respectively weights and nodes of Laguerre polynomials [20, pp. 875-924].

Similarly, integral I_2 can be solved by applying Gauss-Legendre quadrature formulae:

$$I_2 = \left(\frac{\gamma_0 \mu (1+k)}{2\gamma} \right)^{L\mu+i} \int_{-1}^{1} f_2(u) du \cong \left(\frac{\gamma_0 \mu (1+k)}{2\gamma} \right)^{L\mu+i} \sum_{k=1}^{R} B_k f_2(u_k) \tag{1.15}$$

where B_k and u_k, $k=1,2,...,R$, are respectively weights and nodes of Legendre polynomials.

Convergence of infinite series expressions in (1.13) is rapid since we need about 10 terms to be summed in order to achieve accuracy at the 5th significant digit for corresponding values of system parameters.

2.2 Weibull fading channels

Substituting (1.7) in (1.8) integral of the following form need to be solved

$$I = \frac{1}{\ln 2} \int_{\gamma_0}^{\infty} \gamma^{L\beta/2-1} \ln\left(\frac{\gamma}{\gamma_0} \right) e^{-\left(\frac{\gamma}{\Xi\overline{\gamma}} \right)^{\beta/2}} d\gamma. \tag{1.16}$$

After making a change of variables $t = (\gamma / \gamma_0)^{\beta/2}$ and some simple mathematical manipulations, we get:

$$I = \frac{4\gamma_0^{L\beta/2}}{\beta^2 \ln 2} \int_{1}^{\infty} t^{L-1} \ln(t) e^{-\left(\frac{\gamma_0}{\Xi\overline{\gamma}} \right)^{\beta/2} t} dt \tag{1.17}$$

Furthermore, this integral can be evaluated using partial integration:

$$\int_{1}^{\infty} u dv = \lim_{t\to\infty}(uv) - \lim_{t\to1}(uv) - \int_{1}^{+\infty} v du \tag{1.18}$$

with respect to:

$$u = \ln t; \quad dv = t^{L-1} e^{-\left(\frac{\gamma_0}{\Xi\overline{\gamma}} \right)^{\beta/2} t} dt. \tag{1.19}$$

Performing L-1 successive integration by parts [16, eq. 2.321.2], we get

$$v = -e^{-mt} \sum_{p=1}^{L} \frac{(L-1)!}{(L-p)!} \frac{t^{L-p}}{m^p} \tag{1.20}$$

denoting $m = (\gamma_0 / \Xi\overline{\gamma})^{\beta/2}$. Substituting (1.20) in (1.18), we see that first two terms tend to zero. Hence, the integral in (1.17) can be solved in closed form using [16, eq 3.381.3]

$$I = \frac{(L-1)!}{m^L} \sum_{p=0}^{L-1} \frac{\Gamma(p,m)}{p!} \tag{1.21}$$

with $\Gamma(a, x)$ higher incomplete Gamma function [16]. Finaly, $<C>_{opra}/B$ using L-branch MRC diversity receiver over Weibull fading channels has this form

$$\frac{\langle C \rangle_{opra}^{MRC}}{B} = \frac{2}{\beta \ln 2} \sum_{p=0}^{L-1} \frac{\Gamma(p,m)}{p!} . \tag{1.22}$$

3. Constant power with optimal rate adaptation

With ORA protocol, the transmitter adapts its rate only while maintaining a fixed power level. Thus, this protocol can be implemented at reduced complexity and is more practical than that of optimal simultaneous power and rate adaptation.

The channel capacity, $<C>_{ora}$ (bits/s) with constant transmit power policy is given by [1]:

$$\langle C \rangle_{ora} = B \int_0^\infty \log_2 (1 + \gamma) p_\gamma (\gamma) d\gamma \tag{1.22}$$

3.1 κ-μ fading channels

To achieve the capacity in (1.22), the channel fading level must be attended at the receiver as well as at the transmitter.

After substituting (1.1) into (1.22), by using partial integration:

$$\int_0^\infty u dv = \lim_{\gamma \to \infty} (uv) - \lim_{\gamma \to 0} (uv) - \int_0^\infty v du \tag{1.23}$$

with respect to:

$$u = \ln(1+\gamma); \quad du = \frac{d\gamma}{1+\gamma}; \quad dv = \gamma^{p+\mu-1} e^{-\frac{\mu(1+k)\gamma}{\gamma}} ; \tag{1.24}$$

and performing successive integration by parts [16 , eq. 2.321.2], we get

$$v = e^{-\frac{\mu(1+k)\gamma}{\gamma}} \sum_{q=1}^{p+\mu} \frac{(p+\mu-1)! \gamma^{p+\mu-k}}{(p+\mu-q)!} \left(\frac{\gamma}{\mu(1+k)} \right)^q \tag{1.25}$$

By substituting (1.25) in (1.23), we see that first two terms tend to zero. Hence, the integral in (1.23) can be solved in closed form using [16, eq. 3.381.3]. Finaly, $<C>_{ora}/B$ over κ-μ fading channels has this form:

$$\langle C \rangle_{ora} = \frac{B}{\ln 2} \sum_{p=0}^{\infty} \sum_{q=1}^{p+\mu} \frac{\mu^{2p+\mu-q} \kappa^p (1+\kappa)^{p+\mu-q} (n-1)!}{e^{\mu\kappa} \gamma^{-p+\mu-q} \Gamma(p+\mu)p!} e^{\frac{\mu(1+\kappa)}{\gamma}}$$

$$\Gamma\left(-n+p+\mu, \frac{\mu(1+\kappa)}{\gamma}\right) \tag{1.26}$$

On the other hand, substituting (1.4) into (1.22) and applying similar procedure, the expression for the $<C>_{ora}/B$ with MRC diversity receiver is derived as:

$$\langle C \rangle_{ora}^{MRC} = \frac{B}{\ln 2} \sum_{p=0}^{\infty} \sum_{q=1}^{p+\mu} \frac{\mu^{2p+\mu-q} \kappa^p (1+\kappa)^{p+\mu-q} (n-1)! L^p}{e^{L\mu\kappa} \gamma^{-p+\mu-q} \Gamma(p+\mu)p!} e^{\frac{\mu(1+\kappa)}{\gamma}}$$

$$\Gamma\left(-n+p+\mu L, \frac{\mu(1+\kappa)}{\gamma}\right) \tag{1.27}$$

Convergence of infinite series expressions in (1.26) and (1.27) is rapid, since we need 5-10 terms to be summed in order to achieve accuracy at the 5th significant digit for corresponding values of system parameters.

3.2 Weibull fading channels

After substituting (1.6) into (1.22), when MRC reception is applied over Weibull fading channel, we can obtain expression for the ORA channel capacity, in the form of:

$$\frac{\langle C \rangle_{ora}}{B} = \frac{\beta}{2\Gamma(L)(\Xi\bar{\gamma})^{L\beta/2} \ln 2} \int_0^{\infty} \gamma^{L\beta/2} \ln(1+\gamma) e^{-\left(\frac{\gamma}{\Xi\bar{\gamma}}\right)^{\beta/2}} d\gamma. \tag{1.28}$$

By expressing the logarithmic and exponential integrands as Meijer's G- functions [21, eqs. 11] and using [22, eq. 07.34.21.0012.01], integral in (1.28) is solved in closed-form:

$$\frac{\langle C \rangle_{ora}}{B} = \frac{\beta}{2\Gamma(L)(\Xi\bar{\gamma})^{L\beta/2} \ln 2} H_{2,3}^{3,1}\left((\Xi\bar{\gamma})^{-\beta/2} \left| \begin{array}{l} (-L\beta/2, \beta/2),(1-L\beta/2, \beta/2) \\ (0,1),(-L\beta/2, \beta/2),(-L\beta/2, \beta/2) \end{array} \right.\right) \tag{1.29}$$

with:

$$H_{p,q}^{m,n}\left(x \left| \begin{array}{l} (a_1,\alpha_1)....(a_p,\alpha_p) \\ (b_1,\beta_q)....(b_p,\beta_q) \end{array} \right.\right) \tag{1.30}$$

denoting the Fox H function [23].

4. Channel inversion with fixed rate

Channel inversion with fixed rate policy (CIFR protocol) is quite different than the first two protocols as it maintains constant rate and adapts its power to the inverse of the channels fading. CIFR protocol achieves what is known as the outage capacity of the system; that is the maximum constant data rate that can be supported for all channel conditions with some probability of outage. However, the capacity of channel inversion is always less than the capacity of the previous two protocols as the transmission rate is fixed. On the other hand, constant rate transmission is required in some applications and is worth the loss in achievable capacity. CIFR is adaptation technique based on inverting the channel fading. It is the least complex technique to implement assuming that the transmitter on this way adapts its power to maintain a constant SNR at the receiver. Since a large amount of the transmitted power is required to compensate for the deep channel fades, channel inversion with fixed rate suffers a certain capacity penalty compared to the other techniques.

The channel capacity with this technique is derived from the capacity of an AWGN channel and is given in [8]:

$$\langle C \rangle_{cifr} = B \log_2 \left(1 + 1 \Big/ \int_0^\infty \left(p_\gamma \left(\gamma \right) / \gamma \right) d\gamma \right). \tag{1.31}$$

4.1 κ-μ fading channels

After substituting (1.1) into (1.31), and by using [16, eq. 6.643.2]:

$$\int_0^\infty x^{\mu - \frac{1}{2}} e^{-\alpha x} I_{2\nu} \left(2 \beta \sqrt{x} \right) dx = \frac{\Gamma \left(\mu + \nu + \frac{1}{2} \right)}{\Gamma \left(2\nu + 1 \right)} \beta^{-1} e^{-\frac{\beta^2}{2\alpha}} \alpha^{-\nu} M_{-\mu,\nu} \left(\frac{\beta^2}{\alpha} \right) \tag{1.32}$$

where $M_{k,m}(z)$ is the Wittaker's function, we can obtained expression for the CIFR channel capacity in the form of:

$$\langle C \rangle_{cifr} = B \log_2 \left(1 + \frac{(\mu - 1)}{e^{-\frac{\mu k}{2}} \left(\frac{1 + k}{k \overline{\gamma}} \right)^{\frac{\mu}{2}} M_{1 - \frac{\mu}{2}, \frac{\mu - 1}{2}} \left(\mu k \right)} \right). \tag{1.33}$$

Case when MRC diversity is applied can be modelled by:

$$\langle C \rangle_{cifr}^{MRC} = B \log_2 \left(1 + \frac{(L\mu - 1)}{e^{-\frac{\mu k L}{2}} \left(\frac{1 + k}{k L \overline{\gamma}} \right)^{\frac{L\mu}{2}} M_{1 - \frac{L\mu}{2}, \frac{L\mu - 1}{2}} \left(\mu k L \right)} \right). \tag{1.34}$$

Similarly, after substituting (1.5) into (1.31), with respect to [16, eqs. 8.531, 7.552.5, 9.14]:

$$\Lambda(a, x) = \frac{x^a}{a} e^{-x} {}_1F_1(1; 1+a; x) \tag{1.35}$$

$$\int_0^\infty e^{-x} x^{s-1} {}_pF_q(a_1, ..., a_p; b_1, ..., b_q; \alpha x) dx = \Gamma(s) {}_{p+1}F_q(s, a_1, ..., a_p; b_1, ..., b_q; \alpha x) \tag{1.36}$$

$${}_1F_1(a; b; x) = \sum_{k=0}^\infty \frac{(a)_k\, x^k}{(b)_k\, k!} \tag{1.37}$$

expressions for the CIFR channel capacity over κ-μ fading with SC diversity applied for dual and triple branch combining at the receiver can be obtained in the form of:

$$\langle C \rangle_{cifr}^{SC-2} = B \log_2 \left(1 + 1 \Big/ \sum_{p=0}^\infty \sum_{q=0}^\infty f_1 \right)$$

$$f_1 = \frac{\mu^{p+q+1} \kappa^{p+q} (1+\kappa) \Gamma(p+q+2\mu-1)}{2^{p+q+2\mu-2} e^{2\mu\kappa} \bar{\gamma} \Gamma(p+\mu) p! \Gamma(q+\mu) q! (q+\mu)} \tag{1.38}$$

$${}_2F_1\left(p+q+2\mu-1, 1; 1+q+\mu; \frac{1}{2} \right)$$

$$\langle C \rangle_{cifr}^{SC-3} = B \log_2 \left(1 + 1 \Big/ \sum_{p=0}^\infty \sum_{q=0}^\infty \sum_{r=0}^\infty \sum_{s=0}^\infty f_2 \right)$$

$$f_2 = \frac{\mu^{p+q+r+1} \kappa^{p+q+r} (1+\kappa) \Gamma(p+q+r+s+3\mu-1)}{3^{p+q+r+s+3\mu-3} e^{3\mu\kappa} \bar{\gamma} \Gamma(p+\mu) p! \Gamma(q+\mu) q! (q+\mu) \Gamma(r+\mu) r! (r+\mu)(1+r+\mu)_s} \tag{1.39}$$

$$\times {}_2F_1\left(p+q+r+s+3\mu-1, 1; 1+q+\mu; \frac{1}{3} \right)$$

Number of terms that need to be summed in (1.38) and (1.39) to achieve accuracy at 5th significant digit for some values of system parameters is presented in Table 1 in the section Numerical results.

4.2 Weibull fading channels

After substituting (1.6) into (1.31) we can obtain expression for the CIFR channel capacity when MRC diversity is applied in the form of:

$$\frac{\langle C \rangle_{cifr}^{MRC}}{B} = \log_2 \left(1 + \frac{(\Xi \bar{\gamma}) \Gamma(L)}{\Gamma(L - 2/\beta)} \right) \tag{1.40}$$

5. Truncated channel inversion with fixed rate

The channel inversion and truncated inversion policies use codes designed for AWGN channels, and are therefore the least complex to implement, but in severe fading conditions they exhibit large capacity losses relative to the other techniques.

The truncated channel inversion policy inverts the channel fading only above a fixed cutoff fade depth γ_0. The capacity with this truncated channel inversion and fixed rate policy $<C>_{tifr}/B$ is derived in [8]:

$$\langle C \rangle_{tifr} = B \log_2 \left(1 + 1 \bigg/ \int_{\gamma_0}^{\infty} \left(p_\gamma(\gamma)/\gamma \right) d\gamma \right) \left(1 - P_{out} \right). \tag{1.41}$$

5.1 κ-μ fading channels

After substituting (1.2) into (1.40) we can obtain expression for the CIFR channel capacity over κ-μ fading channel in the following form:

$$\langle C \rangle_{tifr} = B \log_2 \left(1 + 1 \bigg/ \sum_{p=0}^{\infty} f_3 \right) \left(1 - \sum_{i=0}^{\infty} \frac{(k\mu)^i}{e^{\mu k} \Gamma(i+\mu) i!} \Lambda \left(\mu + i, \frac{\mu(1+k)\gamma_0}{\bar{\gamma}} \right) \right) \tag{1.42}$$

$$f_3 = \sum_{p=0}^{\infty} \frac{\mu^{p+1} \kappa^p (1+\kappa) \Lambda \left(p + \mu - 1, \frac{\mu(1+k)\gamma_0}{\bar{\gamma}} \right)}{e^{\mu\kappa} \bar{\gamma} \Gamma(p+\mu) p!}$$

Case when MRC diversity is applied can be modelled by:

$$\langle C \rangle_{tifr}^{MRC} = B \log_2 \left(1 + 1 \bigg/ \sum_{p=0}^{\infty} f_4 \right) \left(1 - \sum_{i=0}^{\infty} \frac{(kL\mu)^i}{e^{L\mu k} \Gamma(i+L\mu) i!} \Lambda \left(\mu L + i, \frac{\mu L(1+k)\gamma_0}{\bar{\gamma}} \right) \right) \tag{1.43}$$

$$f_4 = \sum_{p=0}^{\infty} \frac{\mu^{p+1} \kappa^p (1+\kappa) L^p \Lambda \left(p + \mu L - 1, \frac{\mu L(1+k)\gamma_0}{\bar{\gamma}} \right)}{e^{\mu\kappa L} \bar{\gamma} \Gamma(p+\mu L) p!}$$

Convergence of infinite series expressions in (1.42) and (1.43) is rapid, since we need about 10-15 terms to be summed in order to achieve accuracy at the 5th significant digit.

5.2 Weibull fading channels

After substituting (1.6) into (1.41) we can obtained expression for the CIFR channel capacity over Weibull fading channels when MRC diversity is applied in the form of:

$$\frac{\left\langle C\right\rangle_{\text{tifr}}^{MRC}}{B} = \log_2\left(1 + \frac{\Xi\overline{\gamma}\Gamma(L)}{\Gamma\left(L - 2/\beta, \left(\gamma_0 / \Xi\overline{\gamma}\right)^{\beta/2}\right)}\right)\frac{\Gamma\left(L, \left(\gamma_0 / \Xi\overline{\gamma}\right)^{\beta/2}\right)}{\Gamma(L)}. \tag{1.44}$$

6. Numerical results

In order to discuss usage of diversity techniques and adaptation policies and to show the effects of various system parameters on obtained channel capacity, numerically obtained results are graphically presented.

In Figs. 1.1 and 1.8 channel capacity without diversity, <C>$_{ora}$ given by (1.22), for the cases when κ-μ and Weibull fading are affecting channels, for various system parameters are plotted against γ. These figures also display the capacity per unit bandwidth of an AWGN channel, C_{AWGN} given by:

$$C_{AWGN} = B\log_2(1 + \gamma). \tag{1.45}$$

Considering obtained results, with respect that C_{AWGN} = 3.46 dB for average received SNR of 10dB we find that depending of fading parameters of κ-μ and Weibull distribution, channel capacity could be reduced up to 30 %. From Fig. 1.1 we can see that channel capacity is less reduced for the cases when fading severity parameter μ, and dominant/scattered components power ratio κ, have higher values, since for smaller κ and μ values the dynamics in the channel is larger. Also from Fig. 1.8 we can observe that channel capacity is less reduced in the areas where Weibull fading parameter β has higher values.

Figures 1.2-1.4,1.6 show the channel capacity per unit bandwidth as a function of $\overline{\gamma}$ for the different adaptation policies with MRC diversity over κ-μ fading channels. It can be seen that as the number of combining branches increases the fading influence is progressively reduced, so the channel capacity improves remarkably. However, as L increases, all capacities of the various policies converge to the capacity of an array of L independent AWGN channels, given by:

$$C_{AWGN}^{MRC} = B\log_2(1 + L\gamma) \tag{1.46}$$

Thus, in practice it is not possible to entirely eliminate the effects of fading through space diversity since the number of diversity branches is limited. Also considering downlink (base station to mobile) implementation, we found that mobile receivers are generally constrained in size and power.

In Fig. 1.5 comparison of the channel capacity per unit bandwidth with CIFR adaptation policy, when SC and MRC diversity techniques are applied at the reception is shown. As expected, better performances are obtained when MRC reception over κ-μ fading channels is applied.

Figure 1.7 shows the calculated channel capacity per unit bandwidth as a function of $\overline{\gamma}$ for different adaptation policies. From this figure we can see that the OPRA protocol yields a small increase in capacity over constant transmit power adaptation and this small increase in capacity diminishes as $\overline{\gamma}$ increases. However, greater improvement is obtained in going from complete to truncated channel inversion policy. Truncated channel inversion policy provides better diversity gain compared to complete channel inversion varying any of parameters.

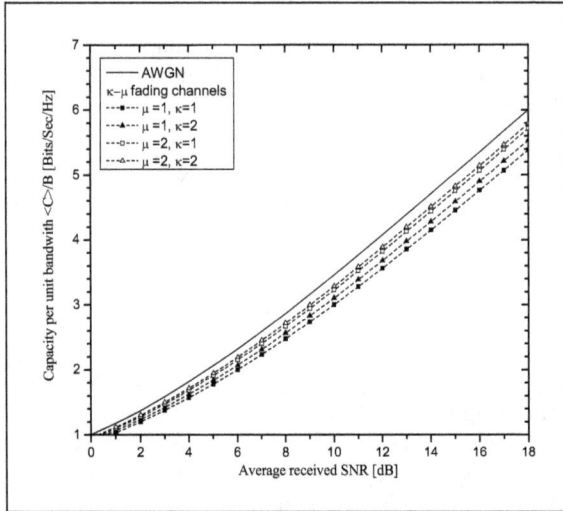

Fig. 1.1 Average channel capacity per unit bandwidth for a κ-μ fading and an AWGN channel versus average received SNR.

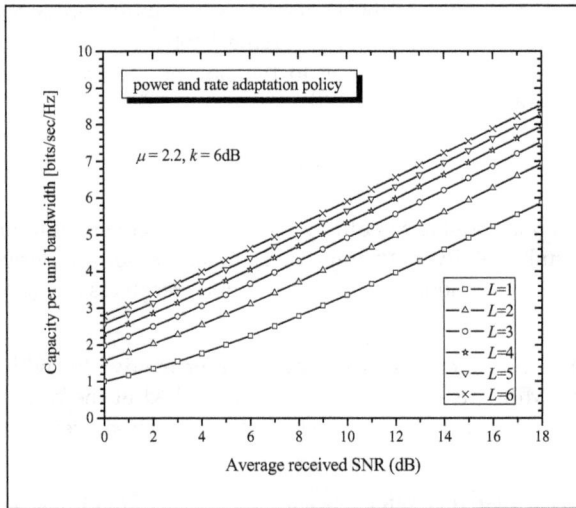

Fig. 1.2 Power and rate adaptation policy capacity per unit bandwidth over κ-μ fading channels, for various values of diversity order.

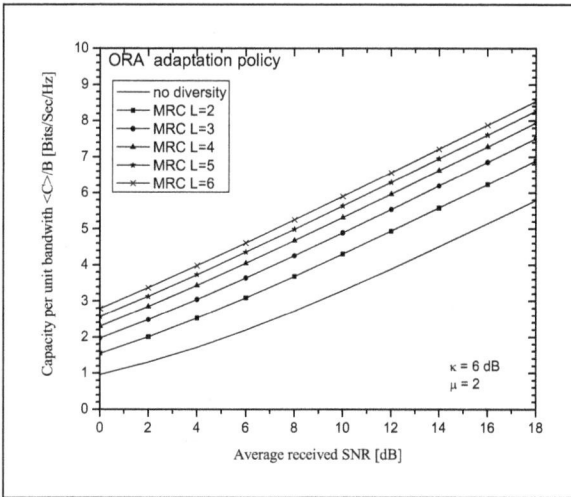

Fig. 1.3 ORA policy capacity per unit bandwidth over κ-μ fading channels, for various values of MRC diversity order.

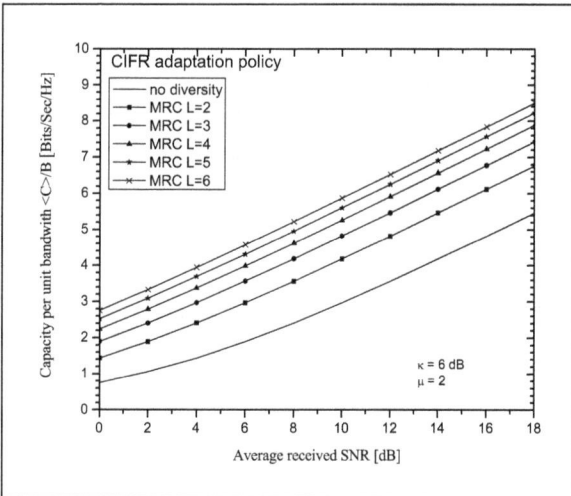

Fig. 1.4 CIFR policy capacity per unit bandwidth over κ-μ fading channels, for various values of MRC diversity order.

Similar results are presented considering channels affected by Weibull fading. Figures 1.9-1.12 show the channel capacity per unit bandwidth as a function of $\overline{\gamma}$ for the different adaptation policies with L-branch MRC diversity applied. Comparison of adaptation policies is presented at Fig. 1.13.

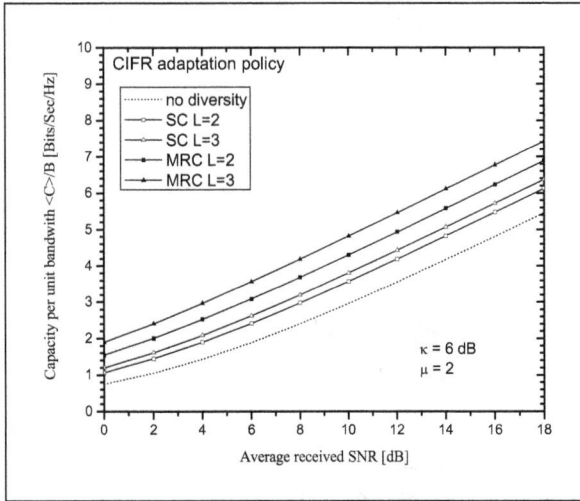

Fig. 1.5 CIFR policy capacity per unit bandwidth over κ-μ fading channels, for MRC and SC diversity techniques various orders.

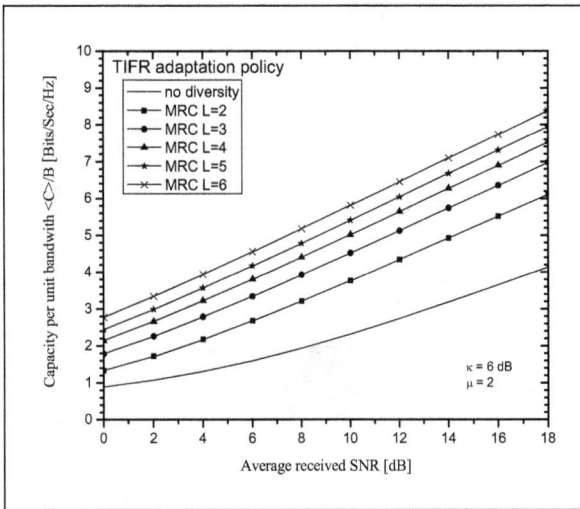

Fig. 1.6 TIFR policy capacity per unit bandwidth over κ-μ fading channels, for various values of MRC diversity order.

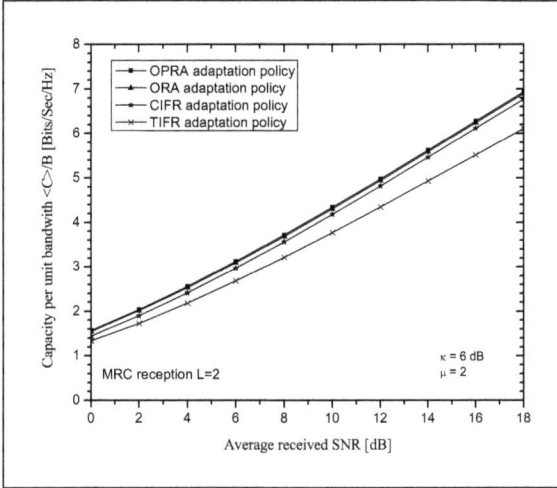

Fig. 1.7 Comparison of adaptation policies over MRC diversity reception in the presence of κ-μ fading.

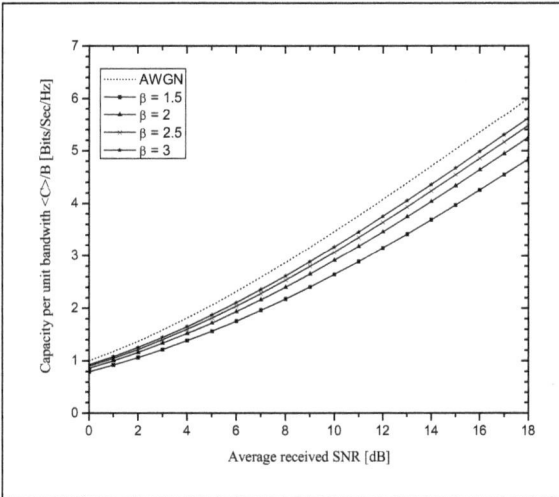

Fig. 1.8 Average channel capacity per unit bandwidth for a Weibull fading for various values of system parameters and an AWGN channel versus average received SNR [dB].

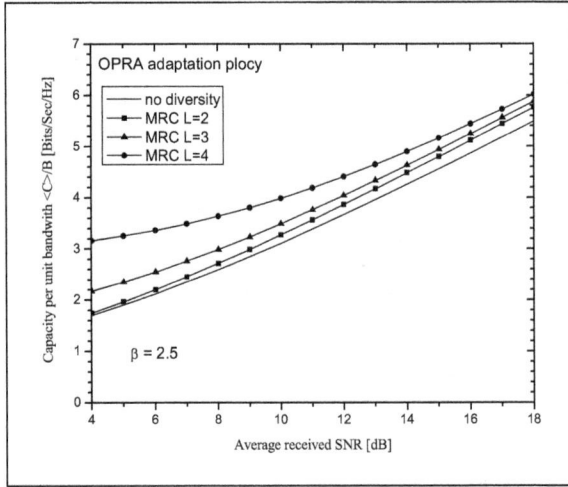

Fig. 1.9 ORPA policy capacity per unit bandwidth over Weibull fading channels, for various values of MRC diversity order.

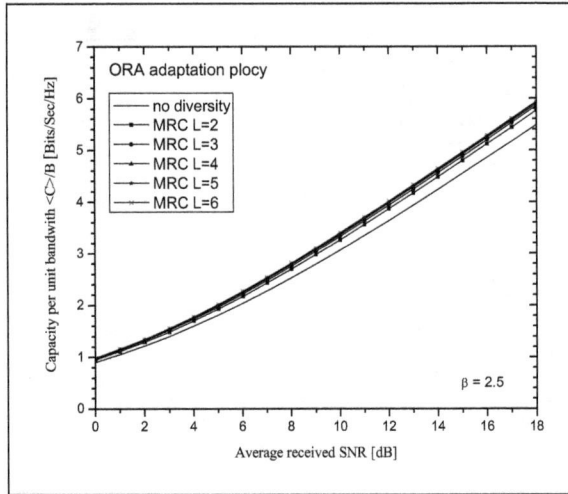

Fig. 1.10 ORA policy capacity per unit bandwidth over Weibull fading channels, for various values of MRC diversity order.

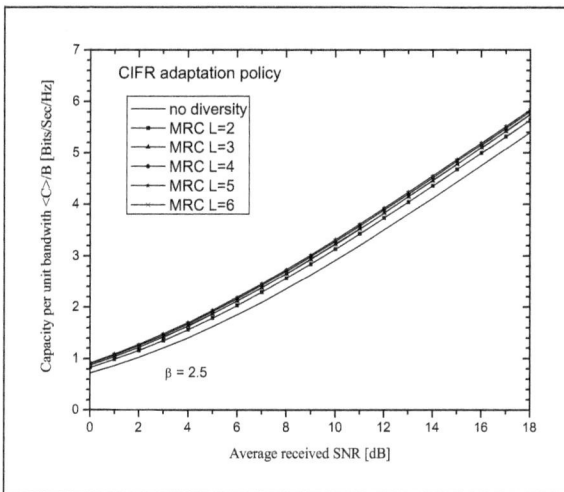

Fig. 1.11 CIFR policy capacity per unit bandwidth over Weibull fading channels, for various values of MRC diversity order.

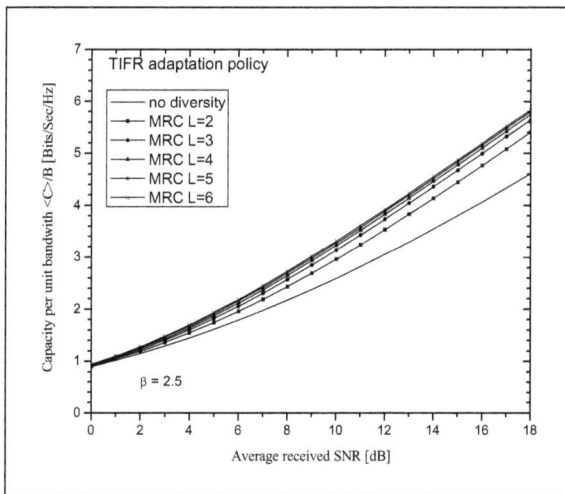

Fig. 1.12 TIFR policy capacity per unit bandwidth over Weibull fading channels, for various values of MRC diversity order.

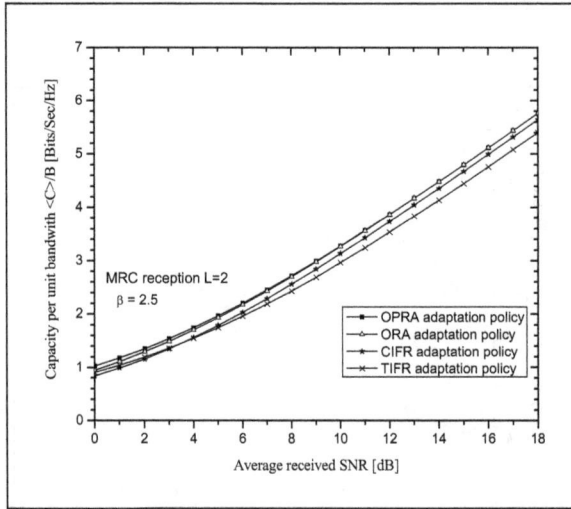

Fig. 1.13 Comparison of adaptation policies over MRC diversity reception in the presence of Weibull fading.

The nested infinite sums in (1.38) and (1.39), as can be seen from Table 1, for dual and triple branch diversity case, converge for any value of the parameters κ, μ and $\overline{\gamma}$. As it is shown in this Table 1, the number of the terms need to be summed to achieve a desired accuracy, depends strongly on these parameters and it increases as these parameter values increase.

Expression (1.38) 6th significant digit		$\overline{\gamma}$ = 5 dB	$\overline{\gamma}$ = 10 dB	$\overline{\gamma}$ = 15 dB
$\kappa = 1$	$\mu = 1$	8	9	10
$\kappa = 2$	$\mu = 2$	15	15	16
Expression (1.39) 6th significant digit		$\overline{\gamma}$ = 5 dB	$\overline{\gamma}$ = 10 dB	$\overline{\gamma}$ = 15 dB
$\kappa = 1$	$\mu = 1$	19	21	24
$\kappa = 2$	$\mu = 2$	23	26	28

Table 1. Number of terms that need to be summed in (1.38) and (1.39) to achieve accuracy at the specified significant digit for some values of system parameters.

7. Conclusion

Cases when wireless channels are affected by general and nonlinear fading distributions are disscused in this chapter. The analytical study of the κ-μ fading channel capacity, e.g., under the OPRA, ORA, CIFR and TIFR adaptation policies and MRC and SC diversity techniques is performed. The main contribution are closed-form expressions derived for the proposed adaptation policies and diversity techniques. Based on them, numerically obtained results are graphically presented in order to show the effects of various system parameters. Since κ-μ model as general physical fading model includes Rayleigh, Rician, and Nakagami-m fading models, as special cases, the generality and applicability of this analysis are more than obvious. Nonlinear fading scenario is discussed in the similar manner, as an analytical study of the Weibull fading channel capacity, under the OPRA, ORA, CIFR and TIFR adaptation policies and MRC diversity technique.

8. Acknowledgment

This paper was supported by the Serbian Ministry of Education and Science (projects: III44006 and TR32023).

9. References

[1] Goldsmith, A. & Varaiya, P. (1997). Capacity of fading channels with channel side information. *IEEE Transactions on Information Theory*, vol. 43, no. 6, (November 1997), pp. 1896–1992.

[2] Freeman, L. R. (2005). *Fundamentals of telecommunications*, John Wiley & sons, Hoboken, New Jersey, 2005.

[3] Sampei, S. ; Morinaga, N. & Kamio, Y. (1995). Adaptive modulation/TDMA with a BDDFE for 2 Mbit/s multimedia wireless communication systems, *Proceedings of the IEEE VTC'95*, pp. 311-315, 1995.

[4] Lee, W.C.Y. (2001). *Mobile communications engineering*, Mc-Graw-Hill, New York 2001.

[5] Ibnkahla, M. (2000). *Signal processing for mobile communications*, CRC Press LLC, Boca Raton, Florida, 2000.

[6] Brennan, D. (1959). Linear diversity combining techniques, *Proceedings of IRE*, vol.47, (June 1959), pp. 1075-1102.

[7] Alouini, M. S. & Goldsmith, A. (1999). Capacity of Rayleigh fading channels under different adaptive transmission and diversity-combining techniques, *IEEE Transactions on Vehicular Technology*, vol. 48, no. 4, (July 1999), pp. 1165-1181.

[8] Simon, M. & Alouini, M. S. (2000). *Digital communications over fading channels*, John Wiley & sons, New York, 2000.

[9] Shao, J.; Alouini ,M. & Goldsmith A. (1999). Impact of fading correlation and unequal branch gains on the capacity of diversity systems. *In: Proceedings of the IEEE vehicular technology conference (VTC-Spring'99)*, Houston, TX, May 1999. pp. 2159–2163.

[10] Bhaskar, V. (2008). Capacity evaluation for equal gain diversity schemes over Rayleigh fading channels, *AEÜ - International Journal of Electronics and Communications*, vol. 63, no. 9, pp. 235-240.

[11] Anastasov, J., Panic, S., Stefanovic M. & Milenkovic V. Capacity of correlative Nakagami-m fading channels under adaptive transmission and maximal-ratio combining diversity technique, accepted for publication in *Journal of Communications Technology and Electronics*

[12] Subadar, R. & Sahu, P. (2011). Channel capacity of adaptive transmission schemes using equal gain combining receiver over Hoyt fading channels, *Communications (NCC), 2011 National Conference on*, 28-30 Jan. 2011, Bangalore, pp 1-5

[13] Sagias, N. (2006) Capacity of dual-branch selection diversity receivers in correlative Weibull fading," *European Transactions on Telecommunications*, vol. 16, no. 1, (February 2006), pp. 37-43.

[14] Yacoub, M. (2007). The κ-μ distribution and the η-μ distribution, *IEEE Antennas and Propagation Magazine*, vol. 49 no. 1, January 2007, pp. 68-81.

[15] Milisic, M.; Hamza, M. & Hadzialic M. (2009). BEP/SEP and Outage Performance Analysis of L-BranchMaximal-Ratio Combiner for κ-μ Fading, *International Journal of Digital Multimedia Broadcasting*, vol. 2009, Article ID 573404, pp 1-8.

[16] Gradshteyn, I. & Ryzhik, I. (1994). *Table of Integrals, Series, and Products*, Academic Press, 5th ed., Orlando, 1994.

[17] Sagias, N.; Zogas, D. ; Karagiannidis, G. & Tombras, G. (2004). Channel Capacity and Second Order Statistics in Weibull Fading, *IEEE Communications Letters,* vol. 8, no. 6, (June 2004) pp. 377-379.

[18] Filho, J. & Yacoub, M. (2006). Simple Precise Approximations to Weibull Sums. *IEEE Communication Letters*, vol. 10, no. 8, (August 2006), pp. 614-616.

[19] Sagias, N. & Mathiopoulos, P. (2005). Switched diversity receivers over generalized Gamma fading channels, *IEEE Communications Letters*, vol. 9, no. 10, (October 2005), pp. 871-873.

[20] Abramowitz, M. & Stegun, I. (1970). *Handbook of Mathematical Functions*, Dover Publications, Inc., New York, 1970.

[21] Academik V. & Marichev, O. (1990). The algorithm for calculating integrals of hypergeometric type functions and its realization in REDUCE system *in Proc. Int. Conf. on Symbolic and Algebraic Computation*, Tokyo, Japan, 1990., pp. 212-224.

[22] The Wolfram Functions Site, 2008. [Online] Available: http://functions.wolfram.com

[23] Prudnikov, A.; Brychkov, Y. & Marichev, O. (1990). *Integral and Series*: Volume 3, More Special Functions. CRC Press Inc., 1990.

Superimposed Training-Aided Channel Estimation for Multiple Input Multiple Output-Orthogonal Frequency Division Multiplexing Systems over High-Mobility Environment

Han Zhang[1], Xianhua Dai[2], Daru Pan[1] and Shan Gao[1]
[1]School of Physics and Telecommunications Engineering,
South China Normal University, Guangzhou
[2]School of Information Science and Technology,
SUN Yat-sen University, Guangzhou
China

1. Introduction

The combination of multiple-input multiple-output (MIMO) antennas and orthogonal frequency-division multiplexing (OFDM) can achieve a lower error rate and/or enable high-capacity wireless communication systems by flexibly exploiting diversity gain and/or the spatial multiplexing gains. Such systems, however, rely upon the knowledge of propagation channels. In many mobile communication systems, transmission is impaired by both delay and Doppler spreads [1]-[7]. In such cases, explicit incorporation of the time-varying characteristics of mobile wireless channel is called for.

The coefficients of a linearly time-varying (LTV) channel can be usually modeled as uncorrelated stationary random processes which are assumed to be low-pass, Gaussian, with zero mean (Rayleigh fading) or non-zero mean (Rician fading) depending on whether line-of-sight propagation is absent or present [1][6]. Recently, the basis expansion models, i.e. the truncated discrete Fourier basis (DFT) models, polynomial models and discrete prolate Spheroidal sequence models, have gained special attentions, especially for the situation that channel is caused by a few strong reflectors and path delays exhibit variations due to the kinematics of the mobiles [1]-[2] [5]-[6] [16] [25]-[28].

In conventional pilot-aided channel estimation approaches, MIMO channels can be effectively estimated by utilizing the time-division multiplexed (TDM) and (or) frequency-division multiplexed (FDM) training sequences [5]-[7] [20]-[23] [25]. Although the channel estimates are in general reliable, extra bandwidth or time slot is required for transmitting known pilots. In recent years, an alternative approach, referred to as superimposed training (ST), has been extensively studied in [8]-[19] [26]-[28]. In the idea of ST, additional periodic training sequences are arithmetically added to information sequence in time- or frequency-domain. The advantage of the scheme is that there is no loss in information rate, and thus enables higher bandwidth efficiency. However, some useful power must inevitably be allocated to the pilots, and thus resulting in information signal-to-noise ratio (SNR) reduction. Meanwhile, the information sequences are viewed as interference to channel estimation since pilot symbols are superimposed at a low power to the information sequences at the transmitter. The existing ST-based channel estimations are mainly restricted to the case where the channel is linearly time-invariant (LTI), where the channel transfer function can be estimated by using first-order statistics [8]-[13] [17]-[18]. In the latest contributions, J. K. Tugnait [16] extended the conventional ST to time-varying environment where the LTV channels are modeled by complex exponential bases. For the issue of training power allocation, the optimal pilot power has been investigated by [24] for different taps of low-pass filter, and then, the optimization of ST power allocation for LTI channel is mathematically analyzed based on equalizer design [15] [19].

In this paper, a new ST-based channel estimator is proposed for OFDM/MIMO systems over LTV multipath fading channels. The main contributions are twofold. First, the LTV channel coefficients modeled by the truncated discrete Fourier bases (DFB), unlike the existing approaches [1]-[2] [5]-[6] [16], cover multiple OFDM symbols. Then, a two-step channel estimation approach is adopted for LTV channel estimation. Furthermore, a closed-form expression of the estimation variance is derived, which provides a guideline for designing the superimposed pilot symbols. We demonstrate by analytical analysis that the estimation variance, unlike that of conventional ST-based schemes [8]-[19], approaches to a fixed lower-bound as the training length increases. Second, for wireless communication systems with a limited transmission power, unlike [10] where the issue of ST power allocation is derived by optimizing the SNR for equalizer design, we provide an optimal solution of ST power allocation with a different point of view by maximizing the lower-bound of channel capacity. Comparatively, the training power allocation scheme [10] can be otherwise considered as a special case compared with the proposed approach. In simulations presented in this paper, we compare the results of our approach with that of the FDM training approaches [5] as latter serves as a "benchmark" in related works. It is shown that the proposed algorithm outperforms that of FDM training, and yields higher transmission efficiency.

The rest of the paper is organized as follows. Section II presents the system and channel models. In Section III, we estimate the LTV channel coefficients with the proposed two-step channel estimation approach. In Section IV, we derive the closed-form expression of the channel estimation variances. Section V determines the optimal ratio of the ST power to the total transmission power by maximizing the lower-bound of channel capacity. Section VI reports on some simulation experiments in order to test the validity of theoretic results, and we conclude the paper with Section VII.

Notation: The letter t represents the time-domain variable and k is the frequency-domain variable. Bold letters denote the matrices and column-vectors, and the superscripts $[\bullet]^T$ and $[\bullet]^H$ represent the transpose and conjugate transpose operations, respectively. $[\bullet]_{k,t}$ denotes the (k, t) element of the specified matrix.

2. System and channel model

2.1 System model

Consider an MIMO/OFDM system of N transmitters or mobile users and a receive array of M receive antennas with perfect synchronization. At transmit terminals, an inverse fast Fourier transform (IFFT) is used as a modulator. The modulated outputs are given by

$$\mathbf{X}_n(i) = \left[x_n(i,0), \cdots x_n(i,t), \cdots x_n(i,B-1)\right]^T = \mathbf{F}^{-1}\mathbf{S}_n(i) \qquad n = 1, \cdots N \qquad (1)$$

where B is OFDM symbol-size, $\mathbf{S}_n(i) = [s_n(i,0), \cdots s_n(i,k), \cdots s_n(i,B-1)]^T$ is the i th transmitted symbol of the n th transmit antenna. $\mathbf{F}^{-1}$ is the IFFT matrix with $[\mathbf{F}^{-1}]_{k,t} = e^{j2\pi kt/B}$ and $j^2 = -1$. Then, $\mathbf{X}_n(i)$ is concatenated by a cyclic-prefix (CP) of length $\bar{L}$, propagating through the respective channels. At receiver, the received signals of m th receive antenna, discarding CP and stacking the received signals $y^{(m)}(i,t)$ $t = 0, \cdots B-1$, can be written in a vector-form as

$$\mathbf{Y}^{(m)}(i) = \left[y^{(m)}(i,0), \cdots y^{(m)}(i,t), \cdots y^{(m)}(i,B-1)\right]^T \qquad m = 1, \cdots M \qquad (2)$$

and the received signals $y^{(m)}(i,t)$ in (2) is given by

$$y^{(m)}(i,t) = \sum_{n=1}^{N} \mathbf{X}_n(i) \otimes \mathbf{h}_n^{(m)}(i) + v^{(m)}(i,t)$$

$$(3)$$

$$= \sum_{n=1}^{N}\sum_{l=0}^{L-1} h_{n,l}^{(m)}(i,t) x_n(i,t-l) + v^{(m)}(i,t) \qquad t = 0, \cdots B-1$$

where $\mathbf{h}_n^{(m)}(i) = \left[h_{n,0}^{(m)}(i,t), \cdots h_{n,L-1}^{(m)}(i,t), 0_{1\times B-L}\right]^T$ is the impulse response vector of the propagating channel from the n th transmit to the m th receive antenna. The channel coefficients $h_{n,l}^{(m)}(i,t), l = 0, \cdots L-1$ is the functions of time variable t which will be defined by (6). The notation $\otimes$ represents the cyclic convolution and $v^{(m)}(i,t)$ is the additive Gaussian noise.

At receiver, an FFT operation is performed on the vector (2), and the demodulated outputs can be written as

$$\mathbf{U}^{(m)}(i) = \left[u^{(m)}(i,0), \cdots u^{(m)}(i,k), \cdots u^{(m)}(i,B-1)\right]^T = \mathbf{F}\mathbf{Y}^{(m)}(i) \qquad m = 1, \cdots M. \qquad (4)$$

From (3) and the duality of time and frequency, the FFT demodulated signals in (4) can be written as

$$u^{(m)}(i,k) = FFT\left\{\sum_{n=1}^{N}\sum_{l=0}^{L-1} h_{n,l}^{(m)}(i,t) x_n(i,t-l) + v^{(m)}(i,t)\right\}$$

$$(5)$$

$$= \sum_{n=1}^{N}\sum_{l=0}^{L-1} FFT\left\{h_{n,l}^{(m)}(i,t)\right\} \otimes FFT\left\{x_n(i,t)\right\} + \overline{v}^{(m)}(i,k)$$

where $\text{FFT}\{\bullet\}$ represents the FFT vector of the specified function and $\bar{v}^{(m)}(i,k)$ is the frequency-domain noise. Compared with the FFT demodulated signals of OFDM systems with LTI channels, the convolution in (5) between the information sequences and the FFT vectors of time-varying channel coefficients may introduce inter-carrier interference (ICI).

2.2 Channel model

As mentioned in [1], the coefficients of the time- and frequency-selective channel can be modeled as Fourier basis expansions. Thereafter, this model was intensively investigated and applied in block transmission, channel estimation and equalization (e.g. [2][5]-[6][16]). In this paper, we extend the block-by-block process [2][5]-[6][16] to the case where multiple OFDM symbols are utilized. Consider a time interval or segment $\{t:(\ell-1)\Omega \le t \le \ell\Omega\}$, the channel coefficients in (3) can be approximated by truncated discrete Fourier bases (DFB) within the segment as

$$h_{n,l}^{(m)}(i,t) \approx \sum_{q=0}^{Q} h_{n,l,q}^{(m)} e^{-j2\pi(q-Q/2)t/\Omega}$$

$$(6)$$

$$= \sum_{q=0}^{Q} h_{n,l,q}^{(m)} \eta_q(t) \quad t=(\ell-1)\Omega,\cdots\ell\Omega , \ \ell=1,2,\cdots$$

where $h_{n,l,q}^{(m)}$ is a constant coefficient, Q represents the basis expansion order that is generally defined as $Q \ge 2f_d\Omega/f_s$ [1], $\Omega > B$ is the segment length and ℓ is the segment index. Unlike [1]-[2] [5]-[6] [16], the approximation frame Ω covers multiple OFDM symbols, denoted by $i=1,\cdots I$, where $I=\Omega/B'$ and $B'=B+\bar{L}$. Since the proposed two-step channel estimation as will be shown in Section III is adopted within one frame, we omit the segment index ℓ for simplicity.

3. ST-based channel estimation

In this section, we propose a ST-based two-step approach for LTV channel estimation. In ST-based approaches [8]-[19], the pilot symbols are superimposed (arithmetically added) to the information sequences as

$$s_n(i,k) = c_n(i,k) + p_n(i,k) \quad k=0,\cdots B-1$$

$$(7)$$

where $c_n(i,k)$ and $p_n(i,k)$ are the information and pilot sequence, respectively. Compared with the FDM/TDM training aided methods [20]-[22], ST requires no additional bandwidth (or time-slot) for transmitting known pilots, and thus offers a higher data rate.

3.1 ST-based channel estimation over one OFDM symbol

For LTV environment where the channel coefficient $h_{n,l}^{(m)}(t)$ is a function of time variable t, the vectors $\text{FFT}\{h_{n,l}^{(m)}(t)\}$ in (5) cannot be approximated as a δ-sequences and, the FFT demodulated signals at the sub-carrier k of the i th symbol is given by

$$u^{(m)}(i,k) = \sum_{n=1}^{N}\sum_{l=0}^{L-1} FFT\left\{h_{n,l}^{(m)}(i,t)\right\}\otimes P_n(i) + \overline{v}^{\prime(m)}(i,k)$$

$$\approx \sum_{n=1}^{N}\sum_{l=0}^{L-1} FFT\left\{\sum_{q=0}^{Q}h_{n,l,q}^{(m)}\eta_q(t)\right\}\otimes P_n(i) + \overline{v}^{\prime(m)}(i,k) \tag{8}$$

$$= \sum_{n=1}^{N}\sum_{q=0}^{Q} H_{n,q}^{(m)}(i,k)\eta_q(t_i)W_q(i,k)\otimes P_n(i) + \overline{v}^{\prime(m)}(i,k)$$

where $\overline{v}^{\prime(m)}(i,k) = \sum_{n=1}^{N}\sum_{l=0}^{L-1} FFT\left\{h_{n,l}^{(m)}(t)\right\}\otimes C_n(i) + \overline{v}^{(m)}(i,k)$ and $t_i = (i-1)B' + B/2$. $W_q(i,k)$ with $k = 0,\cdots B-1$ is the FFT vector of the complex exponential function (CEF) and, can be written as

$$W_q(i,0) = \left[w_q(i,0),\cdots w_q(i,k),\cdots w_q(i,B-1)\right]^T \tag{9}$$

$$= F\left[\eta_q(t_i - B/2)/\eta_q(t_i),\cdots \eta_q(t_i + B/2-1)/\eta_q(t_i)\right]^T .$$

Notice that $W_q(i,k)$ is a cyclic-shifted vector of $W_q(i,0)$ with a shifting length k . On the other hand, ICI introduced by the cyclic convolution $W_q(i,k)\otimes S_n(i)$ depends explicitly on $\eta_q(t), t = 0,\cdots B-1$. When q is not large, the complex exponential functions in (9) are slowly time-varying over an OFDM symbol-duration and, thereby, the principal power or major-lobe of the FFT vector $W_q(i,0)$ may concentrate on its two ends (low frequency tones) with indexes $0,\cdots T$ and $B-T,\cdots B-1$. Using the major-lobe to approximate the CEF vectors $W_q(i,0)$ $q = 0,1,\cdots Q$, we have

$$W_q(i,0) \approx \left[w_q(i,0),\cdots w_q(i,T),0,\cdots\cdots 0,w_q(i,B-T),\cdots w_q(i,B-1)\right]^T \quad q = 0,1,\cdots Q \tag{10}$$

where T is a positive integer.

In general, the FFT vector of the function $\eta_q(t)$ in (10) may have a great side-lobe that results in a great error. For improving the approximation performance, an intuitional idea is to apply a window function to the received signals in order to reduce the side-lobe leakage. The windowed vector of received signals in (3) of the i th symbol is

$$\overline{y}^{(m)}(i,t) = \sum_{n=1}^{N} h_n^{(m)}(i,t)\psi_B(t)x_n(t-1) + v^{(m)}(t)\psi_B(t) \quad t = 0,\cdots B-1 \tag{11}$$

where $\psi_B(t)$ is a time-domain windowing function with a length B . Performing the FFT demodulated operation on the windowed sequences in (10), the demodulated signals, by (10), can be written by

$$u^{(m)}(i,k) \approx \sum_{n=1}^{N}\sum_{l=0}^{L-1} FFT\left\{\sum_{q=0}^{Q}h_{n,l,q}^{(m)}\eta_q(t)\psi_B(t)\right\}\otimes P_n(i) + \overline{v}^{\prime(m)}(i,k)$$

$$= \sum_{n=1}^{N}\sum_{q=0}^{Q} H_{n,q}^{(m)}(i,k)\eta_q(t_i)\overline{W}_q(i,k)\otimes P_n(i) + \overline{v}^{\prime(m)}(i,k) \tag{12}$$

where $\overline{W}_q(i,k)$ is the CEF vector with the windowing function $\psi_B(t)$ as

$$
\begin{aligned}
\overline{W}_q(i,k) &= \left[\overline{w}_q(i,0),\cdots\overline{w}_q(i,k),\cdots\overline{w}_q(i,B-1)\right]^T \\
&= \mathbf{F}\left[\psi_B(t)\eta_q(t_i-B/2)\big/\eta_q(t_i),\cdots\psi_B(t)\eta_q(t_i+B/2-1)\big/\eta_q(t_i)\right]^T \\
&\approx \left[\overline{w}_q(i,0),\cdots\overline{w}_q(i,T),0,\cdots0,\overline{w}_q(i,B-T),\cdots\overline{w}_q(i,B-1)\right]^T.
\end{aligned}
\tag{13}
$$

Compared with (10), the approximation of windowing based vector has a much smaller side-lobe with the same index T. The experiment studies show that by using a *Kaiser* function [5], the approximation in (13) of $T = 2$ may capture almost 99% power of $FFT\left[\psi_B(t)\eta_q(t_i-B/2:t_i+B/2-1)\big/\eta_q(t_i)\right]^T$ for truncated DFBs when $q < B/10$. Substituting (13) into (12), the FFT demodulated outputs can be approximated by

$$
\begin{aligned}
u^{(m)}(i,k) = \sum_{n=1}^{N}\sum_{q=0}^{Q} H_{n,q}^{(m)}(i,k)\eta_q(t_i)\Bigg[\sum_{k'=0}^{T} \overline{w}_q(i,k')p_n(i,k-k') + \\
\sum_{k'=T}^{1} \overline{w}_q(i,B-k')p_n(i,k+k')\Bigg] + \overline{v}^{(m)}(i,k).
\end{aligned}
\tag{14}
$$

The first term of (14) illustrates that $2T+1$ tones, i.e. $p_n(i,k-k'),\cdots p_n(i,k+k')$ should be jointly designed for estimating $H_{n,q}^{(m)}(i,k)$. We refer to such $2T+1$ consecutive pilot tones as a pilot cluster for differentiating from the isolated tones utilized in the LTI channel estimation [19] [22]-[23]. Denote $k_1,\cdots k_\Gamma$ as the pilot cluster indexes located at the τ th pilot symbol and, $p_n(k_\tau-T),\cdots\ p_n(k_\tau+T)$ as the pilot sequences at the pilot cluster k_τ. Since the ST does not entail additional bandwidth, two adjacent pilot-clusters, i.e. k_τ and $k_{\tau+1}$ can be placed closed together. The pilot tone distribution is shown in Fig. 1.

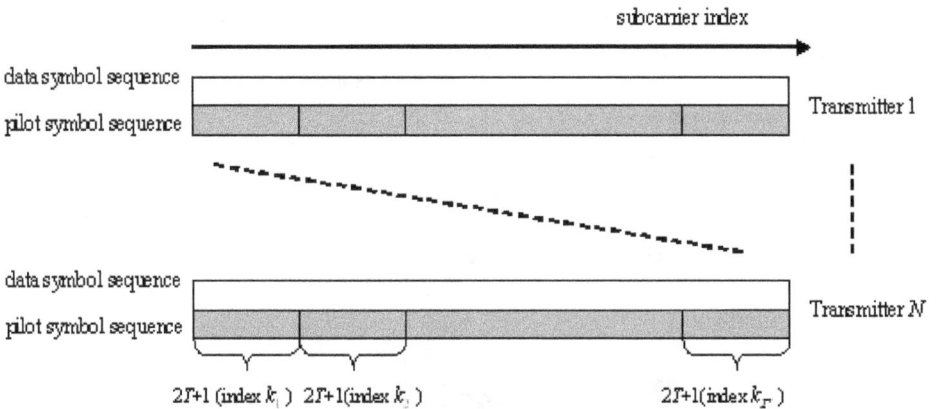

Fig. 1. A typical pilot tone distribution. $2T+1$ consecutive tones are grouped together as one pilot cluster. All pilot clusters are uniformly distributed in frequency domain with each adjacent pilot cluster being closed together.

Then, we focus on ST design. From (14), when the training sequence at each pilot cluster is designed as either a constant modulus sequence, i.e.

$$p_n(i,k_\tau) = p_n(i,k_\tau \pm k') \quad \tau = 1,\cdots \Gamma, \ k' = 1,\cdots T \tag{15}$$

or a δ sequence, i.e.

$$p_n(i,k_\tau \pm k') = \begin{cases} p_n(i,k_\tau), k'=0 \\ 0, \quad otherwise \end{cases} \quad \tau = 1,\cdots \Gamma, \ k' = 1,\cdots T \ . \tag{16}$$

Accordingly, the FFT demodulated outputs at pilot cluster k_τ can be approximated as

$$u^{(m)}(i,k_\tau) \approx \sum_{n=1}^{N} \sum_{q=0}^{Q} H_{n,q}^{(m)}(i,k_\tau) \eta_q(t_i) g_q(i,k_\tau) p_n(i,k_\tau) + \overline{v}^{\,\prime(m)}(i,k_\tau)$$

$$= \sum_{n=1}^{N} H_n^{\,\prime(m)}(i,k_\tau) p_n(i,k_\tau) + \overline{v}^{\,\prime(m)}(i,k_\tau) \tag{17}$$

where $g_q(i,k_\tau) = \sum_{k'=0}^{T} \overline{w}_q(i,k_\tau + k') + \sum_{k'=T}^{1} \overline{w}_q(i,k_\tau + B - k')$ if the training sequence takes value from (15) and $g_q(i,k_\tau) = \overline{w}_q(i,k_\tau)$ for (16), which are all known, respectively. The channel transfer functions $H_n^{\,\prime(m)}(i,k_\tau)$ is given by

$$H_n^{\,\prime(m)}(i,k_\tau) = \sum_{q=0}^{Q} H_{n,l,q}^{(m)}(i,k_\tau) \eta_q(t_i) g_q(i,k_\tau)$$

$$= \sum_{l=0}^{L-1} \sum_{q=0}^{Q} h_{n,l,q}^{(m)} \eta_q(t_i) g_q(i,k_\tau) e^{-j2\pi k_\tau l/B} \approx \sum_{l=0}^{L-1} \hbar_{n,l}^{(m)}(t_i) e^{-j2\pi k_\tau l/B} \ . \tag{18}$$

From (17)-(18), we note that $H_n^{\,\prime(m)}(i,k_\tau), \tau = 1,\cdots \Gamma$ is in fact a LTI system transfer function of which the coefficients are the mid-values of the LTV channel at the i th OFDM symbol interval. As a result, the LTV channel estimation can be approximately reduced into that of the LTI channel [22] and [23] by simply designing the ST sequences as (15) or (16).

Let $\mathbf{H}^{(m)}(i) = \left[\hbar_{1,0}^{(m)}(t_i), \cdots \hbar_{1,L-1}^{(m)}(t_i), \cdots \cdots \hbar_{N,0}^{(m)}(t_i), \cdots \hbar_{N,L-1}^{(m)}(t_i) \right]^T$ be the channel coefficient vector associated with the i th OFDM symbol and stack the FFT demodulated signals at pilot clusters of the i th OFDM symbol to form a vector

$$\mathbf{U}^{(m)}(i,k_1 : k_\Gamma) = \left[u^{(m)}(i,k_1), \cdots u^{(m)}(i,k_\tau), \cdots u^{(m)}(i,k_\Gamma) \right]^T \ . \tag{19}$$

The received signals at pilot clusters can be thus written as

$$\mathbf{U}^{(m)}(i,k_1 : k_\Gamma) = \underbrace{\mathbf{A}(i)\mathbf{H}^{(m)}(i)}_{\text{desired signal for channel estimation}} + \underbrace{\Xi^{(m)}(i,k_1 : k_\Gamma)}_{\text{information interference on channel estimation}}$$

$$+ \ \overline{\mathbf{V}}^{(m)}(i,k_1 : k_\Gamma) \tag{20}$$

where $\bar{\mathbf{V}}^{(m)}(i,k_1:k_\Gamma)$ is the noise vector in frequency-domain, $\Xi^{(m)}(i,k_1:k_\Gamma) = \left[\Xi^{(m)}(i,k_1),\cdots\Xi^{(m)}(i,k_\Gamma)\right]^T$ is the interference vector produced by the information sequences with $\Xi^{(m)}(i,k_\tau) = \sum_{n=1}^{N} H_n'^{(m)}(i,k_\tau)c_n(i,k_\tau)$, $\mathbf{A}(i) = \left[\mathbf{A}(1,0),\cdots\mathbf{A}(1,L-1)\cdots\mathbf{A}(n,l)\cdots\mathbf{A}(N,0),\cdots\mathbf{A}(N,L-1)\right]$ is a $\Gamma \times NL$ matrix with the column-vectors

$$\mathbf{A}(n,l) = \left[p_n(i,k_1)e^{-j2\pi k_1 l/B},\cdots p_n(i,k_\tau)e^{-j2\pi k_\tau l/B},\cdots p_n(i,k_\Gamma)e^{-j2\pi k_\Gamma l/B}\right]^T. \tag{21}$$

Since the matrix $\mathbf{A}(i)$ is known, when $\Gamma \geq NL$, the matrix $\mathbf{A}(i)$ is of full column rank, and the channel coefficient vectors can be thus estimated by

$$\hat{\mathbf{H}}^{(m)}(i) = \mathbf{A}^{\dagger}\mathbf{U}^{(m)}(i,k_1:k_\Gamma)$$
$$= \mathbf{H}^{(m)}(i) + \mathbf{A}(i)^{\dagger}\Xi^{(m)}(i,k_1:k_\Gamma) + \mathbf{A}(i)^{\dagger}\bar{\mathbf{V}}^{(m)}(i,k_1:k_\Gamma)\ m=1,\cdots M,i=1,\cdots I \tag{22}$$

where the superscript '$\dagger$' is the pseudo-inverse operation, and the hat '$\wedge$' indicates the estimation. From (22), the mainly computational effort is directly proportional to the unknown parameter number NL.

Using the specifically designed ST sequences in (15) and (or) (16), the problem of LTV channel estimation for MIMO/OFDM systems can be reduced into that of LTI channel. From (20) and (22), however, we notice that the interference vector due to information sequence can hardly be neglected since the power of data symbol is much larger than the pilot power. For conventional ST based schemes stated in [8]-[13] [17]-[18], first-order statistics are employed to suppress the information sequence interference over multiple training periods in the case that the channel is LTI during the record length. Such arithmetical average process, however, is no longer feasible to the channel assumed in this paper where the channel coefficients are linearly time-variant between consecutive OFDM symbols.

3.2 Channel estimation over multiple OFDM symbols

In this sub-section, a weighted average approach is developed to suppress the abovementioned information sequence interference over multiple OFDM symbols, and thus overcoming the shortcoming of the existing ST-based approach in estimating the time-variant channels.

By (22), the LTV channel coefficients can be obtained following the relationship $h_{n,l}^{(m)}(t_i) = \sum_{q=0}^{Q} h_{n,l,q}^{(m)}\eta_q(t_i)$. Taking the LTV channel coefficient estimation of each OFDM symbol $\hat{h}_{n,l}^{(m)}(t_i)\ i=1,\cdots I$ by (22) as a temporal result, and form a vector as $\hat{\mathbf{h}}_{n,l}^{(m)} = \left[\ \hat{h}_{n,l}^{(m)}(t_1),\cdots\hat{h}_{n,l}^{(m)}(t_I)\ \right]^T$, we thus have

$$\hat{\mathbf{h}}_{n,l}^{(m)} = \boldsymbol{\eta}\hat{\mathbf{h}}_{n,l,q}^{(m)}$$

$$= \begin{bmatrix} e^{j2\pi(0-Q/2)t_1/\Omega} & \cdots & e^{j2\pi(Q-Q/2)t_1/\Omega} \\ \vdots & \ddots & \vdots \\ e^{j2\pi(0-Q/2)t_I/\Omega} & \cdots & e^{j2\pi(Q-Q/2)t_I/\Omega} \end{bmatrix} \begin{bmatrix} \hat{h}_{n,l,0}^{(m)} \\ \vdots \\ \hat{h}_{n,l,Q}^{(m)} \end{bmatrix} \quad n = 1, \cdots N, l = 0, \cdots L-1 \tag{23}$$

where $\hat{\mathbf{h}}_{n,l,q}^{(m)} = \left[\hat{h}_{n,l,0}^{(m)}, \cdots, \hat{h}_{n,l,q}^{(m)} \cdots \hat{h}_{n,l,Q}^{(m)} \right]^{T}$ is estimation of the complex exponential coefficients

vector modeling the LTV channel, $\boldsymbol{\eta}$ is a $I \times (Q+1)$ matrix with $[\boldsymbol{\eta}]_{q,i} = e^{j2\pi(q-Q/2)t_i/\Omega}$. Thus,

when $I \geq Q+1$, the matrix $\boldsymbol{\eta}$ is of full column rank, and the basis expansion model
coefficients can be computed by

$$\hat{\mathbf{h}}_{n,l,q}^{(m)} = \boldsymbol{\eta}^{\dagger}\hat{\mathbf{h}}_{n,l}^{(m)} \quad n = 1, \cdots N, l = 0, \cdots L-1. \tag{24}$$

Substituting $t_i = (i-1)B' + B/2$ into the matrix $\boldsymbol{\eta}$, we obtain the pseudo-inverse matrix as

$$\left[\boldsymbol{\eta}^{\dagger} \right]_{i,q} = e^{-j2\pi(q-Q/2)((i-1)B'+B/2)/\Omega}\Big/I. \tag{25}$$

By (23)-(25), the modeling coefficients (6) can be computed by

$$\hat{h}_{n,l,q}^{(m)} = \sum_{i=1}^{I} e^{-j2\pi(q-Q/2)((i-1)B'+B/2)/\Omega}\hat{h}_{n,l}^{(m)}(i)\Big/I. \tag{26}$$

In fact, (26) is estimated over multiple OFDM symbols with a weighted average function
of $e^{-j2\pi(q-Q/2)t_i/\Omega}\big/I$.

Compared with the conventional ST strategies, the proposed channel estimation is
composed of two steps: First, with specially designed ST signals in (15) and (16), channel
estimation can be reduced into that of LTI channel, and we are allowed to estimate the
channel coefficients during each OFDM symbol as temporal results. Second, the temporal
channel estimates are further enhanced over multiple OFDM symbols by using a weighted
average procedure. That is, not only the target OFDM symbol, but also the OFDM symbols
over the whole frame are invoked for channel estimation. Similar to the first-order statistics
of LTI case [8]-[13] [17]-[18], it is thus anticipated that the weighted average estimation may
also exhibit a considerable performance improvement for the LTV channels over a long
frame Ω.

4. Channel estimation analysis

In this section, we analyze the performance of the channel estimator proposed in Section III
and derive a closed-form expression of the channel estimation variance which can be, in
turn, used for ST power allocation. Before going further, we make the following
assumptions:

(H1) The information sequence $\{c_n(i,k)\}$ is zero-mean, finite-alphabet, i.i.d., and equi-powered with the power σ_c^2.

(H2) The additive noise $\{v^{(m)}(i,t)\}$ is white, uncorrelated with $\{c_n(i,k)\}$, with $E\left[\left|v^{(m)}(i,t)\right|^2\right] = \sigma_v^2$.

(H3) The LTV channel coefficients $\mathbf{h}_{n,l}^{(m)}$ are complex Gaussian variables, and statistically independent for different values of n and l.

From (22)-(26), the mean square error (MSE) of channel estimation is given by

$$
\begin{aligned}
MSE^{(m)} &= E\left\{\sum_{n=1}^{N}\sum_{l=0}^{L-1}\left\|h_{n,l}^{(m)}(i,t) - \sum_{q=0}^{Q}\hat{h}_{n,l,q}^{(m)}\eta_q(t)\right\|^2\right\} \\
&= E\left\{\sum_{n=1}^{N}\sum_{l=0}^{L-1}\left\|h_{n,l}^{(m)}(i,t) - \sum_{q=0}^{Q}h_{n,l,q}^{(m)}\eta_q(t) + \sum_{q=0}^{Q}h_{n,l,q}^{(m)}\eta_q(t) - \sum_{q=0}^{Q}\hat{h}_{n,l,q}^{(m)}\eta_q(t)\right\|^2\right\}
\end{aligned}
\tag{27}
$$

where $\|\ \|$ is the Euclidean norm. In (27), the first error term $\sum_{n=1}^{N}\sum_{l=0}^{L-1}\left[h_{n,l}^{(m)}(i,t) - \sum_{q=0}^{Q}h_{n,l,q}^{(m)}\eta_q(t)\right]$ is caused by the orthonormal basis expansion model in (6), which is referred to as the channel modeling error. The second error term $\sum_{n=1}^{N}\sum_{l=0}^{L-1}\sum_{q=0}^{Q}\left[h_{n,l,q}^{(m)}\eta_q(t) - \hat{h}_{n,l,q}^{(m)}\eta_q(t)\right]$ is duo to the information interference to channel estimation (22) and additive noise. Explicitly, two error signals are mutually independent. Herein, we do not elaborate the topic of channel modeling error and focus on channel estimation error, which is mainly produced by the interference of information sequence. By (H2), the MSE of the estimation in one OFDM symbol can be written as

$$
\begin{aligned}
MSE^{(m)}(i) &\overset{def}{=} \frac{1}{(Q+1)NL}E\left\{\sum_{n=1}^{N}\sum_{l=0}^{L-1}\sum_{q=0}^{Q}\left\|h_{n,l,q}^{(m)}\eta_q(t) - \hat{h}_{n,l,q}^{(m)}\eta_q(t)\right\|^2\right\} \\
&= \underbrace{\frac{1}{(Q+1)NL}\mathrm{tr}\left\{\mathbf{A}(i)^{\dagger}E\left\{\Xi^{(m)}(i,k_1:k_\Gamma)\left(\Xi^{(m)}(i,k_1:k_\Gamma)\right)^{H}\right\}\left(\mathbf{A}(i)^{\dagger}\right)^{H}\right\}}_{\text{estimation variance due to information sequence interference}} \\
&+ \underbrace{\frac{1}{(Q+1)NL}\mathrm{tr}\left\{\mathbf{A}(i)^{\dagger}E\left\{\overline{\mathbf{V}}^{(m)}(i,k_1:k_\Gamma)\left(\overline{\mathbf{V}}^{(m)}(i,k_1:k_\Gamma)\right)^{H}\right\}\left(\mathbf{A}(i)^{\dagger}\right)^{H}\right\}}_{\text{estimation variance due to additive noise}}.
\end{aligned}
\tag{28}
$$

For zero-mean white noise, we have

$$
E\left\{\overline{\mathbf{V}}^{(m)}(i,k_1:k_\Gamma)\left(\overline{\mathbf{V}}^{(m)}(i,k_1:k_\Gamma)\right)^{H}\right\} = \sigma_v^2\mathbf{I}_\Gamma.
\tag{29}
$$

Invoking the assumption (H1), information sequence interference $\Xi^{(m)}(i,k_\tau) = \sum_{n=1}^{N}H_n^{\prime(m)}(i,k_\tau)c_n(i,k_\tau), \tau = 1,\cdots\Gamma$ is approximately Gaussian distributed for a large Γ. Therefore, the channel estimation variance due to information sequence interference can be obtained as

$$E\left\{\Xi^{(m)}(i,k_1:k_\Gamma)\left(\Xi^{(m)}(i,k_1:k_\Gamma)\right)^H\right\} = \frac{\sigma_c^2}{\Gamma^2}\sum_{\tau=0}^{\Gamma-1}\sum_{n=1}^{N}\sum_{l=0}^{L-1}\left|H_n^{(m)}(i,k_\tau)\right|^2 = \frac{\rho_c}{\Gamma}\sum_{\tau=0}^{\Gamma-1}\sum_{n=1}^{N}\sum_{l=0}^{L-1}\left|h_{n,l}^{(m)}e^{-2\pi k_\tau l/B}\right|^2. \quad (30)$$

Substituting (29) and (30) into (28), we have

$$MSE^{(m)}(i) \stackrel{def}{=} \frac{1}{(Q+1)NL}\left(\sigma_v^2 + \frac{\sigma_c^2}{\Gamma}\sum_{\tau=0}^{\Gamma-1}\sum_{n=1}^{N}\sum_{l=0}^{L-1}\left|h_{n,l}^{(m)}e^{-2\pi k_\tau l/B}\right|^2\right)\mathrm{tr}\left[\left(\mathbf{A}(i)\right)^H\mathbf{A}(i)\right]^{-1}. \quad (31)$$

Apparently, the channel estimation performance depends crucially on the matrix $\mathbf{A}(i)$. The optimal estimation or minimum MSE (MMSE) estimation may require $\mathbf{A}(i)^H\mathbf{A}(i) = \Phi\mathbf{I}$ where Φ is a constant. From (21), we adopt the training sequence as $p_n(i,k) = \sigma_p, k = 0, \cdots B-1$ as (15), the above MMSE condition can be well satisfied. We thus have

$$\mathrm{tr}\left[\left(\mathbf{A}(i)\right)^H\mathbf{A}(i)\right] = \Gamma\sigma_p^2\mathbf{I}_{NL(Q+1)}. \quad (32)$$

Substituting (32) into (31), the MSE of channel estimation over one OFDM symbol can be derived as

$$MSE^{(m)}(i) = \frac{\sigma_c^2}{\Gamma^2\sigma_p^2}\sum_{\tau=0}^{\Gamma-1}\sum_{n=1}^{N}\sum_{l=0}^{L-1}\left|h_{n,l}^{(m)}e^{-2\pi k_\tau l/B}\right|^2 + \frac{\sigma_v^2}{\Gamma\sigma_p^2}. \quad (33)$$

It is seen that the first term of (33) is the estimation variance due to information interference, and depends upon the channel transfer functions. We thus define the normalized variance as

$$NMSE^{(m)}(i) = \frac{\sigma_c^2}{\Gamma^2\sigma_p^2}\sum_{\tau=0}^{\Gamma-1}\sum_{n=1}^{N}\sum_{l=0}^{L-1}\left|h_{n,l}^{(m)}e^{-2\pi k_\tau l/B}\right|^2 \Bigg/ \left|\bar{h}^{(m)}(i)\right|^2 \quad (34)$$

where $\left|\bar{h}^{(m)}(i)\right|^2 = \sum_{\tau=0}^{\Gamma-1}\sum_{n=1}^{N}\sum_{l=0}^{L-1}\left|h_{n,l}^{(m)}e^{-2\pi k_\tau l/B}\right|^2 \Big/ NL\Gamma$. Following the definition of (34), we obtain the normalized variance as

$$NMSE^{(m)}(i) = \frac{\sigma_c^2}{\Gamma^2\sigma_p^2}\sum_{\tau=0}^{\Gamma-1}\sum_{n=1}^{N}\sum_{l=0}^{L-1}\left|h_{n,l}^{(m)}e^{-2\pi k_\tau l/B}\right|^2 \Bigg/ \left|\bar{h}^{(m)}(i)\right|^2 = \frac{NL}{\Gamma}\frac{\sigma_c^2}{\sigma_p^2}. \quad (35)$$

From (35), we can find that the estimation variance due to the information interference is directly proportional to the information-to-pilot power ratio σ_c^2/σ_p^2, thereby resulting in an inaccurate solution for the general case that $\sigma_c^2 \gg \sigma_p^2$.

Then, we analyze the channel estimation performance of the weighted average approach over multiple OFDM symbols (the whole frame Ω). Define the vectors $A = \left[\mathbf{A}(1),...\mathbf{A}(I)\right]^T$,

$$\Xi^{(m)} = \left[\Xi^{(m)}\left(1,k_1:k_\Gamma\right),\cdots\Xi^{(m)}\left(I,k_1:k_\Gamma\right)\right]^T \text{ and } \quad V^{(m)} = \left[\bar{V}^{(m)}\left(1,k_1:k_\Gamma\right),\dots\bar{V}^{(m)}\left(I,k_1:k_\Gamma\right)\right]^T,$$

the MSE of the weighted average channel estimator over multiple OFDM symbols is given by

$$MSE^{(m)} = \frac{1}{(Q+1)NL}\text{tr}\left\{\boldsymbol{\eta}^\dagger E\left\{\left\|\mathbf{A}^\dagger\Xi^{(m)} + \mathbf{A}^\dagger V^{(m)}\right\|^2\right\}\left(\boldsymbol{\eta}^\dagger\right)^H\right\} = \frac{1}{I}\sum_{i=1}^{I}MSE^{(m)}(i)\text{tr}\left\{\boldsymbol{\eta}^\dagger\left(\boldsymbol{\eta}^\dagger\right)^H\right\}. \quad (36)$$

Note that the column vectors of the matrix $\boldsymbol{\eta}$ in (23) are in fact the FFT vectors of a $I \times I$ matrix, we thus have $\boldsymbol{\eta}^H\boldsymbol{\eta} = II_{(Q+1)}$ and $\text{tr}\left[\boldsymbol{\eta}^H\boldsymbol{\eta}\right]^{-1} = (Q+1)/I$. Substituting (33) into (36), the MSE of channel estimation over multiple OFDM symbols is given by

$$MSE^{(m)} = \frac{(Q+1)\sigma_c^2}{I\Gamma^2\sigma_p^2}\sum_{\tau=0}^{\Gamma-1}\sum_{n=1}^{N}\sum_{l=0}^{L-1}\left|\hbar_{n,l}^{(m)}e^{-2\pi k_\tau l/B}\right|^2 + \frac{(Q+1)\sigma_v^2}{I\Gamma\sigma_p^2} \quad (37)$$

In (37), the second term is caused by information sequence interference, which may become the dominant component of the channel estimation variance for the general case of $\sigma_c^2 \gg \sigma_p^2$, especially for large SNRs. Therefore, we solely consider information sequence effect. Similar to (34)-(35), we derive the normalized variance due to information interference by removing the channel gain as

$$NMSE^{(m)} = \frac{(Q+1)\sigma_c^2}{I\Gamma^2\sigma_p^2}\sum_{\tau=0}^{\Gamma-1}\sum_{n=1}^{N}\sum_{l=0}^{L-1}\left|\hbar_{n,l}^{(m)}e^{-2\pi k_\tau l/B}\right|^2 \Bigg/ \left|\bar{\hbar}^{(m)}\right|^2 \quad (38)$$

where $\left|\bar{\hbar}^{(m)}\right|^2 = \sum_{i=1}^{I}\left|\bar{\hbar}^{(m)}(i)\right|^2\Big/I$. It follows that

$$NMSE^{(m)} = \frac{\sigma_c^2}{\sigma_p^2}\frac{NL(Q+1)}{I\Gamma} \approx \frac{\sigma_c^2}{\sigma_p^2}\frac{NL(Q+1)}{I\Gamma}\frac{B}{\Omega} = \frac{\sigma_c^2}{\sigma_p^2}\frac{NL(Q+1)}{\theta\Omega} \quad (39)$$

where $\theta = \Gamma/B$ is the training ratio of one OFDM symbol. For conventional ST-based LTI schemes where isolated pilots are exploited for channel estimation [8]-[13] [17]-[18], we have $\theta = 1$. However, for estimating the LTV channels addressed in this paper, Γ pilot clusters, instead of isolated pilot tones, are exploited. Thus, the corresponding training ratio yields $\theta \leq 1/(2T+1)$. From (39), the normalized variance is directly proportional to the information-pilot power ratio σ_c^2/σ_p^2, the training ratio θ and the ratio of unknown parameter number $NL(Q+1)$ over the frame length Ω.

Compared with the variances of channel estimation over one OFDM symbol as in (33)-(35), the estimation variances of the weighted average estimator(37)-(39) is significantly reduced owing to the fact that $I/(Q+1) \gg 1$. Theoretically, the weighted average operation can be considered as an effective approach in estimating LTV channel, where the information sequence interference can be effectively suppressed over multiple OFDM symbols. As stated in conventional ST-based LTI schemes [8]-[13] [17]-[18], channel estimation performance can be improved along with the increment of the recorded frame length Ω, i.e. the estimation variance approaches to zero as $\Omega \to \infty$. This can be easily comprehended that larger frame length Ω means more observation samples, and hence lowers the MSE level. From the LTV channel model (6), however, we note that as the frame length Ω is increased, the corresponding truncated DFB requires a larger order Q to model the LTV channel (maintain a tight channel model), and the least order should be satisfied $Q/2 \geq f_d \Omega / f_s$, where f_d and f_s are the Doppler frequency and sampling rate, respectively. Consequently, as the frame length Ω increases, the LTV channel estimation variance (39) approaches to a fixed lower-bound associate with the system Doppler frequency as well as the information to pilot power ratio. This is quite different from the existing ST-based channel estimation approaches [8]-[19].

According to the theoretic analysis in (37)-(39), the proposed two-step LTV channel estimator achieves a significant improvement over multiple OFDM symbols compared with that of block-by-block process (33)-(35). However, as the frame length Ω is increased, the estimation variances approach to a fixed lower-bound. Further enhancement of the channel estimation should resort to increasing the ST power σ_p^2. For wireless communication systems with a limited transmission power, however, an increased ST power allocation reduces the data power σ_c^2, leading to SER degradation. Accordingly, in the analysis presented in the next section, the ratio of ST power allocation is determined by maximizing the lower bound of the average channel capacity.

5. Analysis of ST power allocation and system capacity

In this section, we consider the issue of ST power allocation where the lower bound of the average channel capacity is maximized and then mathematically derived for the proposed two-step channel estimator.

Define the ST power allocation factor

$$\beta = \frac{E\left[|p_n(k)|^2\right]}{E\left[|p_n(k)|^2\right] + E\left[|c_n(k)|^2\right]} = \frac{\sigma_p^2}{\sigma_p^2 + \sigma_c^2}. \tag{40}$$

For a fixed SNR or transmitted power budget, higher β implies smaller effective SNR at the receiver due to decreased power in the information sequence but higher channel estimation accuracy. Having removed ST sequence, we obtain the received signals in a vector-form as

$$\bar{U}^{(m)}(i) = \left[\bar{u}^{(m)}(i,0), \cdots \bar{u}^{(m)}(i,k), \cdots \bar{u}^{(m)}(i,B-1)\right]^T$$

$$= \underbrace{\sum_{n=1}^{N} \hat{H}_n^{(m)}(i) C_n(i)}_{\text{desired signals: } = \lambda^{(m)}(i)} + \underbrace{\sum_{n=1}^{N} \Delta \hat{H}_n^{(m)}(i)\left[P_n(i) + C_n(i)\right]}_{\text{interference to information signal recovery: } = \mu^{(m)}(i)} + \bar{V}^{(m)}(i) \tag{41}$$

with the received signals $\bar{u}^{(m)}(i,k), k = 0, \cdots B-1$ in (41) as

$$
\begin{aligned}
\bar{u}^{(m)}(i,k) &= \lambda^{(m)}(i,k) + \mu^{(m)}(i,k) + \bar{v}^{(m)}(i,k) \\
&= \sum_{n=1}^{N} \hat{H}_n^{\,\prime(m)}(i,k)c_n(i,k) + \sum_{n=1}^{N} \Delta\hat{H}_n^{\,\prime(m)}(i,k)\big[p_n(i,k) + c_n(i,k)\big] + \bar{v}^{(m)}(i,k)
\end{aligned}
\tag{42}
$$

where $\Delta\hat{H}_n^{\,\prime(m)}(i,k) = \sum_{n=1}^{N}\big[H_n^{\,\prime(m)}(i,k) - \hat{H}_n^{\,\prime(m)}(i,k)\big]$ is the estimation error due to information interference as well as additive noise. Using the proposed two-step estimator (23)-(26), the channel estimation variance can be smoothed over multiple OFDM symbols, and approaches to a small fixed lower bound. The estimated vector $\hat{H}_n^{\,\prime(m)}(i)$ as well as the error vector $\Delta\hat{H}_n^{\,\prime(m)}(i)$, therefore, can be thus approximated to be of the similar characteristics of distribution as that of $H_n^{\,\prime(m)}(i)$. Consequently, following the assumption (H1)-(H3), the interference vector $\boldsymbol{\mu}^{(m)}(i)$ is approximately white for a large symbol-size B, and independent of the noise vector $\bar{V}^{(m)}(i)$. Similar to the procedure of (29)-(30), the covariance matrix of $\boldsymbol{\mu}^{(m)}(i)$ and $\lambda^{(m)}(i)$ can be obtained as

$$
\mathrm{var}\big(\lambda^{(m)}(i)\big) = E\Big\{\big(\lambda^{(m)}(i)\big)^H \lambda^{(m)}(i)\Big\} = \sigma_{\hat{H}}^2 \sigma_p^2 I
\tag{43}
$$

$$
\mathrm{var}\big(\boldsymbol{\mu}^{(m)}(i)\big) = E\Big\{\big(\boldsymbol{\mu}^{(m)}(i)\big)^H \boldsymbol{\mu}^{(m)}(i)\Big\} = \sigma_{\Delta\hat{H}}^2 \big(\sigma_p^2 + \sigma_c^2\big) I
\tag{44}
$$

where $\sigma_{\hat{H}}^2 = \big|\bar{h}^{(m)}(i)\big|^2 = \sum_{k=0}^{B-1}\sum_{n=1}^{N}\sum_{l=0}^{L-1}\big|\hat{h}_{n,l}^{(m)} e^{-2\pi kl/B}\big|^2 \big/ NLB$,

and $\sigma_{\Delta\hat{H}}^2 = \sum_{k=0}^{B-1}\sum_{n=1}^{N}\sum_{l=0}^{L-1}\big|\Delta\hat{h}_{n,l}^{(m)} e^{-2\pi kl/B}\big|^2 \big/ NLB$. Since the ST power allocation factor is derived within each isolated OFDM symbol, we neglect the symbol-index i for simplicity. A lower bound on the OFDM channel capacity with channel estimation error has been derived in [20]-[21] for uniform pilot distribution. Such expression can readily be extended to issue of ST where the pilots are spread over the whole frequency band. Therefore, the lower bound of the average channel capacity for an ST-based OFDM system can be obtained by summing over all the subcarriers, i.e.,

$$
C^{(m)} \geq \bar{C}^{(m)} = \frac{1}{B}\sum_{k=0}^{B-1} E\left\{ \log\left[1 + \frac{\sigma_c^2}{\big(\sigma_p^2 + \sigma_c^2\big)\sigma_{\Delta\hat{H}}^2 \big/ \sigma_{\hat{H}}^2 + \sigma_v^2 \big/ \sigma_{\hat{H}}^2} \right]\right\}
\tag{45}
$$

For the sake of simplicity, we assume the transmission power satisfies that $\sigma_p^2 + \sigma_c^2 = 1$. By (40), we thus have $\sigma_p^2 = \beta$ and $\sigma_c^2 = 1 - \beta$. Considering that the normalized MSE of the proposed two-step channel estimator is sufficiently small and approaches to a fixed lower bound (37)-(39), it allows us to make the approximation of $\sigma_{\Delta\hat{H}}^2 \big/ \sigma_{\hat{H}}^2 \approx \sigma_{\Delta\hat{H}}^2 \big/ \sigma_{\hat{H}}^2 = NMSE^{(m)}$. As a result, $C^{(m)}$ in (45) can be approximated as

$$C^{(m)} \approx \frac{1}{B} \sum_{k=0}^{B-1} E\left\{ \log\left[1 + \frac{\sigma_c^2}{\sigma_{\Delta\hat{H}}^2/\sigma_H^2 + \sigma_v^2/\sigma_H^2} \right]\right\}$$

$$= \frac{1}{B} \sum_{k=0}^{B-1} E\left\{ \log\left[1 + \frac{1-\beta}{(1-\beta)(Q+1)NL/\beta I\Gamma + (Q+1)\sigma_v^2/\beta I\Gamma\sigma_H^2 + \sigma_v^2/\sigma_H^2} \right]\right\} \qquad (46)$$

$$= \log\left(1 + \frac{(1-\beta)\beta I\Gamma}{\beta\left[I\Gamma/\Re_{SNR} - (Q+1)NL \right] + (Q+1)NL\left(1/\Re_{SNR} + 1 \right)} \right)$$

where $\Re_{SNR} = \sigma_H^2\left(\sigma_p^2 + \sigma_c^2 \right)/\sigma_v^2 = \sigma_H^2/\sigma_v^2$. In fact, the averaged channel capacity of (46) is a log-function of β, which is a monotonically increasing function. Therefore, the lower-bound of $C^{(m)}$ with respect to β can be achieved by maximizing the following function

$$\Upsilon^{(m)}(\beta) = \frac{\beta(1-\beta)}{\alpha_1\beta + \alpha_2} = \frac{(1-\beta)\beta}{\beta\left[1/\Re_{SNR} - (Q+1)NL/I\Gamma \right] + (Q+1)NL\left(1/\Re_{SNR} + 1 \right)/I\Gamma}. \qquad (47)$$

where

$$\alpha_1 = 1/\Re_{SNR} - (Q+1)NL/I\Gamma, \quad \alpha_2 = (Q+1)NL\left(1/\Re_{SNR} + 1 \right)/I\Gamma. \qquad (48)$$

Setting the first derivation of $\Upsilon^{(m)}(\beta)$ with respect to β to zero, we obtain (after some manipulations) a quadratic equation in β, i.e.

$$\beta^2 + \frac{2\alpha_2}{\alpha_1}\beta - \frac{\alpha_2}{\alpha_1} = 0. \qquad (49)$$

Consequently, the global maximum of $\Upsilon^{(m)}(\beta)$ can be obtained when

$$\beta = \frac{\sqrt{\left(1/\Re_{SNR} + 1 \right)\left(I\Gamma/NL(Q+1)\Re_{SNR} + 1/\Re_{SNR} \right)} - (Q+1)NL\left(1/\Re_{SNR} + 1 \right)}{I\Gamma/\Re_{SNR} - (Q+1)NL}. \qquad (50)$$

As will be shown in simulations, an increase in the training power allocation factor β does not necessarily improve the overall system performance since a larger β implies a better channel estimation while substantially scarifying the effective received signal SNR at the same time.

6. Simulations

We assume the MIMO/OFDM system with $N = 2$ and $M = 4$. The symbol-size is $B = 1024$ and the transmitted data $s_n(i,k)$ is 8-PSK signals with symbol rate $f_s = 10^7$ /second. Before transmission, the transmitted data are coded by 1/2 convolutional coding and block interleaving over one OFDM symbol. The channel is assumed to be $L = 10$ taps and, the coefficients $h_{n,l}^{(m)}(t)$ are generated as low-pass, Gaussian and zero mean random processes and uncorrelated for different values of n and l. The multi-path intensity profile is chosen to be $\phi(l) = \exp(-l/10)$ $l = 0, \cdots L-1$. The Doppler spectra are $\Psi(f) = (\pi\sqrt{(f_n)^2 - f}\ )$ for $f \leq f_n$, where f_n is the Doppler frequency of the nth user, otherwise, $\Psi(f) = 0$. CP-length is chosen to be 32 to avoid inter-symbol interferences. The additive noise is a Gaussian and white random process with a zero mean.

Test Case 1. Channel Estimation

We run simulations with the Doppler frequency $f_n = 300\text{Hz}$ that corresponds to the maximum mobility speed of 162 km/h as the users operate at carrier frequency of 2GHz. In order to model the LTV channel, the frame is designed as $\Omega = B'\times 128 = (B + \text{CP-length}) \times 128 = 135168$, i.e. each frame consists of 128 OFDM symbols. During the frame, the channel variation is $f_n\Omega/f_s = 4.1$. Over the frame Ω, we utilize truncated DFB of order $Q = 10 > 2f_d\Omega/f_s$ to model the LTV channel coefficients. In order to estimate the MIMO/OFDM channels, the superimposed pilots are designed according to (15) with the pilot power $\sigma_p^2 = 0.2\sigma_c^2$. Fig.2 depicts the LTV channel coefficient estimation over the frame Ω. It is clearly observed that although the channel coefficient is accurately estimated during the centre part of the frame, the outmost samples over the whole frame still exhibit errors. A possible explanation is that as the Fourier basis expansions in (6) are truncated, and an effect similar to the Gibbs phenomenon, together with spectral leakages, will lead to some errors at the beginning and the end of the frame. This may be a common problem for the proceeding literature [1]-[2] [5]-[6] [16] that employing basis expansions to model the LTV channels. To solve the problem, the frames are designed to be partially overlapped, e.g. the frames are designed as $(\ell-1)\Omega - \Psi B' \leq t \leq \ell\Omega$, $\ell = 2,3,\cdots$, where Ψ is a positive integer. By the frame-overlap, the channel at the beginning and the end of one frame can be modeled and estimated from the neighboring frames.

To further evaluate the new channel estimator, we use the mean square errors to measure the channel estimation performance by

$$MSE_n^{(m)} = \sum_{i=1}^{\Omega/B'} MSE_n^{(m)}(i)/(\Omega/B') =$$

(51)

$$\frac{B'}{\Omega}\sum_{i=1}^{\Omega/B'} E\left\{ \sum_{t=0}^{B-1}\sum_{l=0}^{L-1}\left| h_{n,l}^{(m)}(i,t) - \sum_{q=0}^{Q} \hat{h}_{n,l,q}^{(m)} e^{j2\pi(q-Q/2)t/\Omega} \right|^2 \middle/ BL\left| h_{n,l}^{(m)}(i,t)\right|^2 \right\}$$

where $\hat{h}_{n,l,q}^{(m)}$ is the channel coefficient estimation.

We firstly test the two-step channel estimator under the different pilot powers and different channel coefficient numbers to verify the channel estimation variance analysis. The LTV channel is the same as that in Fig.2. As shown in Fig.3, the MSE of the channel estimation approach are almost independent of the additive noises, especially as SNR>5dB. This is consistent with the channel estimation analysis (38)-(39) where the additive noise has been greatly suppressed by the weighted average procedure. Thus, the estimation errors depend mainly on the information- pilot power ratio as well as the system unknowns NL. This is rather different from the FDM training based schemes [20]-[23].

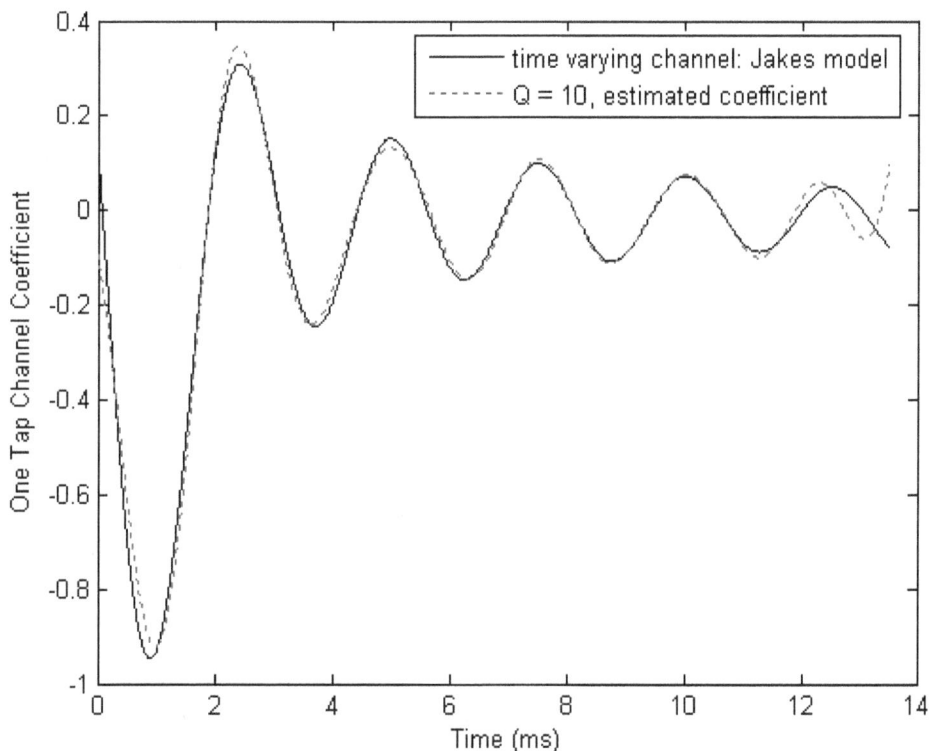

Fig. 2. One tap coefficient of the LTV channel and the estimation over the frame $\Omega = 135168/10^7 \approx 13.52$ ms.

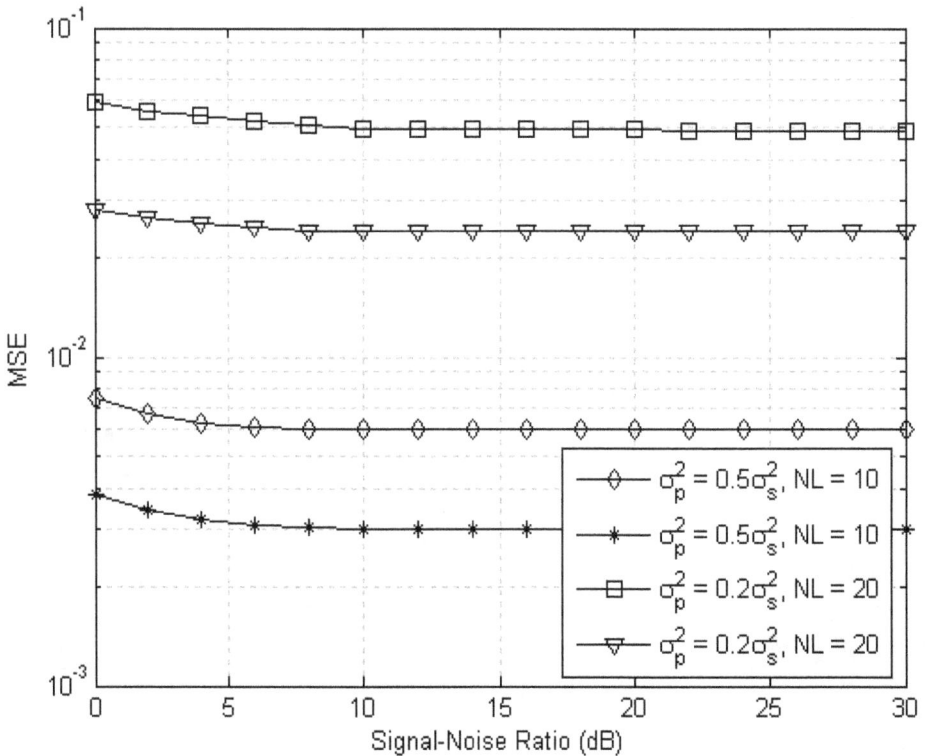

Fig. 3. MSE of the weighted average estimation versus SNR for the LTV channel of $f_n = 300$Hz and $\Omega = 13.52$ ms under the different pilot powers and different unknown parameters.

We then compare the proposed two-step channel estimation scheme with the conventional ST-based methods [8]-[13] [17]-[18] under different Doppler frequencies. In the conventional ST scheme, the LTV channel is firstly estimated from the LTI assumption at each OFDM symbol, and then all the estimations from the frame Ω are averaged to confront the information sequence interferences. It shows clearly in Fig. 4 that for the LTI channel of $f_n = 0$Hz, both the conventional ST and the weighted average estimator exhibit the similar performance. In addition, the estimation performance can be improved with the increment of the frame or average length. However, when the channel involved in simulations is time-varying, the channel estimation performance of the conventional ST-based schemes is degraded extensively. The simulation reveals the shortcoming of the conventional ST in estimating the LTV channels. On the contrary, the MSE level is reduced by the weighted average process (23)-(26) for the LTV channels of $f_n = 100$Hz, 300Hz with $T = 2$ (one pilot cluster is composed of $2T + 1 = 5$ pilots). We also observe that the MSE approaches to a constant as the increment of the frame length, i.e. the lower-bound that associated with the given Doppler frequency.

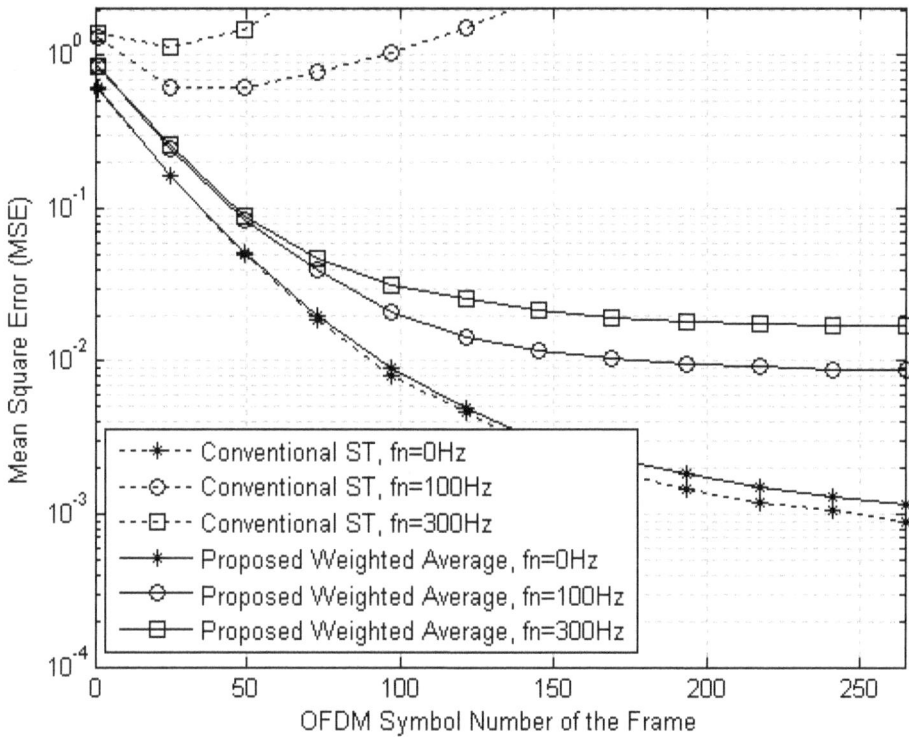

Fig. 4. MSE versus frame or average length under the different Doppler frequencies of the
LTV channel with $\sigma_p^2 = 0.25\sigma_c^2$, SNR = 20dB.

From Fig. 4, we observe that channel estimation performance would be degraded as the
increment of mobile users' speed (or corresponding system Doppler shift). To further
enhance the channel estimation performance of the systems with a limited pilot power while
suffering from a high Doppler shift, an iterative decision feedback (DF) approach can be
adopted at the receiver. Explicitly, the iterative method can be considered as a twofold
process. First, the information sequences are recovered by a hard detector [5] based on the
LTV channel estimation in Section III. Second, the recovered data symbols are removed from
the received signals to cancel the information sequence interference and, thus to enhance the
channel estimation performance. Fig. 5 depicts the performance between the weighted
average scheme and the iterative DF estimator in terms of channel MSE. For a fairness of
comparison, we also simulate the MSE of the FDM training-based channel estimator [5] as
latter serves as a "benchmark" in related works. For estimating the MIMO/OFDM channels,
$\Gamma = 40$ pilot clusters with $\Gamma(2T+1) = 200$ known pilot symbols which are subject to the

proposed pilot specifications in (15) are used in one OFDM symbol. That is, approximately 10% total bandwidth is assigned for pilot tones. Comparatively, as shown in Fig.5, the iterative DF estimation exhibits a more significant improvement than that of weighted average estimation, and outperforms the FDM channel estimator [5] by using a small pilot power of $\sigma_p^2 = 0.25\sigma_c^2$, which conforms that the information sequence interferences can be effectively cancelled by iterative DF procedure. Moreover, it should be noted that since the superimposed pilots are spread over the entire band, the proposed ST-based channel estimator is also feasible to estimate the channel with a very long delay spread, i.e. cluster-based channel.

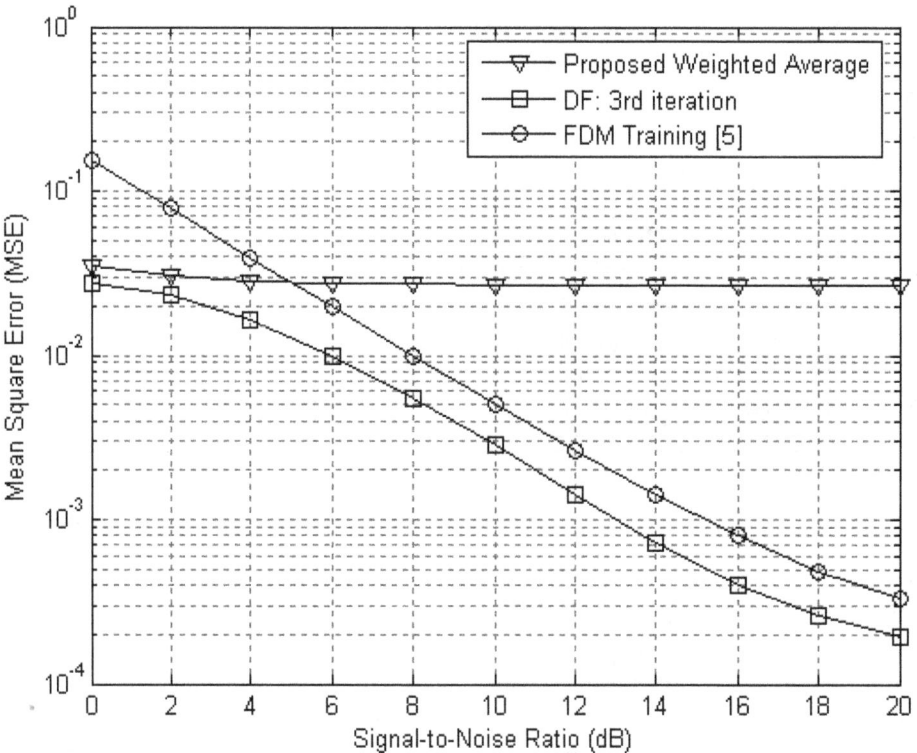

Fig. 5. MSE versus SNR for different estimators for the LTV channel of $f_n = 300\,\mathrm{Hz}$, $NL = 20$.

To further validate the effectiveness of the DF scheme, we also provide the channel estimation MSE of the DF method versus the iteration numbers under SNR = 15dB. Fig. 6 shows that the iterative DF method is feasible for a wide range of system Doppler spreads. Obviously, the enhancement of the iterative DF is at the cost of an increment in computational complexity that is directly proportional to the iteration number. However, as is shown in Fig. 6 that the iterative DF approach converges to the steady-state performance by only a few iterations, the overall computational complexity will be acceptable for many wireless communication systems.

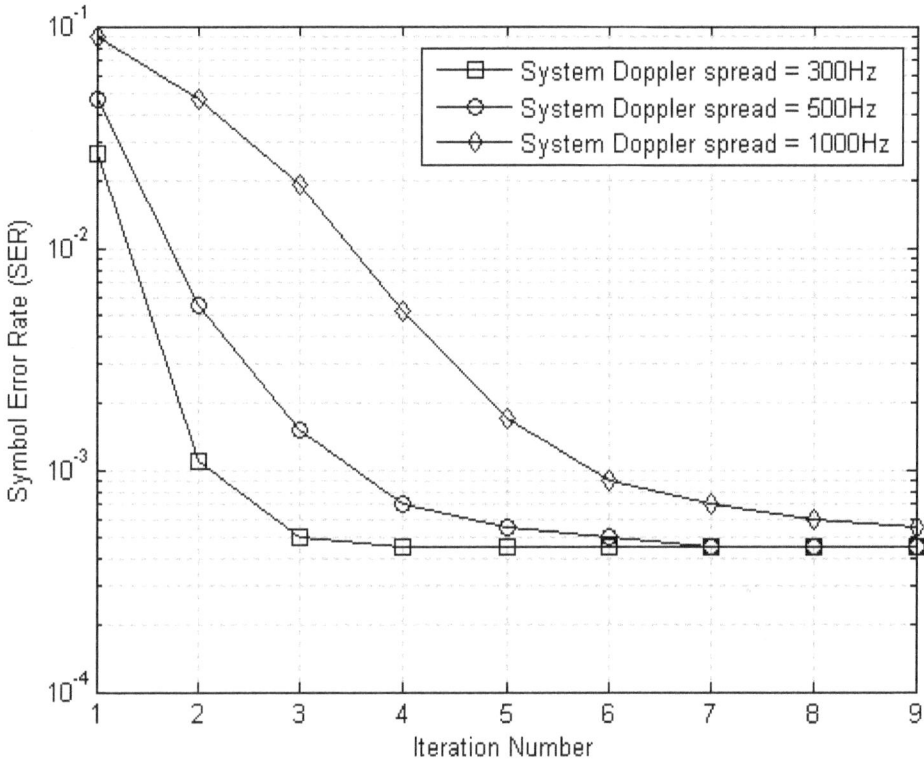

Fig. 6. MSE of the iterative DF channel estimation versus iterations for the LTV channel of
different system Doppler spreads when SNR=15dB.

Test Case 2. Training Power Allocation

As aforementioned, for wireless communication systems with a limited transmission power,
some useful power must inevitably be allocated to the superimposed pilots, and thus
resulting in the received signal SNR reduction. Herein, we carry out several experiments to
assess the effect of ST power allocation factor on the lower-bound of the average channel
capacity for different SNRs.

Fig. 7 shows the effect of different value of training power allocation factor β on the lower
bound of the average channel capacity for received signal SNR = 10 and 20 dB, respectively.
It is seen that the average channel capacity decreases with the increment of β. It reveals that
although higher β implies that higher fraction of transmitted power is allocated to training
leading to more accurate channel estimates, the received signal SNR is substantially
decreased, resulting in potential decrement of the average channel capacity. In addition, we
further simulate the approximated β in order to test the validity of theoretic results in (50). It
can be seen that the approximation of β is almost consistent with that of the actual results.

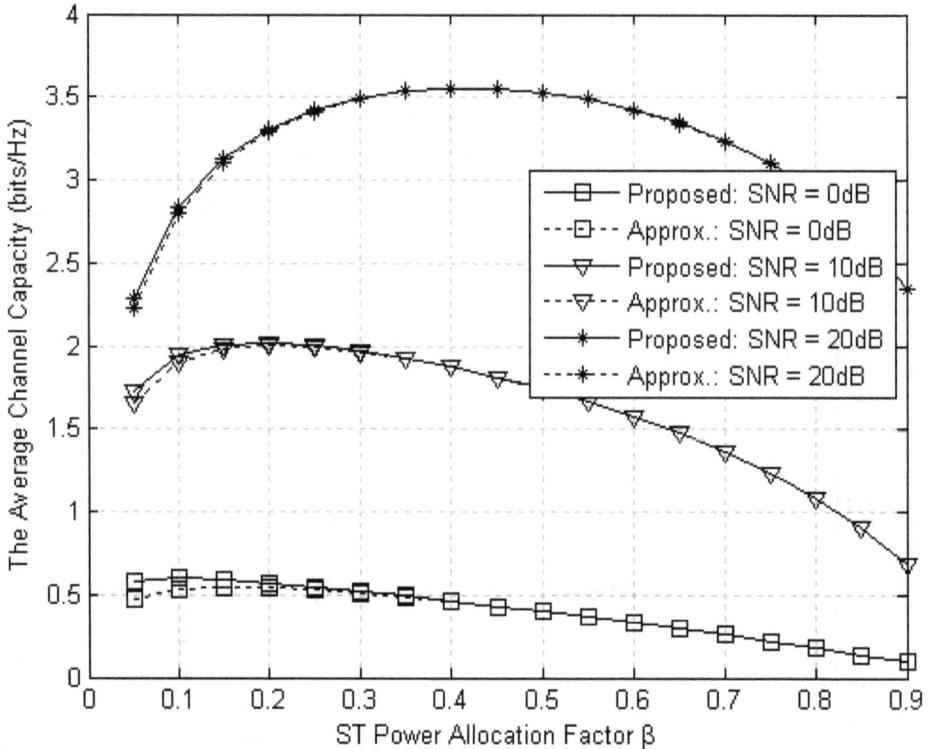

Fig. 7. Lower bound of the average channel capacity versus different values of ST power allocation factor β under SNR = 0dB, 10dB and 20dB, respectively.

Fig. 8 shows the plots of the optimal value of training power allocation factor β versus received SNR for different frame length. It is observed that the increment of SNR leading to a corresponding increase in the optimal ST power allocation factor. This can be easily comprehended that according to (41), the effective interference is composed of two factors, i.e. the bias of channel estimation and the additive noise. That is, for large SNRs, higher β is required to improve the channel estimation performance, thus leading to a reduction of the effective interference. Conversely, when SNR is small, improving the channel estimation accuracy has a small effect in reducing the effective interference. On the other hand, we notice that β decreases as the frame length increases but approximately unvaried when Ω is sufficiently large, i.e. β is almost unchanged when $I \geq 192$. This result arises because we have theoretically analyzed in Section III that the estimation variance approaches to a fixed lower bound that can be only improved by increasing ST power allocation when the frame length is large enough. Therefore, the power allocated to the training sequence can be reduced with no loss in channel estimation performance when the frame length is increased, but finally approaches to a fixed lower bound associate with the channel estimation variance when Ω is sufficiently large. This is somewhat different from those presented in [10].

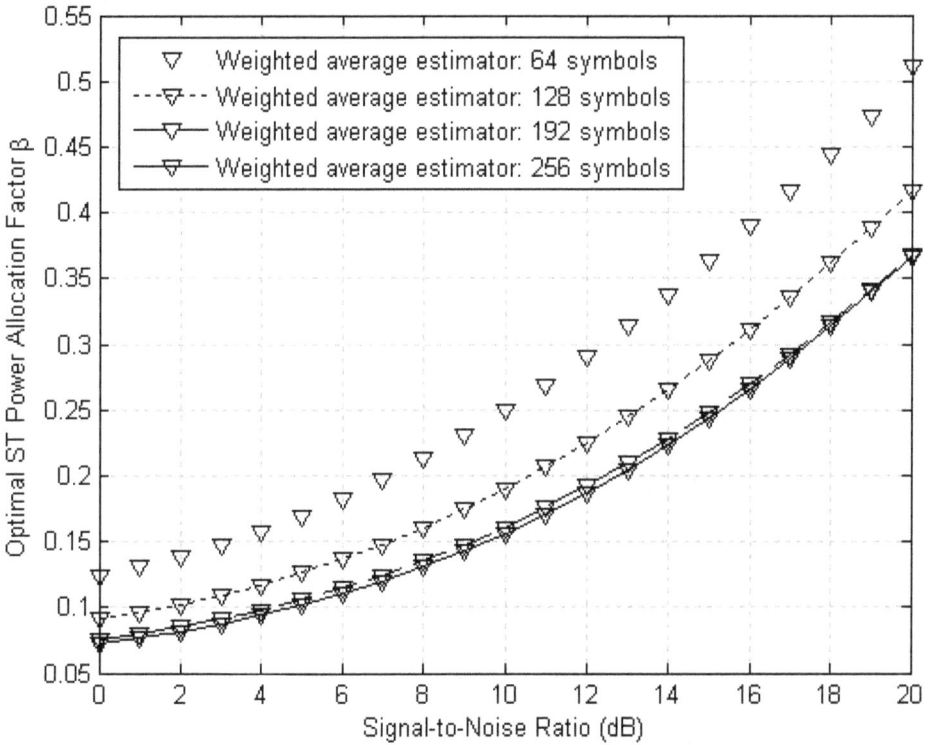

Fig. 8. Optimal ST power allocation factor of the proposed weighted average channel estimator versus SNR for different frame lengths.

7. Conclusions

In this paper, we have a developed superimposed training-aided LTV channel estimation approach for MIMO/OFDM systems. The LTV channel coefficients were firstly modeled by truncated DFB, and then estimated by using a two-step approach over multiple OFDM symbols. We also present a performance analysis of the proposed estimation approach and derive closed-form expressions for the channel estimation variances. It is shown that the estimation variances, unlike the conventional ST, approach to a fixed lower-bound that can only be reduced by increasing the pilot power. Using the developed channel estimation variance expression, we analyzed the system capacity and optimize the training power allocation by maximizing the lower bound of the average channel capacity for systems with a limited power. Compared with the existing FDM training based schemes, the new estimator does not entail a loss of rate while yields a better estimation performance, and thus enables a higher efficiency.

8. Ackownledgment

This work is supported by the national Natural Science Foundation (NSF) of China, (Grant No. 61002012), the Project of NSF of Guangdong Province, (Grant No. 10451063101006074), and also supported by national innovation experiment program for university students, Grant No.C1025338, and Key project of college students' extracurricular science and technology, Grant No. 10WDKB05

9. Reference

[1] G. B. Giannakis, C. Tepedelenlioglu "Basis expansion models and diversity techniques for blind identification and equalization of time-varying channels," *Proc. of the IEEE*, vol.86, pp. 1969-1986, Oct, 1998.

[2] X. L. Ma, G. B. Giannakis and B. Lu "Block differential encoding for rapidly fading channels," *IEEE Trans. Commun.*, vol. 52, no. 3, pp.416-425, March 2004.

[3] H.-C. Wu, "Analysis and characterization of intercarrier and interblock interferences for wireless mobile OFDM systems," *IEEE Trans. Broadcasting*, vol. 52, no. 2, pp. 203–210, Jun. 2006.

[4] X. Dai "Adaptive blind source separation of multiple-input multiple-output linearly time-varying FIR system," *IEE Proc.-Vis. Image Signal Process.*, Vol. 151, No. 4, pp.279-286, August 2004.

[5] Z. Tang, R. C. Cannizzaro, G. Leus and P. Banelli, "Pilot-assisted time-varying channel estimation for OFDM systems," *IEEE Trans. Signal Process.* Vol. 55, no. 5, pp. 2226-2238, May 2007.

[6] X. Dai, "Optimal estimation of linearly time-varying MIMO/OFDM channels modeled by a complex exponential basis expansion," *IET Commun.*, vol. 1, no. 5, pp. 945-953, 2007.

[7] H. C. Wu and Y. Wu, "Distributive pilot arrangement based on modified M-sequences for OFDM intercarrier interference estimation," IEEE Trans. Wireless Commun., vol. 5, no. 6, pp. 1605-1609, May 2007.

[8] G. T. Zhou, M. Viberg, and T. McKelvey, "A first-order statistical method for channel estimation," *IEEE Signal Processing Letters*, vol. 10, no. 3, pp.57-60, March 2003.

[9] J. K. Tugnait and W. Luo "On channel estimation using superimposed training and first-order statistics," *IEEE Comm. Letters*, vol. 7, no. 9, pp. 413-416, Sept. 2003.

[10] J. K. Tugnait and X.H. Meng "On superimposed training for channel Estimation: performance analysis, training power allocation, and frame synchronization," *IEEE Trans. Signal Process.*, vol. 54, no.2, pp.752-765, Feb. 2006.

[11] M. Ghogho, D. McLernon, E. A. Hernandez, and A. Swami "Channel estimation and symbol detection for block transmission using data-dependent superimposed training," *IEEE Signal Process. Letters*, vol. 12, no. 3, pp. 226-229, March 2005.

[12] W. C. Huang, C. P. Li and H. J. Li, "On the power allocation and system capacity of OFDM systems using superimposed training schemes," *IEEE Trans. Veh. Technol.*, vo. 58, no. 4, pp. 1731-1740, May, 2009.

[13] Q. Y. Zhu and Z.Q. Liu, "Optimal pilot superimposition for zero-padded block transmissions," *IEEE Trans. Wireless Commun.*, vol. 5, no. 8, pp. 2194-2201, Aug. 2006.

[14] H. Zhang, X. Dai, "Time-varying Channel Estimation and Symbol Detection for OFDM Systems using Superimposed Training," *IEE Electronics Letter*, vol. 43, no. 22, Oct. 2007.

[15] S. M. A. Moosvi, D. C. McLernon, A. G. Orozco-Lugo, M. M. Lara, and M. Ghogho, "Carrier frequency offset estimation using data dependent superimposed training," *IEEE Commun. Letter*, vol. 12, no. 3, pp. 179-181, Mar. 2008.

[16] R. C.-Alvarez, R. P. -Michel, A. G. O. -Lugo, and J. K. Tugnait, "Enhanced channel estimation using superimposed training based on universal basis expansion," *IEEE Trans. Signal Process.* vol. 57, no. 3, Mar. 2009.

[17] F. Mazzenga, "Channel estimation and equalization for M-QAM transmission with a hidden pilot Sequence," *IEEE Trans. Broadcasting*, vol. 46, no. 2, pp. 170-176, June 2000.

[18] A. Goljahani, N. Benvenuto, S. Tomasin and L. Vangelista, "Superimposed sequence versus pilot aided channel estimations for next generation DVB-T systems," *IEEE Trans. Broadcasting*, vol. 55, no. 1, pp. 140-144, Mar. 2009.

[19] C. -P. Li and W. Hu, "Super-imposed training scheme for timing and frequency synchronization for OFDM systems," *IEEE trans. Broadcasting*, vol. 53, no. 2, pp. 574-583, Jun. 2007.

[20] A. Y. Panah, R. G. Vaughan and R. W. Heath, Jr. , "Optimizing pilot locations using feedback in OFDM systems," *IEEE trans. Vehicular Tech.* , vol. 58, no. 6, pp. 2803-2814, July 2009.

[21] S. Ohno and G.. B. Giannakis, "Capacity maximizing MMSE-optimal pilots for wireless OFDM over frequency- selective block Rayleigh-fading channels," *IEEE Trans. Inf. Theory*, vol. 50, no. 9, pp. 2138-2145, Sep. 2004.

[22] I. Barhumi, G. Leus and M. Moonen "Optimal training design for MIMO OFDM systems in mobile wireless channels," *IEEE Trans. Signal Process.*, vol. 51, no. 6, pp.1615-1624, June 2003.

[23] Y. Xiaoyan, W. Jiaqing, Y. Luxi and H. Zhenya, "Doubly selective fading channel estimation in MIMO OFDM systems," *SCIENCE CHINA Information Sciences*, vol. 48, no. 6, pp. 795-807, 2005.

[24] J. Z. Wang and J. Chen, "Performance of wideband CDMA with complex spreading and imperfect channel estimation," *IEEE Journal on Selected Areas in Commun.*, vol. 19, no. 1, pp. 152-163, Jan. 2001.

[25] J. Z. Wang and L. B. Milstein, "CDMA overlay situations for microcellular mobile communications," *IEEE Trans. on Commun.*, vol. 43, no. 2/3/4, pp. 603-614, Feb/March/April 1995.

[26] X. Dai, H. Zhang and D. Li, 'Linearly time-varying channel estimation for MIMO-OFDM systems using superimposed training,' *IEEE trans. Commun.*, vol. 58, no. 2, pp. 681-693, Feb. 2011.

[27] H. Zhang, et. al., 'Linearly time-varying channel estimation and symbol detection for OFDMA uplink using superimposed training,' *EURASIP J. Wireless Commun. Networking*, vol. 2009.

[28] H. Zhang, X. Dai and D. Li, 'Time-varying channel estimation for MIMO/OFDM systems using superimposed training,' *in Proc. IEEE WiCOM 08*, pp. 1-6, 2008.

Permissions

The contributors of this book come from diverse backgrounds, making this book a truly international effort. This book will bring forth new frontiers with its revolutionizing research information and detailed analysis of the nascent developments around the world.

We would like to thank Dr. Ali Ekşim, for lending his expertise to make the book truly unique. He has played a crucial role in the development of this book. Without his invaluable contribution this book wouldn't have been possible. He has made vital efforts to compile up to date information on the varied aspects of this subject to make this book a valuable addition to the collection of many professionals and students.

This book was conceptualized with the vision of imparting up-to-date information and advanced data in this field. To ensure the same, a matchless editorial board was set up. Every individual on the board went through rigorous rounds of assessment to prove their worth. After which they invested a large part of their time researching and compiling the most relevant data for our readers. Conferences and sessions were held from time to time between the editorial board and the contributing authors to present the data in the most comprehensible form. The editorial team has worked tirelessly to provide valuable and valid information to help people across the globe.

Every chapter published in this book has been scrutinized by our experts. Their significance has been extensively debated. The topics covered herein carry significant findings which will fuel the growth of the discipline. They may even be implemented as practical applications or may be referred to as a beginning point for another development. Chapters in this book were first published by InTech; hereby published with permission under the Creative Commons Attribution License or equivalent.

The editorial board has been involved in producing this book since its inception. They have spent rigorous hours researching and exploring the diverse topics which have resulted in the successful publishing of this book. They have passed on their knowledge of decades through this book. To expedite this challenging task, the publisher supported the team at every step. A small team of assistant editors was also appointed to further simplify the editing procedure and attain best results for the readers.

Our editorial team has been hand-picked from every corner of the world. Their multi-ethnicity adds dynamic inputs to the discussions which result in innovative outcomes. These outcomes are then further discussed with the researchers and contributors who give their valuable feedback and opinion regarding the same. The feedback is then collaborated with the researches and they are edited in a comprehensive manner to aid the understanding of the subject.

Apart from the editorial board, the designing team has also invested a significant amount of their time in understanding the subject and creating the most relevant covers. They scrutinized every image to scout for the most suitable representation of the subject and create an appropriate cover for the book.

The publishing team has been involved in this book since its early stages. They were actively engaged in every process, be it collecting the data, connecting with the contributors or procuring relevant information. The team has been an ardent support to the editorial, designing and production team. Their endless efforts to recruit the best for this project, has resulted in the accomplishment of this book. They are a veteran in the field of academics and their pool of knowledge is as vast as their experience in printing. Their expertise and guidance has proved useful at every step. Their uncompromising quality standards have made this book an exceptional effort. Their encouragement from time to time has been an inspiration for everyone.

The publisher and the editorial board hope that this book will prove to be a valuable piece of knowledge for researchers, students, practitioners and scholars across the globe.

List of Contributors

Zhao Wang, Eng Gee Lim, Tammam Tillo and Fangzhou Yu
Xi'an Jiaotong - Liverpool University, P.R. China

Yue Li, Jianfeng Zheng and Zhenghe Feng
Tsinghua University, China

Onofrio Losito and Vincenzo Dimiccoli
Itel Telecomunicazioni srl, Ruvo di Puglia (BA), Italy

Hamsakutty Vettikalladi, Olivier Lafond and Mohamed Himdi
Institute of Electronics and Telecommunication of Rennes (IETR), University of Rennes 1, France

Antonio F. Mondragon-Torres
Rochester Institute of Technology, USA

Han-Nien Lin
Feng-Chia University, Taiwan, R.O.C.

Suramate Chalermwisutkul
The Sirindhorn International Thai-German Graduate School of Engineering, King Mongkut's University of Technology North Bangkok, Thailand

Hui Yu and Xi Chen
Shanghai Jiao Tong University, China

Mihajlo Stefanović and Jelena Anastasov
Faculty of Electronic Engineering, University of Niš, Serbia

Stefan Panić
Faculty of Natural Science and Mathematics, University of Priština, Serbia

Petar Spalević and Ćemal Dolićanin
State University of Novi Pazar, Serbia

Han Zhang, Daru Pan and Shan Gao
School of Physics and Telecommunications Engineering, South China Normal University, Guangzhou, China

Xianhua Dai
School of Information Science and Technology, SUN Yat-sen University, Guangzhou, China